D0065491

YET BEING SOMEONE OTHER

YET BEING SOMEONE OTHER

Laurens van der Post

WILLIAM MORROW AND COMPANY, INC.

New York   1983

Library of Congress Cataloging in Publication Data

Van der Post, Laurens.
    Yet being someone other.

    1. Van der Post, Laurens.   2. Ocean travel.
3. Travelers—South Africa—Biography.   I. Title.
G540.V33   1983   910.4′5   82-20824
ISBN 0-688-01843-2

Printed in the United States of America

FIRST U.S. EDITION

1 2 3 4 5 6 7 8 9 10

To
Captain Katsue Mori, Master of the *Canada Maru*,
Japan's senior marine officer and a great sailor,
with gratitude and affection.

"So I assumed a double part, and cried
And heard another's voice cry: 'What! are *you* here?'
Although we were not. I was still the same,
Knowing myself yet being someone other –
And he a face still forming; yet the words sufficed
To compel the recognition they preceded.
And so, compliant to the common wind,
Too strange to each other for misunderstanding . . ."

T. S. Eliot, 'Little Gidding'

CONTENTS

Acknowledgements

5

CHAPTER

1 The Ambivalent Cape

9

2 Return to the Sea

27

3 The Singing Whale

61

4 The Ship and the Captain

107

5 Full House

183

6 The Shadow in Between

256

7 The Sword and the Flower

301

ACKNOWLEDGEMENTS

I owe a great deal to my wife, Ingaret Giffard, for editing my
manuscript, to The Hogarth Press, and especially to Norah
Smallwood who, with the late Ian Parsons, took over where
Virginia and Leonard Woolf left off, and has been the best of
publishers and the most generous of friends.

The title of this book is taken from the poem *Little Gidding* by
T. S. Eliot and my grateful thanks are due to Mrs T. S. Eliot
and Messrs Faber & Faber for permission to quote on page 6
eight lines from that poem.

YET BEING SOMEONE OTHER

The Ambivalent Cape

UNTIL it was all over, I had no idea how deeply I would be affected by my last voyage in the last liner from the Cape. Although it is several years now since I made that round voyage in the *Windsor Castle* from Southampton, I have not yet made my peace with the event and doubt if I ever shall. Whenever I return to the Cape, as I do regularly, and look down on Table Bay from the slopes of the mountain which gives it its name, it looks like a ghost harbour. It is still full of shipping but not the kind which made it, for me, one of the most exciting harbours in both the world and history. Unless there happens to be one of the rare cruise liners which treat the sea and its ports as sites for floating Butlin camps, I see only tanker after tanker, one more gigantic and unprepossessing than the other, interspersed with the proliferating classes of container ships, even uglier in line and general appearance than the greasy giant oilers.

The classical cargo ship, with its funnel amidships and tall masts fore and aft, is almost as rare as the cruising, pleasure-bound passenger ships, and about to vanish altogether. But what makes the scene so haunting is that the regular passenger ships which called in at the Cape for some four centuries and made the harbour a place of beauty, grace and colour every single day of the year, including Christmas and Sundays, have gone forever.

The passenger liner, bound on a distant mission and performing a regular service of great cultural value in a world scattered, separated and far-flung over seven seas, has ceased to exist. Its service on the run between the United Kingdom and the Cape lasted longest, as became a service that was the first. It tried desperately to maintain its regular if reduced function long after passenger services in the rest of the world had accepted defeat and vanished, but in the end it, too, had to surrender, and the surrender was fulfilled with an immense dignity and unforgettable grace by the *Windsor Castle* on its final voyage.

The absence of these ships is what makes Table Bay such an empty and reproachful scene for me now. The loss of the beauty itself is hard enough to bear because the ships of the line, of which the *Windsor Castle* was the last, were perhaps the most beautiful ever built. From the moment they first appeared in the blue South Atlantic water, they caught the imagination of all and held it attentive until the end. To this day the Cape-coloured people, who have an acute sense of the sea and an unfailing eye for what is appropriate on one of the most unpredictable and stormiest of oceans, still sing songs about ships of the past. One of the most haunting of their songs is a ballad about the confederate cruiser *Alabama* which was the *Emden* of the American Civil War and which will be remembered in music as long as ships retain their meaning in the imagination of living men. Another celebrates *The Scot* built in the nineties of the last century and one of the loveliest of ships, which represented the 'Union' part of the 'Castle' line and held the record for the fastest run between Britain and the Cape until the thirties of this century. Accordingly, the ships of the combined Union Castle line, which carried the Royal Mail, were treasured to the end.

The ships themselves were invariably of a clean and graceful design and always managed to be contemporary, incorporating all that was most modern in nautical evolution without breaking with what was best in the traditional architecture of the past. Moreover, no ships took to the seas half as well-dressed as they; their svelte hulls were of a lilac colour, just showing the red of their petticoats as the dancing sea of the Cape, excited by their beauty, swung them higher than perhaps it should have done. Their decks and masts were immaculately white and their funnels scarlet, with a band of black around them at the top to prevent stain of smoke and fume of oil and to hold them clean and clear against the sky.

Hard to bear as the loss of such beauty is, yet harder still is the knowledge that they were the end not merely of a line but of a long and significant era of history, leaving us deprived and unprepared on the hazardous frontier of an undiscovered world of mind and spirit and universe. That last voyage compelled me to look, as I had not previously looked, into the paradox implicit in the fact that someone born so far in the interior of Africa as I was should be so deeply concerned over the event, and prompted me to uncover in myself a sense of history and the sea which was in many ways not only a dominant factor in the evolution of my own life and spirit but of some importance for understanding what happened, is happening and is about to happen in our time.

I begin therefore in Africa and with this unusual sense of sea and ships. No matter how far we pushed north towards the unknown heart of the darkest of continents, this sense of the sea followed us about like our own shadows. Even when our creaking ox-wagons were probing the catchments of Crocodile, Elephant, Buffalo and Pafuri rivers, nearly two thousand miles from the Cape, it was so much in the minds of all those who led that march up-country that it compelled them to seek places with access to the sea and its traffic on which they had resolutely turned their backs. More significant still, when it was not a part of conscious reckoning, it was there deeply embedded as some reflex of the spirit which surprised everyone by the violence of its reaction when touched on by some external event such as the ending of the Union Castle line.

I say this with certainty, because in the isolated little Voortrekker community beyond the Orange (or Great) River, as the aboriginal Bushmen called it, which was a thousand and one miles from the sea itself, I and most of my friends had this compulsive interest in anything to do with the sea. Our interest attracted knowledge just as a magnet attracts minute filings of iron. The largest section of my father's immense library which was set apart with books for his many children, consisted of stories about the sea. *Robinson Crusoe, Masterman Ready, Typee* and *Moby Dick* were all there, also Dana's *Two Years before the Mast*, Bullen's *The Cruise of the Cachelot*, Kingston's *The Three Midshipmen* and its sequel. There were numerous parallels from Dutch and French literature. But one book in particular, Ballantyne's *Coral Island*, had been read and re-read so much by my elder brothers and had such a wide circulation among their friends, that when at last I, the youngest, was able to read it alone and unaided, the copy was nearly falling to pieces.

Usually I would read perched as high as possible on a branch of an enormous Imperial Chinese mulberry tree, a pagoda of Pekin green; high, wide and awesome. It stood at the centre of our garden and, securely hidden by its enormous leaves, I read *Coral Island* from cover to cover. I had to climb down many times from my exalted perch to retrieve loose pages which, in the excitement caused both by the story and by the discovery of a great new world of the imagination, had slipped through my fingers and fluttered down on to the purple mulberry-stained ground. I would look out over that crowded garden and see the blue of the day breaking over it in the long swell of the African sun, and see it as surf and foam in an ocean of light. So intense was this experience that it filled my senses, as did the swish of the sea in a large mother-of-pearl shell which we often held to our ears. A fancy more real than reality, that I was on some island in the unknown Pacific, would overcome me.

It would take the spire of the village church, and the cock on the weather vane crowing at the village below, to remind me that I was landlocked in Africa and some thousand miles from a sea I had not yet seen. Somehow, however, I knew that it was the sea and its role in the evolution of history which had produced so great a hold on me. This was confirmed when I came to read the sequel to *Coral Island*, called *The Gorilla Hunters*. In this book the same splendid threesome of Victorian manhood, Jack, Ralph and Peterkin, were sent out on yet more dangerous adventures in my beloved Africa. But for all that, the reading was for me tame by comparison with the first book simply because the element of the sea was lacking. It was the sea too which, in a version simplified for the young, established the *Odyssey* firmly at the centre of my emerging imagination. There, to this day, I see it as a dominant metaphor of the profound pattern of departure and return which encompasses the human lot on earth.

There were also the happenings on the long journeys between farm and village in the horse-drawn carriage reserved for my mother. She knew what a trial so long a journey in the heat and dust was for her younger children and so would invariably begin the journey by telling us a story, maintaining it with effortless art until she could bring it to an end just as we topped the pass in the line of red hills that encircled the village. From there we could see the church with houses huddled around it, in Anatole France's incomparable simile: 'like a hen and her chickens'. From beginning to end she told the story with a sense of the importance of detail, and also the authority of a fastidious memory which had been encouraged in her by the indigenous people of the land who had been nurses and companions to her from birth. These people lived without ever losing their central conviction that a story truly told is a kind of religious experience without which life itself is diminished in colour and meaning. Their primitive capacity for recollection, which, in the absence of pens, paper, notebooks and other civilised means of recording, was dedicated to precision and totality as a matter of survival, is still sustained by a passion to avoid error and serve truth to a degree undreamt of in our computer philosophies.

My mother's memory was not only of the stories, myths and legends of Africa but also of long and complex European novels. When in later years I came to read Dickens, who was a particular favourite of hers, I was amazed to discover how little of the essential story my mother had left out. But on the privileged occasions when I made the journey with her alone and was asked what kind of story I would like, I invariably chose one connected with the sea. I remember in particular an account of *Robinson Crusoe*. From the moment we set off I did not notice anything on the long journey until Crusoe was safely restored to the world and I was called back to what passed for reality by the fact that the horses' hooves were suddenly pounding the hard surface of the main street of the village itself. And the sound amplified by the white walls of the gabled houses which lined it was, I imagined, like the beat of a retreat by the marines at the end of the day on the afterdeck of a ship of the line.

I could multiply the examples almost indefinitely but it is enough perhaps to add only that as today other boys play trains and cars, aeroplanes and spacecraft, so in those days I played ships and made models of sailing vessels to launch on the dams and streams of the interior which, inevitably, like the craft in the fiction and history which had inspired me, suffered shipwreck in their turn.

There was confirmation of all this too in the impact of my first vision of real ships at sea. My father had taken me along with him to the Cape where he had to attend a gathering of the National statesmen summoned to deliberate the consequences of the act of Union of the four provinces which constituted the South Africa of today. I have forgotten almost everything about that fateful journey and the details of the crossroad experience of history with which it was concerned. The company of men like the ex-President of my native Orange Free State, Marthinus Theunis Steyn, and Generals de Wet and Herzog, and the subsequent visits to our hotel of Botha, Smuts, John Xavier Merriman and Abraham Fischer, made almost no impression on me.

I remember mainly the sea and my first view of it at midnight. Tired as I was by the long journey behind us, I persuaded my father to take me down to the foreshore. I was in the grip of a feeling of urgency that I cannot explain to this day. It was as if I had a personal appointment of the utmost immediacy and nothing, not even the sleep which I badly needed, should be allowed to make me late for it. I have no clear factual remembrances of the occasion but a great many feelings, most of them indescribable. The sea itself was at first invisible. One might not have known it was there were it not for the music with which it announced its presence. There was no wind, but some other power seemed continually to be shaking it and compelling it to move out of the darkness to break into a dazzle of white at our feet. Looking back on it all, it was perhaps the ideal introduction because that was how it was in my unformed mind: something coming out of the dark of the beginning before recorded time towards the "here and now" in which we find ourselves.

I cannot say that I found pleasure in that first meeting because it was so overwhelming. There seemed far more to it than physical sensation and excitement. There was something so awesome about it as to constitute almost an intimation, if not a revelation, of a super-natural kind. Particularly was this so when I looked up from the dazzle of foam at our feet and into the midnight sky of the Southern Hemisphere, packed and stacked with stars in a way that the northern sky can never be. I noticed how this white fringe of the South Atlantic seemed to correspond to the Milky Way, and its spume of unformed stars lacing the tidal wave of darkness arched high and unimpeded as it bore down on the Southern Cross.

All at that moment seemed but foam and spray of one immense great ocean of darkness, and the experience did not become tangible until the following day when I saw the ships lying in the dock and riding at anchor in the dark blue Roadstead of Table Bay. With the morning light over all that golden scene, they seemed to figure as in a sort of dream which comes only to those who are still innocent and at one with life. I was fearful at that first moment of daylight and reluctant to look away and up at my father who was telling me about the harbour, and pointing out the mail-ship *Balmoral Castle* as being the most important vessel among the shining concourses. I feared they might vanish in the seconds that it would take me to look up at him and back at the Bay. It all seemed too beautiful to be true and not only enchanted my eyes but made a kind of royal music on the wind of morning in my ears. Indeed, from that moment I took to waking up early in the dark, apprehensive that I might have imagined all that I had seen the day before, and that the next morning would expose the brilliant illusion. At first light I would be at the window of my room and, as the dawn came up, it was as if the Bushveld were on fire behind the navy blue Hottentot's Holland mountains to the north of Table Bay. I would watch some of that flame flicker and flare on the red funnels of a ship far out in the Bay, burnish the tips of the slender masts of another ship moored deeper in the shadows, and then turn into a gush of blood the wound of rust in the flank of a tramp-ship until finally the crowded maritime scene was illuminated and brought out of the darkness in coats of many colours and the whole vision of the evening before reconfirmed in every detail.

The great Cape itself, like the grit in the oyster on which the mystery of the pearl turns, was fixed firmly in my imagination at first sight and then transformed into a permanent point of growth for my love of the sea. My first view of it was towards evening from among the milk-bushes and dunes beyond Blueberg Strand. To look at Table Bay and Mountain from that point and at that hour was part of a solemn ritual of my father's life. He would always say it was the nearest he could get to reliving his first view of the Cape from the deck of a sailing ship some fifty-five years previously after a voyage of seventy-nine days from Europe. It had been visible like a smudge of cloud on the horizon of the blue sea on that morning so long ago. But then on a day made unbearably long, he emphasized, by an extreme hunger for land after so many confined days at sea, he and the whole crew had had to watch it acquiring substance with ponderous slowness until it stood in front of them tall and majestic, and much as we were seeing it at that moment. The view was more beautiful than anything the sailors had foreseen, and for him lifted the experience to an almost religious dimension.

The discovery that all this was an immediate part of my own family story, as well as the history of my native country, subsequently had a great deal to do with the impact of the experience. My impressions too were inflamed by the nature of that particular hour.

There is always for us who are truly born of the earth of Africa a kind of drama in the sunset hours, which no contrived theatre can equal. They are not merely remote impersonal moments confined to the external world. They are events joining the wheeling systems of a universe beyond even the light-year limits of our discerning, in a single totality because they correspond to a pattern within the human spirit, charged with a natural mythological import as of participation at that precise moment in some kind of cosmological sacrament. I remember that even my father's voice as he talked to me about it was not quite normal. It had something 'congregational' about it, like that of some devout being repeating his responses to an ancient lesson in the temple. So much of this indescribable flow of the experience came to dominate my imagination that my father, always so clear, near and dear in my memory, is fixed like a fly in amber in that strangely translucent moment; an image of some initiate contained in a world beyond any immediate knowing, and listening forever to some inaudible murmur of the universe made audible by the miracle of the hour.

When he did speak again it was to stress how uncanny it was that he could show me the view exactly as it had been that evening, fifty-five years before, when the ship was preparing to beat up and down the Roadstead stained with sunset so as to be ready for disembarkation at dawn the next day. Devil's Peak, which he pointed out to me on the left of the central mountain, the Lion's Head on its right, falling away into its rump at Signal Hill, they all stood just as boldly then as they did now, except that Table Mountain itself, contained as it was in its own halo of light, seemed perhaps higher, wider, more authoritative and imperious than it had been in the past. And so much did he appear to relive the hour that it was as if the years in between had become a weight of some kind which caused his shoulders to sag whenever he pointed out the scene. And that alarmed me. But the difference, he feared, was subjective, because there was still no doubt in his mind that the objective message was the same. The empire of the sea ended there and an imperative of land as great, of which the mountain was the plenipotentiary, began right there where it stood. His voice, still in the grip of the original emotion, deepened as he spoke, and indeed every inch of the mountain's height confirmed what its name implied – The Cape of Good Hope – of great and new good hope for all of them perhaps. He understood why his own father had referred to David the psalmist, who had once called on his own fatigued self in another land of promise to lift up his eyes to the hills from whence would come his help. How much those on board were about to need help from beyond themselves, and how unprepared they were for life in the world of which that Cape was ambassador, was illustrated the next morning, by something considered trivial and rather funny at the time.

They were watching numbers of tall, broad-shouldered, muscular, dark men wading out through the surf towards the ship's boats in which the passengers had been lowered, in order to carry them ashore in their arms. My father's youngest brother, seeing what was happening, cried out in terror at the black faces that he had never seen before, "Father, please don't deliver me to the devil!"

It was just as well to remember, my father reminded me, history being such an unpredictable and contradictory process, that not everyone who had seen the Cape in the past had thought of it in terms of hope. The first name bestowed on it by Europeans was the Cape of Storms. There was no doubt that where the Cape drove a wedge into the sea for some ninety miles between the Indian and Atlantic Oceans, the greatest storms of wind and water of which nature was capable were constantly born and reborn. Looking at the cold Atlantic, utterly resolved and at peace in the bay as if already curled up for sleep in the curve of the arm of land which kept it from the warm Indian Ocean only some thirty miles to the east, such duplicity seemed unbelievable.

But our farewell visit provided a disturbing confirmation of what my father meant. At first glance Table Bay was tranquil and full of light. But as the sun began to go down, the green-blue sky turned sulphur and suddenly, between it and the mountain, we noticed a strange kind of murk, half mist and half cloud, smoking over the sea towards the south before it fanned out and came racing towards us. A moment before, the long level light had so enveloped the mountain that it seemed not lit up from without but rather glowing from within like a lamp from its own flame. Transformed, too, was the flat marble top of the mountain itself which had led man to name it 'Table'.

My father had already made it clear to me how much he disapproved of the name. He thought it was an unpardonable euphemism, diminishing the mountain's reality and typical of the arrogant way in which Europeans had taken liberties with the essentially wild and pagan phenomenon of Africa. It was too good an opportunity for him to miss another homily on the absurdity of this European attempt to domesticate a great mountain by calling the advancing cloud its table-cloth. Impelled by immense forces, it was already fluttering in rags and tatters torn of itself over rocks and cliffs, as it pressed down on to the mountain. We had hardly observed this sudden change in the scene when we heard one of a group of Malay fishermen working on their nets nearby exclaim, "Allah! World! People! Look at that old mountain, thinking up his own weather again!"

After the untroubled calm and sense of magisterial benevolence which had been my experience until then, the transformation was frightening. For me the Cape, in an instant, had become a living and vengeful gigantic presence. The impression went so deep that when, many years later, I was to encounter its personification as the resentful Titan, Adamastor, in the *Lusiad* of Luis de Camões, I instantly understood what had inspired him.

By the time we arrived back at our hotel a south-east gale was racing down the face of the mountain, pouring through the spaces between Devil's Peak and the Lion's Head, and testing every anchor of that line of ships that we had seen riding so confidently in the darkening Roadstead as they waited for room and shelter in the crowded dock at the far end of the Bay. Before long it turned into one of the most fearful of south-easters ever unleashed over the Town and Bay. We heard its music riding high and wild all through the night. The thick walls of the building shook, shuddered and at times felt as if they were reeling like a ship under the blows of the wind. Such sleep as I had was troubled perhaps by the first inkling that life was not made with the clear-cut simplicity for which one longed, but that the reality might be profoundly ambivalent and the manifest appearance only skin deep.

There is not another cape that has so dominated not only my own but also, I believe, the imagination of the world. Alone among the great capes of the world, the Cape of Good Hope seems to have been dressed to perfection by both earth and fate for the two-sided reality that it was destined to play in our history: at one moment an image of hope, at another the ultimate symbol of storm.

When the great Bartholomew Diaz returned to Lisbon in 1488, the few survivors among the many who had accompanied him on that brave voyage down the west coast of Africa gave such a harrowing account of the storms they had encountered off the Cape, and how they had beaten about and fought seas and winds for thirteen days before they discovered that they had rounded the southern-most point of Africa and were at last heading north, that they induced a vast majority of the Portuguese to think of it as they did, the Cape of Storms. But Diaz himself, who had in vain tried to make them sail on further north to what he was certain was a new way to the East, added to the ambivalence when on reflection he called it the Cape of Good Hope. Columbus himself was present at the court in Lisbon when Diaz later came to report to his King, and there is to this day a copy of the Imago Mundi with a note in Columbus's own hand, recording that Bartholomew Diaz on that occasion referred to it as the Cape of Good Hope.

His discovery was the culmination of more than a century of endeavour by the Portuguese to find an alternative route to the East. Columbus and many others had set sail with much the same intent to probe the unknown oceans westward towards the Americas. Historically, the explanation has always been a purely rational one, that it was lust after the spice, silk and treasure which the great city states of the Mediterranean like Genoa and Venice had for so long monopolised. For me, however, there has always seemed to be more to it than this plausible one-dimensional explanation: it was part of one of the most significant compulsions of the Renaissance spirit.

If it had been merely a matter of materialistic greed, I do not believe that the sailors of Spain, Portugal and the rest of Europe who soon followed would have overcome the superstitions of centuries and ventured out of sight of land and sail over great oceans which had been an inviolate frontier of the western world for so long. Laughable as superstitions may always appear in hindsight, they were powerful enough in an age when only the esoteric few believed that it was the earth which moved around the sun and not the other way round. The rest believed firmly that the earth itself was flat and stood on pillars, as not only many mythologies emphasized but also one of the most famous psalms in the Bible. Convinced as they were that the earth was flat, they had no doubt that if one sailed far enough one would come to the edge of the world and drop off into a bottomless abyss, inhabited by unimaginable monsters and evil spirits. They felt this not only about the seas but also about the land which surrounded them.

For centuries no-one living among the Alps of Europe ever dreamed of climbing them out of fear of the supernatural forces which their beliefs invested in their summits. Indeed C. G. Jung, one of the greatest explorers of the European spirit in a truly twentieth-century way, told me that if one wanted to fix a precise moment at which the Renaissance began, it would be the day when the Italian poet Petrarch decided to defy all superstition and climb a mountain in the Alps, just for the sake of reaching its forbidden summit. Perhaps this was therefore the first tangible evidence of a profound shift in European spirit, the first sign that, after centuries of introspection, the imagination of the West was swinging over into another principle of itself, and beginning to focus on an external world for too long underrated.

Accordingly this eruption of a passion for exploring the unknown world beyond the seas took on a far greater meaning for me than the plausible explanation which conventional history allowed. It had always seemed axiomatic to me that faith, in the sense of an intimation of evidence of things not seen and yet to come, can only be replaced by faith of a greater and more contemporary kind. Somehow in this profound shift of focus in the area of the human spirit where faith arises, greed found the power great enough to demolish the impediments which had confined the imagination of the West for over a thousand years in an overwhelming concern for the world within.

In the part of Africa where I lived, concepts such as whether the earth was round and rotated round the sun or not were still subjects of passionate argument and sometimes violent debate at school and in our churches. My father, who had twice heard the President take an uncompromising stand on this issue against distinguished international personalities, had often spoken rather whimsically about these occasions. One was the meeting with Mark Twain who, because of this, subsequently had much characteristic fun at Kruger's expense. The other was a meeting between the President and the lone American sailor, Joshua Slocombe who, I believe, was the first man to travel single-handed round the world. On his arrival at Port Natal, Slocombe took time off to call on Kruger, the most accessible of Presidents. Asked what he was doing, Slocombe answered innocently enough that he was sailing around the world. To his amazement he found himself instantly and sternly rebuked for making so erroneous and blasphemous a claim.

My father's account of those two incidents was a source of great amusement in our family circle. But outside I knew it was no laughing matter and indeed was so dangerously controversial a subject that I soon learned it was best left to our teachers of geography and physics. It was rumoured that Kipling wrote 'The Village that voted the Earth was Flat' after a brief visit to us during the Boer War.

As a result, from an early age I was receptive to the achievement of the great Spanish and Portuguese sailors who had discovered our Cape and the Americas. What started as an instinctive feeling became in time an informed conviction that, although material gain played a significant part in the matter, a far more powerful urge of the human spirit had come to the fore. This urge was for me the real heart of the great reawakening of Western man that we call the Renaissance. The great new activity of man and ships demonstrated that there can be no thrust into what is unknown in the external world without being accompanied, if not preceded, by an equal and opposite thrust into the world of the spirit.

I do not pretend that any of this was consciously present in my mind then. It was more an instinctive concern for the meaning of history, sustained by the voyage ten years after Diaz of perhaps the greatest and most ruthless Portuguese sailor of all: Vasco da Gama. Following in the wake of Diaz, he rounded the Cape in the calmest of seas, seeing it, I believe, much as I had first seen it myself with my father in that translucent, sunset moment, locked within the stream of my memory like a tranquil lagoon. The feeling of hope induced in da Gama by the strangely passive sea of the Cape was reinforced by his progress north and up along the inhospitable and harbour-less coast of Africa to Kilindini, which is Mombasa today, and thence to India and the Far East.

This achievement, which was a breakthrough as much in the mind as the physical world, was enough for all the world to take it at face value and to confirm until this day the vision of it as a Cape of Good Hope. However, when I came to learn at school that the Portuguese word for storms was 'tormentos', another emphasis was added to my own private preference and the sense of the ambivalence of experience in which it was rooted. Tormentos suggested both storm and torment to me. As a result I came to amend my own name for the Cape by borrowing from Camões and thinking of it always as a 'far-flung and much tormented Cape'.

The sense of ambivalence that I felt about the Cape seems to have been shared by the sailors who faced the immense hazards of the voyages between Europe and the Far East. The hope of gain and riches, hitherto beyond the dreams of a poor country, without doubt sustained Portugal in reinforcing an increasingly disastrous traffic of ships to the Far East. But the sailors, soldiers, administrators and their families who lost their lives in unbelievable numbers through shipwreck and malnutrition – even the survivors could never participate to any appreciable extent in the gain that was expected with such high fantasy – must have been motivated by something else or they could not possibly have endured and perished as they did. My own hunch was that as ordinary people they lived not by ideas so much as by an instinct that flesh and blood was created precisely in order to serve this other meaning of history that I have mentioned. Much as those in command of the world exploited the energies of simple men and women who were motivated by an instinctive belief that they were promoting a continuation of purposeful history and, above all, that the society and culture to which they belonged was an expression of great meaning, there was nothing that they could not endure. For instance, I was convinced that the people who conducted the traffic by sea between Europe and the Far East, from the greatest captains to the most humble sailor, could not have done what they did if they were not possessed by the feeling that somehow they were serving the ultimate reality they called God.

It was significant that almost all the Portuguese ships had religious names, unlike their English, Dutch and French counterparts, who were soon to follow and add a ruthless, bloody rivalry to hazards of the sea. In some four hundred years of sailing between Portugal and the Far East, only one hundred and forty-one out of one thousand, three hundred and forty-four of their greatest ships, and many lesser ones, had names devoid of religious associations. Wherever they landed and established ports, trading centres and ultimately colonies in Africa and, above all, India and the Far East, they put the conversion of the indigenous peoples to Christianity before trade and profit. In this they displayed a fanaticism of faith that was both great and terrible. Were it not for the horrors of the two World Wars, and particularly the concentration camps and gas chambers of Germany, one would be tempted to deny any claim of humanity in the Portuguese of that day. Vasco da Gama himself thought nothing of setting fire to the ships he encountered in Far Eastern waters and then watching them burn, without compassion for the women who thronged the burning decks and held up their babies and children in an appeal for mercy. At Calicut he cut off heads, hands, ears and noses of eight hundred prisoners before throwing them in the hold of a ship which was allowed to drift ashore while all night long, the historians tell us, the beaches were crowded with weeping men and women searching for corpses of murdered friends and relations. And Vasco da Gama was by all accounts the personification of what passed for a Portuguese gentleman and nobleman in attendance at Court. For those considerations he was especially chosen to lead the first Portuguese venture into the forbidden and unknown world beyond the Cape. Pedro Cabral, Francisco d'Almeida and Alfonso d'Albuquerque, also nobles and viceroys who followed da Gama, were just as cruel and militantly zealous in their pursuit of what they regarded as their Christian duty to convert the unbelievers of the Far Eastern world. The editor of the account of the voyages of the great French sailor/explorer, François Pyrard, tells us that the outstanding characteristic of the Portuguese during the first fifty years of their penetration into Asia was their fanaticism. Indeed, Pyrard remarks: "Throughout the whole period the religion of Christ was invoked for the sanction of acts of cruelty and oppression hardly surpassed even in the annals of the East".

There were, it is true, great and honourable exceptions, particularly among the Jesuits, who were the most zealous of missionaries. The outstanding example of course was St Francis Xavier. By force of his example and passion for the cause of his Church, he alone established Christianity in Goa and the south of India so firmly that Roman Catholicism to this day is still a significant reality in the lives of millions, despite the teeming Hindu context of their world.

Even more impressive was Francis Xavier's achievement in Japan. He converted the Japanese to Christianity in such numbers that the Tokugawa Shogun who ruled the country banished Xavier's successors and all the Portuguese from the country. It is interesting how different was the behaviour of their successors and rivals, the Dutch, who persuaded the Shoguns to let them take the place of the over-zealous Portuguese Christians. Greed alone seems to have been their motive, as their murdering of the English traders in the island of Amboina in 1623 had already suggested. It was a passion almost as ruthless as the religious obsessions of the early Portuguese. The Dutch established what they called a 'factory' at Doshima, an island off Nagasaki, and were allowed to send a strictly limited number of ships to Japan, but on condition that, twice a year, the commander of the factory and his principal officers allowed themselves to be paraded through the streets of Yedo (as Tokyo was then called) dressed like clowns and other comic figures and generally subjecting themselves to public humiliation. Above all, once a year they had to trample on a wooden cross in front of a Japanese official. Yet for centuries the material gain was thought well worth the depravity implied in the process so totally without reservation that Toynbee in his *Study of History* comments that the reader of the story "is left at a loss between admiration for the tenacity and disgust at the servility with which these Nordic Protestant occidentals held their ground and made their money".

Coming as I do from Dutch and French Huguenot ancestors in almost equal proportions, this aspect of history in which the Cape of Good Hope was to play so crucial a part became a salutary corrective to any tendency towards excessive nationalism that I may have had. It also allowed me to view the Portuguese excesses in the context of a period when for example the tortures of the Inquisition could be accepted without any of the horror which would be felt today. At the same time I was able to picture the heart-rending sight of Portuguese Armadas setting sail from Lisbon with, so the historians tell us, vast crowds of weeping townsmen of all classes led by priests carrying crucifixes, following the sailors down to the shore to pray for their safe return.

No wonder Lisbon five centuries ago had coined its own proverb for these occasions, "Deos as leva, Deos as traz" (God carries them out, God brings them back). For them it was clearly in every way more of an affair with God in the hands of God than a vulgar commercial transaction. Already in Diaz's day Portugal was singing a form of Fado, a 'song of fate' which remains a popular art form to this day, hinting at all those other greater non-rational and subliminal urges in the Renaissance spirit which ultimately tend to redeem somewhat the increasingly sordid affair between Portugal and Africa and the Far East. It went something as follows:

> *I saw, my mother, the ships,*
> *sail into the sea,*
> *and I died of love.*
>
> *I went, mother, to see*
> *the ships on the Strand*
> *and I died of love.*
>
> *The ships on the sea*
> *for them I went to wait*
> *and I died of love.*
>
> *For them I went to wait*
> *and I found I could not*
> *and I died of love.*

Shipwreck, almost unendurable hardship at sea, and the constant and mysterious disappearance of vessels became so normal a part of Portuguese experience that it inspired a special literature of its own. Ordinary Portuguese men and women had their imagination so inflamed by what was increasingly a national horror story that they acquired an insatiable appetite, not just for factual records of what happened at sea but for fiction about the sea, ships and the men who sailed in them. It was called Literatura de Cordel, loosely translated as 'string literature'.

It was given this name because so many stories of this kind came from the pens of popular Portuguese writers that they were rushed into print in a glorified pamphlet form and displayed all over Lisbon, strung up on string and hung up outside shops like some new sort of salami of the imagination, pre-cut for instant consumption. This phenomenon, as much as any other, illustrates what a platitude of everyday existence tragedy at sea was in the life of post-Renaissance Europe, and how the storms which Shakespeare used in *Pericles* and most of all in the greatest of his plays, *The Tempest*, were not products of far-fetched imagination but valid and even commonplace occurrences of the times. As if this were not tragedy enough, disease, malnutrition and general hardship made the already hazardous voyages seem as if those fleets were visited by all the plagues of Egypt. One Portuguese flag ship, for example, lost nine hundred persons out of eleven hundred men, women and children. Even at the end of the sixteenth century, after a hundred years of experience, forty-seven percent of the ships which left Lisbon never returned. One historian commented that if tombstones could be placed in the sea to mark the places of all the dead thrown overboard, the route from Lisbon to Goa would look like one vast graveyard.

Even the pragmatic Dutch, following on the Portuguese and with access to a century of their experience, suffered greatly from the same illnesses and causes of malnutrition. When their first fleet of four ships sailed from Holland for the Far East in April 1595, the voyage to the Cape took even that efficient fleet four months. But by then so many had died that it was necessary to burn one of their remaining ships and distribute the crew members between the remaining three ships. Similar misfortunes continued to cause losses far too great for the liking of the Lords Seventeen who came to rule over the Dutch East India Company, the infamous John Company of our popular and not unblemished history. As their ships continued to arrive at various ports of call with such water as was left already turned into a kind of porridge of worms, and with the last biscuits more weevil than cereal, they too searched desperately for a remedy. In the process they could not fail to notice that the Portuguese by this time had partially solved the problem by maintaining firm bases at Luanda, Mozambique and Mombasa. Accordingly the Dutch tried to take Mozambique by force and, when they failed, took more notice of the fact that their loss of life might have been even greater had they completely shunned the Cape as the Portuguese had done.

In their far more extravert and pragmatic approach, the difference now made by the replenishment of supplies picked up during improvised calls at the Cape from time to time, was significant. From then on, calling in at the Cape became a regular feature, so that it was used even as a kind of post office. Outward-bound ships left letters and reports, carefully wrapped against damp, under great stones that were inscribed to serve as letter boxes where they could be collected by homeward-bound vessels, and new reports left.

It took the wreck in Table Bay of just one ship, the *Haarlem*, to bring about a solution to the problems caused by storms and malnutrition. Those of the officers and crew of the *Haarlem* who managed to get ashore not only survived at the Cape for nearly a year until they were picked up by a fleet on its way back to Holland in 1648, but they managed to make gardens near streams of fresh water and grow vegetables of all kinds from seeds they had brought ashore. They grew pumpkins, water melons, cabbages, carrots, radishes, turnips, onions and garlic and, more significantly, they established some relationship of trust with the copper-coloured and cattle-owning Hottentots of the Cape.

Up to that time the Hottentots had acquired a reputation for treachery and militant brutality so great that the Portuguese who should, by natural right, have been the first to explore this potential of the Cape, had dismissed it as a port of call and base for the future. This had begun with the murder at the Cape by Hottentots of one of Portugal's viceroys, Francesco d'Almeida, together with many of the nobles and soldiers in his company. But the survivors of the *Haarlem* had proved also that the antagonism could be overcome and so finally prompted the East India Company to build a fort at Table Bay in 1652 and to establish a permanent base for supplies and fresh water. This was the moment when the concept of our role in life in Africa as that of halfway house to the East was born, notwithstanding the Cape's modest beginnings as a glorified kitchen garden. None of the Lords Seventeen in Holland ever wanted it to become a colony. But they were soon to find, as men still do everywhere, that when they underrate the submerged and non-rational motivations of the human spirit, which the ambivalent Cape came to symbolise for me, supposedly new situations quickly develop a mind and will and character of their own.

The first gardens had hardly been planted, the vineyards laid down, the fort they called a castle built, when a number of men appointed to work and manage the gardens, as well as obtain cattle and sheep from the Hottentots, rebelled against the Dutch company. One fine day they walked out of their contracted roles and insisted and ultimately succeeded, in spite or resistance from governors and the Lords Seventeen, to set up in the market gardening and farming business on their own account. Before long these men, against all restrictions and discouragement, indeed against all reason, started to push steadily inland. However much both the Dutch in Holland and their British successors were to try and discourage this thrust into the remote and harsh interior of Africa, the great and ever-increasing pull towards the heart of the country was too much for them. The fact that men whom Holland and Britain had tried to hold back had committed their lives to Africa, and that their children and future generations were to be born in Africa, subtly transformed them into something new which not only had never been in the reckoning of those who brought them there, but also transcended their own conscious intentions. However much they acknowledged their debt to Europe and continued to think of themselves as of Europe, insisting always, as they do even today, that they are still European promoting European civilisation in a dark, pagan continent, they were transformed into something which was neither totally European nor totally of Africa but had undergone, as it were, a deep-sea change into something new and strange. And this something came to have a dynamic of its own which has yet to be determined and defined, and given a contemporary expression which also would have to be valid for all the yellow, copper, chocolate and remarkable black peoples of southern Africa who were drawn into their lives by this strange and irresistible march up-country.

Meetings between different cultures, particularly of so-called civilised and so-called primitive men, are events of the most traumatic and fateful consequences, as immense as they must be unforeseen end irreversible. Neither 'civilised' nor 'primitive' men can ever be the same again once they have met one another and started the process of being a part of one another's daily lives. No matter how much the exponents of one culture would like to limit participation, interpenetration or even partnership with other cultures, these are matters to which no human or conscious limits can be set. Those who led the Great Trek of the descendants of the first kitchen gardeners deeper and deeper into the interior, like the members of my mother's family who have taken a leading part in these events for nearly three hundred years, far from leaving their problems behind soon began to discover, as they are still discovering today, the appalling responsibilities imposed on them by that immense, new human encounter in southern Africa which was brought about in so apparently nonchalant a manner by the founders of the first settlement at the Cape.

And it was not in Africa alone that similar irreversible processes were set in motion in the human spirit. As a result of the movement started by the Portuguese, the Spaniards, Dutch, French and English all moved along the coasts of Africa and into the remote interior of the continent. There was the expedition of the brother of Vasco da Gama, hastening to help the Emperor of Abyssinia in his war against the Muslims. And it continued even into India, Ceylon, Indonesia, South-East Asia, the Philippines, Formosa, China and Japan. Millions of human beings of contrasting and conflicting cultures were inflicted with the seeds of great unrest. This was due to a stark necessity now suddenly imposed on them, namely to change and renew themselves and the world which they inhabited. It was to be the same in the Americas and Australia and, as this empire of the West expanded over the centuries for the first time in the history of mankind, the whole of the world was caught fast in this over-riding necessity which they could deny at their mortal peril, as is so amply witnessed and confirmed by the social unrest, the wars and the impetuous withdrawal from Asia, the Pacific, Africa, Caribbean and New World by European powers themselves confused and shaken by the encounter.

There is proof enough for me therefore in history that wherever we went in Africa from 1652 onwards, it was as if an invisible Cape presided over our imagination and destiny. Whether as a cape of storms or of hope, or for that matter both, there was total agreement on both the conscious and the instinctive levels that we owe our origin at that Cape to its being the perfect half-way house between East and West – not only in the physical world but also in whatever those two opposites represent in the world of the spirit. For me this made our own little history potentially more meaningful than most, and out of imponderables such as these was born the improbable conviction that sometime my own life would not be content with a provisional, half-way condition, but would insist on completing, as best it could, the whole of the journey implied in our unique beginning.

CHAPTER TWO

Return to the Sea

ONE of the disadvantages of being the thirteenth of a family of fifteen children was that I arrived on the stage after many of the principal actors had already made their exits. I was thus too late to know my father's father and his French wife and to hear her sing in what was, by common family consent, the most beautiful of voices which had delighted the Court in Holland. However, much as I loved music, what I regretted perhaps most of all was that I could not hear my grandfather himself add to the account of the voyage which my father had given me during those moments of ritual I had been privileged to share with him at sunset on the beach of Blueberg Strand.

I had accordingly to depend on other experiences to support the reality of this immense but indefinable sense of the sea I seemed to have had from birth. One of the first of such experiences was the departure of an elder brother for further study in Holland and America. This became a subject of prolonged and anxious deliberations in the family circle. I was amazed by the knowledge of the ships that sailed between South Africa and Europe which came to the surface in the ensuing process of argument and counter-argument. This took place not only with my father, mother, elder brothers and sisters, but even with my mother's father, who had never even seen the sea. He was a member of a family which had always been in the forefront of the strange, irresistible movement from the Cape deeper and deeper into the interior. He had grown up with only the great African 'sea of land' around him. Yet it now appeared that he too had the keenest interest, as well as the least flexible views, on the whole matter. Between them all they seemed to know intimately the ships and their performances; all about the character and reputation of the men who commanded them, and the companies who owned them. This was perhaps not so surprising because I learned that, ever since the beginning of traffic between Europe and the Cape by sea, the names and records of the captains were matters of extreme personal and private concern to almost everyone in southern Africa. As time went on their names became household words. In the Cape itself, the growing numbers of the spirited communities of coloured peoples, with their own spice of Malay, Indian Pakistani and Hindu ingredients, also had their own direct and long connections with the sea and would name houses and children after ships and captains who had most excited their imagination.

But it had never occurred to me that in our isolated community my own family's knowledge of these things was almost as great. Once the comparisons had been made between the various ships and lines, it was unanimously decided that my brother could not possibly sail in a ship of any line but the Union Castle Company, whose Royal Mail ships conducted, among other enterprises, a regular weekly service between the Cape and Southampton. One of their ships, the *Balmoral Castle*, was already prominent in my imagination: it was like a lighthouse to me, because this was the first mailship which my father had pointed out to me from our hotel window overlooking the crowded maritime scene in Table Bay.

Although the ships which had to perform this service for the family were never in doubt, some of their masters were. I was startled to hear that one ship had to be dismissed because its captain was reputed to like his liquor too much; another because he was so determined to maintain the run of his ships to time that he shortened the length of his voyage by cutting corners, and it was whispered that he often scraped the bottom of his ships in the most dangerous fashion. "Et ça finira mal", my father would remark, quoting a favourite expression of his mother. It always sounded most ominous to me, being in incomprehensible French. Sooner or later, my father maintained, he was going to run his ship aground.

Another captain was dismissed because he was held to be too casual and did not maintain sufficient discipline among his officers and crew. Another, and this was held to be perhaps the most serious disability of all, was reputed to be an atheist who was bound, sooner or later, to incur the displeasure of the Almighty at sea. This was an argument of which my devout grandfather was the keenest exponent. Yet another captain was held to be too fond of socializing and flirting with the ladies; dancing and playing cards in lounges and smoke-rooms when he should have been doing his duties on the bridge. Only one ship was dismissed because of suspect sea-keeping qualities. It was said to roll dangerously in bad weather, and, ever since the disappearance of the *Waratah*, there was a fear deeply embedded in the minds' of all who might have to face a journey by sea. Finally it was decided that a ship called the *Kenilworth Castle*, and its captain, had all the appropriate qualities to carry my brother safely to his destination.

But, even so, such was our family's respect for the irrevocable and unpredictable nature of voyages by sea that, on the Sunday after his departure, prayers were said for his safety in our local church, much to my own embarrassment. In my view it was quite unnecessary, because I never doubted that the *Kenilworth Castle* would perform its duty with the utmost despatch and safety, because it happened to be a sister-ship of the *Balmoral Castle*, already immortal in memory. Yet these prayers continued every Sunday until, at last, a cable arrived to say that the ship had reached its destination without mishap.

Other examples of how the sea continued to excite our interest were provided by my grandfather himself. I was staying with him once at his farm. This was the first of all European farms to be established north of the Orange River. Its pioneering pre-eminence was always confirmed for me by the evocative name bestowed on the farm – Boesmansfontein – the Spring of the Bushman. The Bushmen, of course, were the unique original Stone Age inhabitants of my native country, and ever since my childhood they have continued to play an important role in my own evolution.

I remember vividly the day when my grandfather's favourite monthly magazine arrived with the weekly mail. It was printed in high Dutch and was called *The Church Messenger*. It had a bright yellow cover which I found repulsive with its title and sub-titles printed in a massive and inelegant black type. But it did have one virtue in my young eyes – its illustrations. In this particular issue there was one of what I thought to be the most beautiful ships I had ever seen. However, to my amazement, my grandfather seemed to be far from impressed by the beauty of the vessel. Pointing at it with a finger trembling with indignation and speaking in a voice which sounded to me like an Old Testament prophet, he resounded, "Laurens Jan, do you know what they say about this ship you are looking at? They say it is unsinkable. And mark my words, the Almighty will have something to say one day about so arrogant and blasphemous a boast."

This observation was typical of him. It provided an illuminating glimpse into the devout nature of his upbringing and character. There was nothing that his own life and the love of his family and ancestors had taught him to fear more than lack of humility and respect for the power and glory of his God. Young as I was, I knew that he had already found himself on occasions painfully wanting in his own lack of the 'fear' of God required by the Old Testament that he loved. I knew from my mother, who told us the story as a mark of the devout spirit of our grandfather, that on the birth of his first son he had him baptized in his own name. Three months later this son died. A second son was christened likewise after him. Not long after the second son died as well. A third son suffered the same fate. My grandfather decided that clearly these were signs that he had sinned against the Holy Spirit, and, after months of contrite and sombre self-examination, he concluded that he was egotistically trying to secure some kind of vicarious immortality for himself by imposing his own will and personal predilections upon the workings of Providence. Accordingly, when the fourth son was born, he gave him names totally unconnected with him, his family or his ancestors. And his son lived. Sadly, this son too came to a tragic end, for he was killed fighting against the British with my father in the Boer War.

To return, however, to the afternoon of the arrival of the *Church Messenger* and to my grandfather's rejection of the arrogance of men who could claim that they had built an unsinkable ship. The lesson learnt was to go deeper when, in a later edition of the magazine, the same illustration was reproduced with the news that the ship had sunk after a collision with an iceberg in the North Atlantic, together with a monstrous loss of life. The name of the ship, of course, was the *Titanic*.

There was yet another reference to the sea among the many made by my grandfather which I was to remember when reading an account of the devastation caused, not only at sea, but among the vineyards of the Cape, by a black south-easter; the rarest and most violent of the winds produced by the mountainous Cape and the sea between them. The report, in the manner of those days, had allowed itself considerable license in the description of the power of the winds and the sea, and in particular the height of the waves hurled against the shore. My grandfather read the report very slowly, because he had been born in an ox-wagon; his youth had been spent on trek with his parents, and so he had never had any formal schooling. Such education as he had received had been from his own parents, and almost his only text-book had been the Bible. I remember many occasions on which he had to compose complicated letters to officials about his large farm. My mother, who loved him very much and understood the difficulties far better than we could, would command us, "Children, do be silent. Can't you see that grandfather is writing a letter?"

At the end of this long report, reading by the light of a large oil-lamp, he looked up and away as if seeing through the windows in the dark night something not visible to the eyes of ordinary men. He looked for so long that I was sure he had totally forgotten me, and his promise that night to tell me more of his own life as a boy. In the silence I remember how unusually loud and portentous was the sound of the bleating of his sheep in the kraals nearby, the lowing of the cattle similarly protected against the wild animals, the call of the night plover, the 'commando' birds as we called them. Finally there was the mournful bark of some jackal setting out on its prowl for food through the night. But my grandfather had not forgotten me. He had merely been absorbed in some profound picture evoked by the news which he had just read and the inevitable religious conclusion to which he had been led.

"Laurens Jan", he told me at last, "I know that we all are always in the hollow of the hand of the Lord, but it must be awful to feel his fingers moving underneath one as they seem to do at sea."

From that time on my own personal relationship with the sea became rooted less in books and history and more in my own experience of external realities through our vacations on the coast. There I made myself most unpopular because I preferred to be taken to the docks and shipyards to see the ships riding at anchor or lying in their berths along the quay, rather than going to collect shells and play on the shore as others did.

But, alas, this period in my life did not last long because, when my father died, almost on the day of the outbreak of the First World War, we found ourselves unexpectedly deprived of ready money. We had land, sheep and cattle in plenty; in various parts of the country we owned more than five hundred thousand acres. We were almost self-sufficient in food, clothes, leather for veld-shoes, candles and soap, and seemed to have very little need of buying what the world outside had to offer, except 'luxuries' like sugar and matches, and the sort of things needed only for weddings and funerals. But we were utterly dependent on my father for money. He had had one of the largest legal practices in the country, so he had earned substantially. Alas, royalties from his books and the money he earned in the course of his public life had been ours. But this source of wealth suddenly vanished and there was no means of replacing it in great measure. Because we lived so far away from the sea, the bales of wool, the skins, bones and even the wheat and maize despatched to the coast were sold there for little more than it had cost for their transport.

As a result we had only one more vacation by the sea, a memorable one for me because it provided another demonstration of the importance of the Cape to the sea, and of both to the world at large. I remember on that holiday the harbour crowded with ships as I had never seen it before: ships which at morning and sunset seemed to be encircled in a halo of gold. Most of them were troop-ships carrying men, many of whom were my relations, who were afraid that they would not get to Europe in time for the war and would arrive to find the fighting over. Everyone around me, except perhaps the weeping women and children waving these ships out to sea, was obsessed with feelings of romance about war. It made no difference, as yet, that casualty lists were growing longer by the day, and that more realistic accounts of the terrible losses experienced all over Western Europe and the Middle East were beginning to circulate in the newspapers. In the two months we spent by the sea it was noticeable how the traffic of ships carrying troops from India, Australia, New Zealand and other parts of the British Empire, of which we had become a part, increased by the week. Also there were the hospital and other general ships, so that one outstanding fact of this general romanticized interest in the process of the war soon became apparent to me. It was simply that, concerned as everyone was with the fighting on land, it was the war at sea that really mattered most to us all.

Proof of this, as well as the first indication that war was no longer the romantic affair with fate which people in the beginning had believed it to be, became apparent through something which occurred in the war at sea. I was in Cape Town the day the news hit the city: the sinking of the great Cunard liner, the *Lusitania*, a name which already had such important associations for me with the Portuguese, who had discovered the Cape, and which Camões had taken as the title of his epic. It was the first sinking of a passenger ship by a German submarine, and was the signal to the world that war was never again to be just an affair between soldiers and other professionals, but was a grim matter of killing as many people as possible, including civilians, and women and children. The sense of outrage produced among the people of the Cape was terrifying. Those kindly, hospitable and respectable people of a community which regarded itself as the most civilised in the land, went on a mindless, indiscriminate rampage. Riots broke out spontaneously all over the Cape, and the worthy citizens set about smashing the windows and setting fire to the stores and property of anyone of German origin, indeed of everyone suspected merely of having German connections. No lives, I believe, were lost. But the destruction was on such a scale that when calm returned, a sense of shame afflicted the whole community, and I was convinced that it was the fact that it had happened at sea and not on land which had roused such people to behave in a manner of which even wild animals would be incapable.

What made this message even more pertinent and poignant was that during my vacation I had formed a friendship among the children whom I met along the beach with a young German boy of German parents. The sinking of the *Lusitania* caused the South African Government to react vigorously and with indecent haste. It rounded up the German community all over the country. I had to watch my new young friend, himself totally unaware of the causes of the sudden new hostility against him and his family, being escorted by the police to the station and put on a train for an internment camp up-country. Whether he himself shared the internment of his parents I was never to know. My last picture of him was looking infinitely forlorn and desperately over his shoulder at me as he was led off, too stunned to respond to my frantic signals of farewell. After our own family returned to the interior I was never to hear of him or his parents again. From that moment on our annual visits to the coast came to an end. I was not to see the sea again until the age of seventeen, when I left school to go into the world to find work.

One of the most marked characteristics of my own countrymen in the interior was their love of learning and respect for education. Their main aim was to give their children as much as possible of the best schooling available, which the necessities of a pioneering life had denied them, even at elementary levels. My own family was typical in this respect. Despite the loss of income caused by my father's death, my mother had somehow managed to continue to send all her children to the best

available schools, and from schools to universities. It was assumed that
I would continue in the same way. But although I had enjoyed myself
enormously at school, my last two years there had seemed to me to be
an increasing waste of time. These years had continued to give me
much of lasting value, through personal relationships with other boys
and masters, but in no other way administered to what I was convinced
I sought. I felt it as singularly pointless to go on studying Jane Austen's
Pride and Prejudice for two years when I had read it with fascination in a
single night. Apart from the scientific subjects that I was compelled to
take and for which I uncovered a totally unsuspected liking, particularly in mathematics and physics, I took so little interest in what
happened in classes that I barely passed my school leaving examina-
tion. Indeed, if I had not been good at games, which were a source of
great prestige, I might have got into serious trouble.

I did not know precisely what I was going to do, although I always
knew I wanted to write. I had already written a great deal of poetry in
the High Dutch which was taught at my school and also spoken on all
official, formal and religious occasions. But I also knew that, although
writing was my principal aim, I did not want to be *just* a writer.
Somewhere in my nature there was something at work akin to what I
came to call an 'Elizabethan pattern of purpose'. Often at night or even
day-dreaming in classes I had visions of myself going to sea. But one
thing was clear in my mind: I was not going to continue this
nonsensical process of an elongated slow-motion education by going to
university. It was an immense shock to my family when I told them
this. I was the first of my family to have taken such a line and became
the only one not to go to university. This particularly distressed my
mother. I remember her saying to me rather sadly that she would not
force me to do anything I did not want to do, but I had to realise that
the choice was clear: I had to go to university or out into the world. So
into the world I went.

The manner of my going was sudden, uncalculated and unforeseen.
At the time it looked like a haphazard product of chance. Yet I was to
discover as the years went by that chance seemed to know more about
my own inner needs than I did myself. I had not been at home on the
farm for long, yet it felt like a kind of no-man's land. I was conscious of
having irrevocably broken out of the protective cover of an institution
which had given me much, and also had become separated from friends
who were dear to me, few of whom I was ever to meet again. And in the
unknown wider world I was expected to assume total responsibility for
finding my own way. There was anxiety and a certain sense of dismay
in this powerful yet totally untried impulse from within which made me
go against the traditions of my family and particularly my mother, to
whom I was devoted, and whom I still regard as the most remarkable

woman I have ever known. Then suddenly I received a letter from an elder brother who was particularly close to me.

I owe him much in all sorts of ways. Although he was to die young, without leaving any so-called 'mark' upon the life of his time, his going meant more to hundreds of anonymous people than the death of many of the famous at that time. He knew all about my love of writing; in fact he was the only member of the family to whom I had shown my poetry. Because he had handwriting of classical form, in contrast to mine, which has always been abominable, he had copied the poems into a red leather-bound book. In his letter was an advertisement from a Port Natal evening paper calling for applications for the post of what was then called a probationer on its editorial staff. He suggested that I should hasten to Port Natal and apply personally for the position.

He stressed that the odds against my getting a job were very great. Durban is the city which grew up around Port Natal. It was the most 'English' part of South Africa, and my brother told me that in the year he had been working there he himself had not met anyone who spoke anything but English. Also, they were people who had the greatest reservations about those of us who lived in the interior, and they had strong feelings of superiority, even scorn, for our supposed deficiencies. Yet he had a hunch that I should, as he put it, have a go at it.

I caught the next train to Durban, encouraged that I was going, as it were, back to the sea. But the editor of the newspaper agreed to see me with reluctance, warning me on the telephone that my chances were remote. It was essentially, he stressed, an English-speaking South African whom he sought and not someone who came from the 'backveld'. All this increased, to an almost unbearable degree, the tensions which I felt before going to see him, and I hardly slept the night before. I arrived at his office in a highly nervous state.

Yet when I actually met him I was aware that I was in the presence of a singularly humane, impulsive and unworldly human being. He was perhaps too human for the extremely difficult task of making a modern national newspaper out of the ruins of a parochial journal which had just been bought by the largest newspaper group in the country. Twenty-four hours after I left him he sent for me and told me that he was no longer going to consider further applications; the job was mine if I still wanted it. He added that, on reflection, it might be extremely good for the newspaper to have on its staff someone who could speak for the 'backveld'.

I was not aware at the time of how brave a decision this was. The whole of his staff consisted of men who had been recruited in England and who had all their experience and training like himself on English papers. So much was the 'Englishness' valued and recognised that part of the terms of their employment stipulated six months long leave in

Britain every two years at the company's expense. I was the first South African to be taken on the editorial staff—a staff, moreover, which had a Kiplingesque sense of empire and its mission. Had I been born of English-speaking parentage and brought up to share in their own imperial vision, perhaps they might have taken more kindly to this revolutionary appointment. But as men who continued to talk of England as 'home', they resented the decision and protested bitterly against my appointment. But, as I was to discover, they had their full share of fairness and respect for the point of view of others, particularly others against whom they had fought in battle, which was so marked a feature of their country. Within a few weeks they accepted me and could not have been kinder and more anxious to help.

I was installed in the editor's office, on the opposite side of a large desk from him. To everyone's surprise, he had taken the unusual step of making himself responsible for my newspaper education. He had done it, he told me, because in my preliminary interview, when he asked me why I wanted to work on a newspaper, I had replied that I regarded newspaper work as a gateway to writing, and possibly even an immediate form of literature itself. Happily this coincided with his own conception of his profession, and from then on I was involved in an intensive and exacting course of classical newspaper training. I was sent to night school every weekday evening to learn shorthand and typing, which I thought would drive me insane. I was also made to study the laws of libel, and the famous libel actions of legal history. The most fascinating of all for me turned out to be the one which was brought by Whistler against Ruskin, when Ruskin called him, among other things, a "cockscomb who had the effrontery to charge excessive sums of money for throwing pots of paint in the public's face". The case ended in an award of a farthing's worth of damages.

When my shorthand achieved the minimum speed of ninety words a minute I was sent to practise objective and accurate reporting in the magistrate's and high courts of the city. Trying as I found this work, I never regretted it because it taught me how to observe carefully and to respect the vital distinction between news and comment.

My 'release from bondage' came about when the Prince of Wales was on a prolonged tour of South Africa. Inevitably he came to Durban and, not surprisingly, this most English of cities was determined to give him the greatest of receptions. My own newspaper decided to devote the entire newspaper to the royal visit. One of the senior men, who had been delegated to report on a particular aspect of the visit, went ill with fever the day before the Prince's arrival and was confined to bed. Inexperienced though I was, the editor instructed me to take his place. This I did, but because of the speed with which I had attempted a kind of reporting I had never done before, and the rush with which it had to be set up and published, I felt sure that I had done extremely badly. Yet the next morning on my arrival at 7.30 am the editor and his staff congratulated me on what they said was the most memorable bit of reporting of the day. Privately I remained unconvinced, but it had important consequences. The editor thought so highly of my work on that and other royal occasions that followed, that he made me responsible for one of the most important pages in the paper, called 'Point and Shipping'.

The Point was the long dockyard salient of the harbour; the shipping needs no explanation. It is hardly necessary to explain the delight with which I now found myself in close and intimate contact with things of the sea.

Moreover, in the months of training that had gone before, the feeling I had that our country owed everything to its being a half-way house to the East had been broadened and intensified in Port Natal. This feeling had developed during my most impressionable years through our early contact with the Cape. Cape Town was still the place where South Africa experienced the greatest impact of the western world, but Durban had long since replaced it as the point where we had the most immediate contact with the East. Even the composition of peoples inhabiting Durban and its neighbourhood seemed to keep this fact alive in my imagination.

The very location of the offices of the newspaper seemed by the day to inflame my sense of history. In those days it was called *The Natal Advertiser* and was the forerunner of the influential *Daily News* of today. It was located in a sleazy, ramshackle building so out of date that, when the newspapers were 'put to bed' in the afternoons and the presses started to roll, walls and floors trembled and shook so much that it felt as though all would fall in pieces to the ground. The building was in an untidy, dilapidated side-street called Grey Street, on the fringes of the Indian Market and commercial quarter.

In the interior, where I had spent my childhood, there had been very few Indians. It is true that there were Indian communities, both Hindu and Moslem, in the Cape, but they were small compared with those of Natal. In the Cape, the East was more amply reflected in the remarkable Malay community of the Western Province. The Malays had been brought by the early Dutch as slaves from Malaya, Sumatra and Java to work at the Cape, mostly as skilled craftsmen and artisans. They had contributed immensely to what there was of grace and elegance in the metal-work of the early and elegant Cape furniture, and unique style of architecture. They had branched out as the tailors and merchants, and, as master-fishermen, were to pass on their skills to the Cape-Coloured community. The latter evolved as a result of mixed marriages of European, Hottentot, Bushmen, Indian, Malays and an incredible assortment of other races. Although the Malays retained their language, the Moslem faith and a great many of their traditions, they had in many other ways cut the ties which bound them to their places of origin, and committed themselves to the soil and needs of Africa. Most of what is good and unique in the way southern Africans cook is to this day due to the Malays and their translation to life in Africa.

The kind of India I now came to know in Natal revealed how much of a halfway-house to the East we still were. Not a day passed in which my new environment did not remind me how close remained our relationship with the East. I had only to walk out of the entrance of the *Natal Advertiser* to be in a world that was as much of the East as of Africa. Since it was an evening paper, we started work early, before the municipal scavengers had been round to 'clean up the streets', and I used to pick my way through the discarded skins of bananas, pineapples, loquats, mangoes, mandarins, custard apples, persimmons, grenadillas, guavas, papayas and other fruit native to the East but assiduously cultivated by skilful Indian market gardeners. In fact the fruit and vegetable part of the Indian Market in the light of an early sub-tropical morning, with piles of polished fruit in pyramids on their stalls, would glow like treasure and never ceased to delight and excite my eyes.

If my assignment was not pressing I would take a longer route through the market. The variety, brilliance and totally un-African colour of the scene would be reinforced by the saris of the Indian women, the silk trousers and tunics of their Moslem fellow-countrymen, who now pass under the name of Pakistanis. But in those simpler and more naive days they were all at one in India under the rule of the British, whose politicians called it the brightest jewel in the Crown of the Empire. Indeed for me there was jewellery in the mere colour of their dress, and a totally new idiom of beauty in their physical appearance and bearing. From the smallest child to the oldest man or woman, I rarely saw a single being who was not comely or of the most marked beauty. Years of service in the old British Indian Army later on in my life, and the extensive knowledge of India that I acquired, have confirmed and enlarged this impression which, at the age of just over seventeen, had an overwhelming impact on me.

Perhaps even more than the colour of the people, and the sound of Urdu, Hindustani, Gujurati and many dialects of Indian speech of which I understood not a single word, this living feeling of the East, which Renaissance man had suffered so much to reach, was reinforced by the evocative scent of mixed spices.

Smell has its own language, with no need of words to express it. Scientists tell us that the nose in the animal is the oldest of the organs of our senses and, therefore, retains an unrivalled power to evoke what is inexpressible. This was my experience in Port Natal. I only had to be aware of the subtle smell of a complex of spices and the whole of our history seemed to come alive in me. Even when I returned late in the day to write against time, with a messenger standing beside me to take my report away sheet by sheet, the scent of the spice and exotic fruit came drifting in through the window and joined with the unforgettable smells of our archaic office – compounded by whiffs of copy paper, printer's ink and other chemical emanations from the liquid metal waiting to be transformed into print by the machines of the lino-typists next door.

As exciting as the Indian quarter and market was the presence everywhere of the indigenous people of Africa. They were still very much as they had been before we had, in a sense, spoilt their world. These were of course the Zulus, who did most of the hard work in the city, its rapidly expanding industries and in the harbour. They scorned to join in the cultivation of the land and left the Indians to maintain the large sugar plantations that were becoming so dominant a part of the economy of the Province. Yet the Zulus were not only to be seen in the cities. There was hardly a household, however modest, which was not staffed with one or more Zulus, if only for general work about the house. It was astonishing to me how so proud a people with so long and great a military tradition, the greatest and most sustained of all the Bantu nations in the whole of Africa, had also an astonishing gift for domesticity and above all for cooking. Even the largest homes of the rich citizens of Durban had Zulu butlers, cooks, footmen, waiters, young house-boys, and Zulu gardeners. And for all the paternalism of the process, which has become such a term of abuse today, their employers in the main had a close, affectionate and not un-worthy human relationship with their Zulu servants, whom they regarded as part of the family. If it had been in any way an undignified relationship I do not think that the Zulus, for whom honour is their highest value, could have given forth so great a feeling of self-respect and maintained such grace and confidence of being as they mingled with the more sophisticated peoples of India and Europe in the markets and streets of the city. They seemed to walk through life as if instinctively aware of being a kind of royalty within themselves. They had always attracted me more than any other of the Bantu nations among whom I had grown up, and as a result I had a keen feeling of kinship with them. This respect and affection, going so far back into my childhood, deepened until it became a dominant experience of those remote days, and it still stands today as fast as ever among my own feelings for my native country.

Even more striking than the men were the women and girls who would be sent from far out in the country to what had become for them a legendary city, in order to buy necessities of life and the rare luxuries they could barely afford to take back to their kraals in Zululand. They would walk down the main streets of the city, crowded as it usually was in the morning with European women and their daughters dressed in their idea of the latest fashions, tripping their way to the most fashionable department stores. But somehow it was these Zulu women who seemed to be the real ladies to me. Dressed only in beaded leather aprons, and adorned with glittering copper or metal bangles round their ankles, beaded necklaces designed always with a taste that was fastidious as it was bright, dangling between their full, bare breasts, they would carry themselves as if they were the only women in the streets who really knew how to dress in that hot, vivid climate. Used as I was to this phenomenon up-country in the backveld, I was at first amazed by noticing how both the Indians and Europeans were disconcerted by the ladies of Zululand and the furtive glances of admiration drawn from them in passing. The Zulu men, however, the moment they saw one of their country-women whom they thought exceptionally beautiful, would call out, irrespective of lurching trams, cars and Indian vegetable carts, in deep, resonant voices to those they admired most. It was always remarkable how delicate was the response evoked when this uninhibited, eloquent and innocent masculine praise rose above the noise of the traffic of a large city. A smile of great delicacy would appear on the faces of the young women, and a graceful hand would be quickly lifted to hide a shy smile of gratification.

All these complex effects of colour, fruit, scent, delicate products of the East and the clothes of their purveyors, all on fire under the sun, as well as daily contact with the vivid people of Africa, were reinforced by a new kind of music which was special to Port Natal. Of all the arts, perhaps music goes deepest and elicits more responses beyond reason and logical explanation. This brilliant intermingling of colours and what was oldest from the East and Africa, had music to accompany the impact they made on my senses. Most of the Europeans who could afford to in that sub-tropical world lived on the hills called the Berea which encircled the harbour. It was cooler and healthier there than at sea level. Often I would walk down towards the sea at night after visiting friends who lived there. I would pass through the Indian Quarter, past houses where Indian music was being played either on gramophones or on the instruments themselves. It was sensitive and aquiver with an undertone of something akin to pain. Even the most resolved melodies sounded as if they might have come not from man-made strings but from the living nerves and tissues of the music itself. I had never known music like it and thought it the most mythological sound on earth, charged with an uncompromising nostalgia for the ages of the great Hindu gods, trembling in a moment of revelation that was neither past nor future.

As I grew to know this music better, I came to recognise specific themes. I remember moments when even the heaviest and darkest of sub-tropical Zulu nights and all Africa were abolished by these sounds as of shot silk, and I would find myself translated to some desolate foreshore on the southernmost tip of India beside Krishna himself, the Hindu god made man, who had become as close a companion in my imagination as Odysseus. There the music would induce such a stillness of grief that I would step out quietly not to miss the least nuance of sound, picturing him looking across the star-sown water which separates India from what was Ceylon, while lamenting the loss of his love, Sita, who had been stolen from him by the diabolical king of the Singhalese island. A kind of cosmic grief would come over me as I listened thus to Krishna's lament coming as it were alive and bleeding from the strings of his own being. I would be moved almost to tears at the most tender passages as, for instance, when the music recalled the moment when a squirrel appeared at his side to comfort him. Krishna, greatly moved by the little animal's compassion, would stroke him with his fingers so ardently that the signs of the god's gratitude were imprinted for ever in the parallel white lines of hair which the Indian squirrels carry, like bars of music, to this day.

Then, as I moved out of the Indian quarter, music of another kind would weave into the fading pattern of Oriental sound. Although it lacked the richness and complexity of the Indian music, it was even more profound in its implications. It was the music played on guitars by Zulus on their way down from the Berea where they worked to close by the sea where they lived. Released from work they would set out, each one playing a tune of his own invention as he marched with a loose stride, similar to that of a legendary hunter, from the hill to the sea. At first all the tunes sounded alike but one soon learned to distinguish the variations. Even if their composition were not a broad highway of music as we and the Indians had evolved, it was like their tentative footpaths, bravely made by vulnerable bare feet of countless generations of anonymous and forgotten men and women through the tangled bush and veld of Africa as they wound their way up one hill and down another and through plains and valleys, out of the night of our aboriginal beginning into the light of our modern day.

After music of this kind the European music coming from cafés and hotel terraces of the city itself was almost intolerable. The last thing I wanted was to be exhorted, as one of the most popular tunes of the day was urging everyone in Port Natal, to say "No! No!" to someone called "Nanette" and then go on to "Tea for Two".

But once on the beaches I would find a place to sit alone on the sand, just below the last line on the summit of dunes where the indigenous trees and bush of Africa, which still surrounded and penetrated the city, made Port Natal one of the loveliest harbours in the world. There I would find company in the music of the Indian Ocean which, even on the calmest of days, pounded the shore like the great horses of the Homeric god of the sea, their long white phosphorescent manes streaming and fine necks arched over polished hooves braced for their impact on star-silver sand.

For all the loneliness implied in moments such as these, there was neither regret nor hurt in them, except a certain discomfort from a growing longing that seemed to have always been there, to sail those seas. Accordingly I look back on countless moments like those without regret or even nostalgia, but only with unqualified gratitude to life for giving me so privileged a chance of communion with the sea and its meaning, both in the dimension of the here and now of daily life as in the depths of the spirit where, through the symbolism of the external world made manifest, we are in touch with all that has been and all that is to come.

Part of my new responsibilities was to be the gatherer of special news from the port and the ships which moved endlessly in and out of it. As far as the East was concerned, I found that there were weekly sailings between Port Natal and India. Every week the ships left crowded with passengers and cargo, and another would arrive to take its place. An even greater number of irregular arrivals and departures came to swell the rich traffic between southern Africa and the East. I was amazed also by the vast numbers of ships that arrived from other parts of the world – Japan, Australia, New Zealand, South America and the United States. I soon realized that what the Portuguese had so bravely and improbably started, had become the busiest and most crowded ocean route in the world.

The role of royalty, as it were, among all this shipping, was played by the mail ships of the Union Castle Line. This Royal Mail had appropriate courtiers in the mixed passenger and cargo ships of the line in its intermediate branches, which maintained four regular sailings a month: two from Port Natal up and along the coast of Africa, through the Suez Canal and on to England; then back again by way of the Outer Islands like Ascension and St Helena. The other two sailings were bound on the same voyages in reverse. The mail ships arrived punctually once a week from England by way of the Cape and sailed just as punctually once a week back. The pre-eminence of these ships and my relationship with them became so significant that they will need a book of their own. However, important as they were, they were by no means the most interesting or romantic ships calling at Port Natal. It was almost as if their regularity and conformity to duty imposed by the economic establishments of Africa, gave them a kind of commuter character. For all the beauty of their lines and colour, and great ships as they were, my appreciation was for the moment diminished by a sense of rebellion against what was established and socially pre-ordained. They took on for a while a respectability too great for my liking, as if travelling in them would merely be an enlarged kind of coming and going between one suburb and another, and I tended to be far more fascinated by other lines which ventured on other and more remote seas. Ships like those of the

Blue Funnel Line that went out to the East, Antipodes and back, the Red Star liners which sailed regularly between New Zealand, Australia, South Africa and Europe and the P & O ships which made a speciality of carrying immigrants and cargo between Australia and England. There was the German Africa Line too, sailing regularly between Hamburg, Southampton, the Cape and East African ports like Dar-es-Salaam (The Haven of Peace, as this ancient Arabic name would have it), the first harbour of what was then still Tanganyika and as a result of the First World War, mandatory territory entrusted to Britain. There were ships of the Holland-Africa Line, small but fastidious, circumspect models at sea reflecting the strong sense of civic and public duty, and love of good husbandry, of their country of origin. They were solidly built, often manned and served by outstanding crews and officers of unassailable respectability. The ships' decks, cleaned and scrubbed even behind the ears of their funnels, were as bright as the house of any burgher of Amsterdam. And they kept one of the best furnished and most appetizing tables of any ships afloat. I privately thought of the men who manned these ships more as burghers of the sea.

There were also regular sailings of the ships of the great Japanese ocean-going concerns: the Nippon Yusen Kaisha and Osaka Shoshen Kaisha, carrying Japanese emigrants on their way to Brazil, adding to the nucleus of their countrymen who had settled there and were to turn it into one of the largest, most industrial and important of the foreign colonies established in that vast projection of Portugal on the continent of South America. Even had I not been there to witness the arrival of the Japanese ships, I would know that they had docked, because the streets of the city would soon be filled with Japanese men, women and children who had been deprived of their own colourful national dress and sent out into the world in the uniform of European respectability that struck dismay in my heart. All the clothes seemed made to the same design of the drab and cheap material of a lustreless grey, so unlike the dress of the Indians, always clad in colourful silk.

This teeming ocean traffic was made even more vivid by the names of the ships themselves. They indicated not only the high-standing role of ships in the long history of man but something of the romance of the breakthrough they must have been far back in unrecorded time. There were the ships of the Blue Funnel Line, for instance; their hulls painted a true black, with immaculate white decks, and their names in glittering white letters. They bore names such as *Anaeus, Nestor, Agamemnon* and so on to the ships of the Red Star Line, determined to go one better. Their hulls were a pure sea-green with spotless white decks but with buff funnels and names like *Diogenes, Sophocles, Demosthenes*. The great Cunarders also drew on the myth and legend of Classical Europe as the only source capable of doing justice to all the imponderables provoked by ships in human imagination. They had lovely ships with names like the doomed, glorious *Titanic*, the tragic *Lusitania* and their happier successors like the *Aquitania, Berengaria* and *Mauretania*. I was struck, too, how often the British Navy called its ships by names that had meant much to me in my reading of the Classics – *Dido, Iphigenia, Leander, Penelope, Arethusa, Thetis* and the *Bellerophon* which had carried Napoleon on his first stages of exile in St Helena and so brought the Corporal-Emperor within the orbit of our halfway-house history.

Indeed, the fundamental conviction that these names had to express the high status of ships in the spirit of man ruled everywhere, so that other nations than the British used names associated with their highest values. The Italians, for instance, names their greatest ships *Dante Alighieri, Michelangelo, Leonardo da Vinci* and so on. The French, in their post-Revolution dedication to the ideological and rational in the human spirit, would go for what seemed to them their finest conscious purposes and made their ships express ideas rather than history in names like *Liberté, Fraternité* and so on, and the *Ile de France* was exceptional perhaps because such a name alone could evoke the high status of the island of culture and civilization that the French have been in European beginnings.

Much could be written on what one can deduce, from the names of these ships, about the national character of the countries who had launched them. The Dutch, for instance, seemed totally untroubled then by history, and would give their ships names most likely to flatter the countries with whom they were designed to promote trade, and chose towns and cities of my native country like *Bloemfontein, Klipfontein* and so on. The Germans, on the other hand, in those between-the-wars days when they were still smarting from the loss of their own empire in Africa, showed how impelled they were unconsciously by this profound, new-old spirit of the great Teutonic tribal surges which had brought them out of Asia to overrun Europe, with names like *Ubena, Usambara* and *Watussi*.

Perhaps the most prosaic names of all came from the United States. On the western ocean they did, it is true, make concessions to history and an overall feeling of pride by calling their finest ships names such as *Washington* and *United States*.

What struck me forcibly at the time was that all the ships whose names expressed the greatest sense of continuity of past and present were those carrying cargo as well as passengers, while the more functional the ship, in the modern meaning of the term, the wider the break in names with history. I realize now how grave a warning this was of a withdrawal of imagination from so ancient a form of commerce and communication. It should have alarmed me as much as the psychiatrist is alarmed when he encounters in his consulting-rooms the equivalent spirit of the human being which he identifies with the terrifying term of 'dissociation of consciousness'. But like everyone else it was only with hindsight that I was to realize how such changes in fashion of ships' names were evidence of a profound shifting of the spirit of the time, and how the names still alive in my memory were a selection of patches of a Joseph's coat of many colours. They suggested how diminished already was the great Renaissance dream which they and their kind had served for five centuries. Ships came and went from Port Natal, each with a uniform of its own, bearing such names as *Volturno, Atlantic, Vascomar, Travessa, Statesman, Clan McGillivray, Saxon, Umvoti, Umgeni, Bosphorus, Burgermeister, Chicago-Maru, Mohamedi, Barrabool, Beethoven, Angola, Jason, Florida, Franconia, Britain, City of Nagpur, Strathmore, Mohican* and *Finlandia*. The list of Union Castle Line ships is too long for inclusion here, but during my days in the harbour I saw them all, from the *Arundel, Dunnottar, Dunvegan*, up to the more famous *Edinburgh, Warwick, Carnarvon, Gloucester Castles* and the *Windsor Castle II*, the elegant predecessor of the flagship of this illustrious fleet and last childless survivor of the proud and ancient line which was to end some five years ago. Even this list is the merest sample, for those were the abundant years destined to culminate in a moment when Nöel Coward could be certain that everyone in England and America would enjoy the joke when, in *Present Laughter*, his main character exclaimed: "It's been a good day for the Union Castle Line".

Many of the ships I had known by sight and perhaps been inside when seeing off friends and acquaintances, but I had not yet got to know them as well as I began to now, doing my early morning rounds as the shipping correspondent of my paper.

I would start my day calling in at the signal station on the Point, as it is still called today. The Point was a long, narrow spit of sand stretching far out to sea, where the great curve of land north of Durban came to an end to face a narrow opening of sea which led into the vast, natural harbour of Port Natal. There it was paralleled by another projection of land, the high ridge of an elongated hill called the Bluff. The Bluff was covered then with indigenous forest and bush, full of little apes, monkeys and birds of all kinds and colours which, seen from the Point, seemed to burst and fall like showers of confetti over the sparkling leaves. On the tip of the nose of the Bluff, where it fell away almost perpendicularly into the uninhibited swell of the Indian Ocean, stood erect as a Roman sentinel a tall, white lighthouse of classical design. This first lighthouse, unlike its successors, was as elegant and appropriate on the nose of the Bluff as any altar candle ever lit in a church. By day it was white in the Zulu sun that smarted in one's eyes; by night, brushing the dark with long wings of light. There was something indeed sanctified about it. Immediately behind it was an even taller white flatmast securely held in position by long hawsers of steel embedded in concrete. The moment a ship was sighted and identified, the appropriate flags, to convey not only the identity of the ship but its needs in the harbour, would be run up the mast in bright, sparkling, coloured flats, inducing immense feelings of excitement and expectation.

Both the Bluff and the Berea were covered with thick, dark indigenous bush. The houses built on it had, as yet, done nothing to violate the unique African reality of the great ridge, so that its contribution to the natural beauty of the setting was still important. In summer the dark surface of the bush took on a shine of jade which would be set even more aglow by the scarlet flame of flamboyants or look hushed and blessed under the subtle, ethereal purple mist of the flower of jacaranda. The bush was still so much a part of the harbour and the city that in the Edward, then as now my favourite hotel, sitting on the verandah at night drinking coffee, I would notice with delight how, at the far end where the verandah and bush met, apes and monkeys would vault out of the night on the tables and quickly empty the sugar basins, embarrassing the Indian waiters who had an almost religious dedication to their duties.

What added to the excitement the port generated was the fact that it was still a harbour in the making. In the centre of the lagoon which lay behind the Point, dredgers were at work day and night, removing deposits of sand that the rivers from the interior had left in the tidal waters of the Bay. A channel had been opened and was being widened between the Point and the innermost reach of waters at the far end. This was called Congella, and more than a hundred years before my family had participated there in a battle against the English sent from the Cape to conquer them or drive them out of this new home they were making for themselves in what Vasco da Gama had named 'Natal'.

On the way to Congella, where ships of all nations ploughed daily through the blue water of this locked lagoon, there was an island called Salisbury Island. It was a miniature of the Africa before either the Zulus or we ourselves had come to disturb it. It was given over entirely to the trees, plants, birds, butterflies, insects and apes with silver hair and black faces, to sit like burnt jewellery of Java on the tangled branches of Zulu green. It epitomized for me all these contrasts and contradictory features of Port Natal, making it a kind of script of hymn ancient and modern, never ceasing to stir and excite one's senses. Only the Point where the signal station stood had lost all its allegiance to the past. The land there was covered with a long line of quays to which the ships could be tied. The Point had even been extended by one of the longest breakwaters in Africa to protect the ships coming through the narrow entrance to the Bay from a natural bar of sand thrown up by the pull of the tides and the thrust of the water from within. And even with this breakwater, the task of captains and pilots responsible for steering their ships to shelter within the Bay was one of the most difficult in the world. I myself was to see how dangerous their work remained, despite all the dredging and lengthening of breakwaters to improve the entrance, when on a still and early winter's afternoon of unstained blue, and with the Indian Ocean breathing deeply in a rhythmical swell as if enjoying a Sunday after-dinner sleep, a ship of some fifteen hundred tons was caught in a cross-current as it emerged from the entrance, wrenched sideways on to the sea, turned over and slowly sunk.

No ship's captain, however experienced and however familiar with this narrow entrance and its problematic bar of sand, enigmatic currents and cross-winds, was ever allowed to bring his ship in on his own. This was the responsibility of an expert band of pilots in the employ of the harbour authorities who, every time a ship signalled, hastened out in one of the powerful tugs to board it in the Roadstead. There, he would take over from the captain on the bridge and pilot the ship to its allotted berth. In time I came to know those pilots well, but not before I had learned much about their personal histories, lives and characters from an education I received at the signal station on the Point.

The station was a tallish square tower of red brick, manned by two ex-Yeomen of Signals from the Royal Navy: a fact which introduced a slight, and to me, rather endearing element of superiority, if not snobbery, in their attitude to mere ships of commerce which they had been called upon to serve. I knew them just as Mr Clark and Mr White. They must, of course, have possessed Christian names, but I never discovered what these were as they never referred to each other by name, close shipmates though they had been. All Clarks in the British Services, I was to discover later in a decade of duty in the British Army, were, for some unfathomed reason, called 'Nobby' and all Whites 'Knockers'. And that is how these two unfailingly referred to each other.

Knocker White, who had, I suspected, a slightly cleft palate, because of a Churchillian slur to his voice, was the senior. Out of this sense of naval pre-eminence he imposed a class if not a Hindu caste system in his attitude to the multitude of ships which came and went through Port Natal. The ships of commerce which came closest to evoking appreciation in the two men whose life and imagination, I was to discover, hardly had room in it for anything except ships, were, of course, the Royal Mail ships of the Union Castle Line. Mr White's satisfaction, which was silently shared by his mate Nobby, was evident when a Royal Mail steamer, coming up from the dangerous waters off Agulhas and the stormy Cape, rounded the Bluff early on a Sunday morning and he nodded as he glanced at their clock which was as delicately tended and accurate as any naval chronometer. That the Royal Mail was on time to the second had an element of relief in it that always made me smile because of a concern so living and delicate. Other passenger ships came a close second-best because, for both of these men, although they had both long since retired from a Navy that had made them Yeomen of Signals, liners, as for Kipling, were ladies. But, thereafter, a nice progression of class distinctions came subtly and automatically into operation. Cargo ships were craft of humbler origin and occasion: they were the tradesmen of the seas. If Knocker and Nobby had had their way, I sometimes thought that they would have seen to it that this harbour which they served so well would have a tradesman's entrance for ships of this kind. Cargo ships of regular lines were, it is true, somewhat redeemed by the fact that their schedules also demonstrated a certain discipline and dedication to service which they could appreciate. But the tramps, of which there were a great number, were a different case altogether. They were an inferior form of ship life comparable, on this Indian gateway, to untouchables in the Hindu caste system; and of an order they were almost ashamed to serve.

I was to see Mr White often put to his eye the kind of telescope which I am certain had not changed since Nelson's day, and focus it on a smudge of smoke coming over the horizon from somewhere out of the East. So well did he know his ships that it only needed the colour on the tip of the funnel and suggestion of masts and line to tell him what kind of ship it was. The moment the vessel was identified he would hand the telescope to Mr Clark saying:

"Nobby old chap, clap your eye on that there creature and tell me what you make of it."

I could tell from the tone of his voice immediately whether it was a lady, tradesman, tramp or worse.

Mr Clark, who was of a serious turn of mind and not given either to irony or sarcasm, would take a look that was as long as it was thorough, before remarking with solemn deliberation:

"Something, Knocker old man, uncommonly high in the water; I dare say another of *those* in ballast from Singapore or somewhere further east, looking for employment."

"Something, is hardly the word", Mr White would reply slowly. "If you ask me, another of those promiscuous hussies coming to solicit, uncalled for like, in a respectable harbour."

"Hussies", "wenches" and "broads" and, at the worst, "tarts", were common ingredients in his vocabulary for tramps, particularly those under Latin flags who were 'no better than they should have been'. Brazen, shameless, saucy, loose, promiscuous-like, were adjectives joined to them according to the degree of denigration allotted. For, even on this far perimeter of their preferences, the discipline which had become second nature to them insisted on law and order in selection.

Over the years their knowledge of the ships which used Port Natal, and of their captains, crews and even owners, though at times apocryphal, was almost as great. I was always in danger of staying in the signal station much too long, because I found their conversation and repartee as interesting as it was enchanting. If Dickens had ever inserted ocean-going characters into any of his great novels, I am certain they would have been among the first, because there was something truly and nobly Dickensian about this pair. What made the temptation to stay even more difficult to resist was the fact that they were the two greatest and most inspired drinkers of tea I have ever met among the people of a great tea-drinking nation. They had it on tap, and I was lucky if I could get away without having two or three cups of 'cha' and ship's biscuits with these two most generous and hospitable of men.

Moreover, they knew the people who served the harbour even better than the ships. They knew the masters of the ocean-going tugs, which were a speciality of the port: tugs built to withstand the stormiest of weather and equipped to go and salvage or rescue ships in danger anywhere on the African coast, as they sometimes did. Indeed, I know one tug and its master who had sailed nearly two thousand miles to salvage a ship which had run itself on to a coral reef at Mikandani on the East Coast.

As Nobby and Knocker took me into the privileged circle of their friendship, they thought it their duty to inform me of any detail about the character of each master of the tugs, and more often of the ships of the port, captains and pilots, who obviously could be sources of news to me. I was amazed at the accurate way in which they could tell, just from the way in which a ship was being lined up in the Roadstead outside, which of the pilots was in charge on the bridge. They were two of the most astute observers of men and their ways that I had encountered. As a result of all they taught me, I too could tell before long what line a ship belonged to by the colours of hulls, funnels and masts. Most important of all, they taught me the best way of approaching ships' masters and pilots, of whom no two were alike, and each of whom had his own propensity to some extreme of sensitivity, if not touchiness, produced by deep-sea loneliness and lives spent in dealing with the unpredictabilities of both sea and weather, and the caprices of chance, which a mere long-shoreman could never understand.

What made Mr White and Mr Clark such valuable friends and pilots to me, in this new world, was in no small degree due to the fact that familiarity with the sea had never created, as familiarity was supposed to do, any grain of contempt, but raised, to an extraordinary height, their sense of respect for the difficulties of men and ships at sea, bordering, at times, on acute apprehension for the tasks facing them. As a result, what tended to make me late in the pursuit of my duties, even more than the cups of sickly, brewed, sweet tea and the talk, was the drama of seeing ships first being lined up out in the Roadstead by the pilot, and then making a dash, at full speed, for the narrow entrance into the Bay. There was always one of these powerful tugs in attendance to correct and steady them in case one of the unpredictable factors like current, swell, wind or a combination of all three would make a ship swerve from its proper course. There were occasions, indeed, when their prompt action alone saved ships and, incidentally, spared pilots and masters an official enquiry and possible loss of tickets.

I would be appalled by the speed with which the greatest of ships hurled themselves at the entrance, speed so high that two more tugs waited inside the harbour mouth to help them brake once across the bar. My two teachers, however, would dismiss my fears with their own approval, saying that a ship at full speed was much easier to keep on course than a slow one. Yet, they would then go on to reawaken my fears with too many qualifications to mention here.

The only exceptions to the rule that all ships had to be brought in by pilot into Port Natal, were ships of the Royal Navy. On these rare occasions when cruisers and sloops of the African station from Simonstown at the Cape called in at Port Natal, the delight of Mr White and Mr Clark was unlimited. It was external confirmation of the first article of their faith: there were no ships like the ships of the Royal Navy, or sailors like those who manned the British ships of war. I never witnessed a naval ship crossing the dangerous bar without one of them remarking to the other, with a touching satisfaction, that they had not ever seen it done better. Then they would both instantly hope and pray that all the assembled pilots in the Port Captain's office had been watching and taking note of how these things should be done.

The three of us often watched this sort of thing happening like people in the stalls of a theatre with all attention caught up in the climax of a play. There was a day, for instance, when one of the great Cunarders, the *Franconia*, and the largest ship which had as yet attempted to visit Port Natal, crowded with wealthy American passengers on a world cruise, had been lined up and aimed like an arrow at the entrance. It was not the best of days for such a venture, with a great swell sweeping over the breakwaters of the Point and a strong wind blowing across the bar. Conscious of all the elements and high drama involved, Mr White and Mr Clark had telescope and binoculars trained on the ship. Together we watched it speeding, bows pitching, for the middle of the entrance. Hardly had its nose found the entrance when even to my naked eye it seemed to veer sharply to port. Immediately the tug in attendance appeared to jump the swell with the thrust of its powerful screws as it answered the full-speed ahead signal from the *Franconia*'s bridge. Mr Clark let fall his binoculars on his chest and went to put his hand on the emergency telephone. Mr White, who seemed to know what his companion had done without looking, kept his telescope steadily on the Cunarder. He said in the quietest of voices, which years of naval training and experience in a World War had taught him was essential in the emergencies of his vocation:

"Hold it, Nobby old man, just hold it a tick or two."

Perhaps thirty seconds later he added with a relief and emotion that was obviously not prescribed in naval regulations:

"I think she's going to be all right, Nobby," paused, and a few seconds later, "She's done it . . . she's done it! Trust Harry England to do the necessary."

Harry England was one of the senior and favourite of their pilots, not only because he was extremely capable, but also because he had served as a Royal Naval Volunteer officer in the World War just behind them. I was to know him well and like him enormously because, in spite of a certain sharpness and quickness of temper, not uncommon among men small in physical stature (for he was a mere five foot six), his was a most generous and chivalrous spirit. Indeed, he was due to die from drowning while trying to save the life of some unknown visitors who had got into trouble in the Indian Ocean surf that always boiled and bubbled along the sands to the north of the Point.

This tragic event revealed aspects of the character of both Mr White and Mr Clark I might never have come to know otherwise. I had long before realised their delicacy of feeling for all the matter-of-fact face they chose to present to the world in order to create an image of men in such control of their emotions that a casual observer could have overlooked this part of their personalities to the point of suspecting them of having either very little feeling, or feelings weak enough to be subject to so unruffled a command in the routine of their work. But on this occasion, deeply shaken, they had a dialogue with each other so personal, private and important that I might not have been there with them.

It started by Mr White asking me, as I appeared at the top of the ladder, whether I had heard the news. At my 'Yes', both he and Mr Clark went silent. They busied themselves with preparing tea and putting out their supply of biscuits. I knew instinctively, however, how upset they were and myself remained silent as well. We were on our second cup of tea before Mr White remarked, "All the same, Nobby old man, I still think he did wrong ever to learn to swim."

The 'he', of course, was Harry England.

"I know what you mean, Knocker old man," Mr Clark replied thoughtfully, "but I still hold that if he hadn't learned to swim we might never have known what manner of man he was, though we knew he was man enough."

This reply seemed to bring some meaning stirring deep down in Mr White to the boil, for he answered with an emphasis unusually sharp for him.

"But he didn't have to prove himself that way. He didn't carry those ribbons and have those gongs pinned on him by the King himself at the Palace for nothing, him as wrote his name on all the seven seas, navigating our ships as well as liners, tramps, and the like. He proved himself in the war along with the best of the officers you and I served under for thirty years or more. He was sailor enough to have known that it was somehow wrong to have been one of these new kind of matelots who go swimming around."

It was a subject which had been hinted at before in their conversation, but I had never had the courage to ask them to elaborate for I knew that it was connected with matters of almost religious importance to them and, therefore, not something open to public discussion. It was just that, as far as the Navy in which they had served for so long was concerned, very few sailors could swim and, for that matter, thought it right to swim.

Mr Clark, who was always somewhat more eclectic in his approach to his calling, and out of attachment to Mr White, was anxious that he should not express his feelings aloud in a manner which he might regret afterwards before someone like myself who was not within the circle of immunity of their profession. He pretended to a calm which he obviously did not altogether feel himself.

"But Knocker old man," he said soothingly, "you know even better than I do; it takes all sorts to make a world, even one of real ships as ours is. If there's a sort among them that wants to be different, as we've known some of the best, and sets out to teach himself to swim, he's as much right to it as those of us who don't think it natural in a sailor. You know, Harry England couldn't help being different from the others. You would be the first to admit that if this difference made him want to swim, not even the likes of us have a right to criticise him for it."

Far from calming Mr White, this agitated him so much that I could see Mr Clark's concern increasing to a definite alarm.

"You've got me wrong," Mr White replied ominously, dropping his usual 'Nobby' and 'old man'. "You should be the first to know I'd never criticise such as Harry England. It may well be, as you say, right for him to be different – even to the point of learning to swim, but I maintain that it was a tragedy, nonetheless, that his difference took him that way. If ever a man was a born sailor, he was one. And born sailors must know that either through what those coroner chaps call death by natural causes or death in action, the proper place for a sailor to have done with his life is in the sea. And the sooner he goes into the sea, once his name is on the slate, the better. It just isn't dignified for a sailor to dispute his last order of all and argue with his lot by swimming around in the drink when his ship has gone down under him."

And, feeling that he had perhaps been rather vehement with a shipmate and friend, he added: "Wouldn't you say, Nobby old man, that a sailor must trust the sea as his natural element and most likely to do what's right by him and not just use it for pleasure by swimming about in it as all these land-lovers do here for their promiscuous-like indulgence?"

He paused, and then repeated his argument with greater precision. "He must know, as all the thousands of sailors before us came to know, that when his name, rank and number are called by Providence, the sooner he goes to his home with respect and at the double, the better it must be."

"Speaking as someone who himself never took to swimming, or holds with it in any particular way," Mr Clark replied, "I must admit, Knocker old man, that you have a point, but . . ."

I was never to know what the 'but' implied; I guessed it had some qualification still lingering over a hint of what might have appeared a censure of Harry England in Mr Clark's mind, because he realised fully now that it was sheer grief at England's going which had made Mr White suddenly deliver himself of so forthright a revelation of his own philosophy as a sailor. So he contented himself with, "Swimmer or not, he's a great loss to us all. I've never known a pilot trained under the red-duster as good as he, and I don't expect we shall see another."

"Aye aye, you're right, and it should have been enough for me," Mr White admitted, not looking at either of us. Then he muttered, for my benefit, "But your cup must be empty, young lad. We're forgetting our manners, Nobby old man. Give our friend here another cup of cha."

Even so, there was enough of a sense of trouble and real grief at work within the hearts of these two good men, who themselves had experienced and suffered so much, that I felt an intruder for the first and only time in their tower.

I would have left then and there if Mr Clark, at that moment, had not spotted a smudge on the horizon of a morning of unbelievable calm, like a gleaming circle of Euclidian precision around a sea unusually dark with the blue of a sky overflowing with light. His reflexes on such occasions were as fast as they were relevant. The smudge was immediately in the focus of his telescope. He studied it for perhaps a minute in silence and then exclaimed, "The *Khandalla*, I think."

I knew enough of the ways of the port by then to recognise the ship as one of the British India Line, which maintained a weekly service between Port Natal, the East Coast ports of Africa and Bombay.

"And the *Khandalla* in trouble, if I'm not mistaken. She's in need of a Port doctor, I think." Mr White also took a look and confirmed that it was indeed the *Khandalla* and that she was flying the yellow flag to show that she had some infection on board.

"I wouldn't be surprised if she comes to us rotten with Smallpox or even the Pest," his favourite term for the Bubonic Plague, which had once produced the Black Death in Europe and still erupted from time to time in India. "You can never tell what those black and yellow ladies" – the colours of the ship of the British India Line – "will pick up in those contaminated waters."

Starting with this introduction to the world of ships which Mr White and Mr Clark between them gave me, my circle of friends and acquaintances rapidly expanded. Like almost everyone else I was to meet at that time, the oddity I presented as a product of the remote interior, or the 'backveld' as they had learned to call it with some denigration, far from turning out to be an impediment, was transformed by the instinctive sense of chivalry which invested the natural character of people of British origin at the time into a key which opened many official doors that had been closed to other newspaper men before me. Even today, I have only to think of the great kindnesses and generosity shown to me by everyone to feel my own sense of gratitude and privilege come as alive and active as a kind of eruption in my spirit.

From the Port Captain, who seemed to me appropriately called Shepherd, to the pilots, ships' agents, chandlers, berthing masters and tug captains, I seemed to have a new world of friends almost without end. The pilots, who had to pass exacting examinations in the craft, if not art, of handling ships in the most difficult circumstances, would now telephone me with points of interest about the ships they brought into harbour. They would often invite me to accompany them on the bridge of the more regular ships, whose masters they knew well, to observe them taking the ships out to sea.

When the moment came for the ship to drop the pilot, now that it was safe and heading well out into the open Roadstead, I would always follow him down the ladder of the heaving flanks of the vessel, and onto the decks of the tug which had accompanied us, with intense feelings of regret that I could not have stayed on the ship we had just abandoned.

Then, there were the masters of the large ocean-going tugs which had accompanied us and who were even more accommodating in some ways than the pilots. One in particular whom I was to know well was a certain Captain Coulthurst. Although he had for years possessed a master's certificate of his own and was well qualified to be one of the pilots senior in rank to himself, he resolutely refused promotion and stuck to his function as mere tug-master. I think this was because he was by nature a small-ship man. All sailors I have ever known naturally divide into two categories: those who preferred small ships and the greatly increased feeling of companionship and common identity which they gave to all who served in them; and those who would select the biggest ships of both Navy and Merchant Marine, perhaps, because they found there a kind of anonymity which enabled them, in the midst of many, to be more their own men than they could have been locked into the close family circle of those who served in the smallest ships.

Captain Coulthurst, I was to discover, valued personal relationships with the men he commanded more than other considerations. Moreover, the smaller ships gave him an independence of command and freedom of movement away from overt authority almost to the point of being imperceptible, as his small tug often was in so great a port. So as a senior tug-master he continued happily to the end of his days.

When he was on duty, knowing how in the afternoons, when my paper had 'gone to bed', as a stranger in a new city I was inevitably often alone and with time on my hands to spare, he would keep a look out for me somewhere along the quays of the harbour. All I had to do was to wave at his tug and he would turn its nose in my direction, pick me up deftly and let me spend the rest of the afternoon watching him carrying out his multitudinous duties. In the long periods between manoeuvring a ship away from the quay and conducting it out into the Roadstead and bringing another one back, he would lie somewhere with just enough turn on the screws of his tug to keep it in position against the press of the tides. We would then go and wait in his cabin. Over tea and biscuits we would talk about things of the sea and books (he was extremely well read) or, perhaps, listen to some of the records he played on a gramophone that was always of the latest make – for his love of music, as I discovered, was even greater than his love of the sea.

Before long I was asked to musical evenings in his flat in the newest, most luxurious of establishments built especially for single or childless people. It, too, had a wonderful view of the inner harbour and the heights of the Bluff beyond, but what distinguished it far more was the fact that the walls were lined not with books but with records. In fact the drawing-room was designed for listening. Although he was always talking of spending his long leave doing a round of concert halls and opera houses of Europe, it never came about. I suspect it was because any spare money he had went to still his most immediate hunger for music. The few friends that he had all shared his love, and we would spend memorable evenings listening to his record recitals.

His musical taste, however, had one element which I could not altogether share; he had an inordinate love for the music of Wagner. I came to understand this many years later when I made the acquaintance of more masters of ships and discovered that among the musical ones they too had a marked preference for Wagner. I was to wonder whether this was not due to the deep-sea nature of the imagination of the composer who sent his music itself to bear down on his listeners in waves like those of the swell which announces the approach of typhoon and hurricane over a deeply troubled ocean. It was like a premonition of evil to come which, even at his sub-tropical windows, made me shiver inwardly.

But for the moment, Captain Coulthurst had no doubt that it had much to do with the nature of the storms encountered between Port Natal and the Cape and that they alone could have awakened the *Flying Dutchman* theme in the imagination of the world. These storms, he said, and he had experienced several, produced at their climax something so excessive and incomparable that sailors in his day were as overawed as they were afraid. How much more so, therefore, in our Portuguese beginnings, had they to be attributed to supernatural, even diabolical, causes which only legends and myths could express, and, for him, the greatest of these was *The Flying Dutchman.*

I remember as he spoke thinking it ironic, perhaps even a form of poetic justice determined by the sense of outrage at the universe's rule of proportion, that according to our histories the birth of the legend appears to be associated with the death of the great Bartholomew Diaz himself, who first experienced the storms at the Cape. This great sailor seems to have been unafflicted by any of the arrogance to which lesser beings might have succumbed as the result of so epoch-making an achievement. He did not hesitate to sail as a mere ship's captain under Pedro Alvarez Cabral, who followed immediately on Vasco da Gama after his first round voyage to the Far East. On this particular journey what heightened considerably the feelings of awe and superstitious fear of sailors, whose vocation from the earliest times tended then to cultivate an inordinate respect for the caprices of chance and the intrusion of abnormalities of nature in their lives at sea, was the fact that when Cabral's fleet of fifteen ships and twelve hundred soldiers and sailors reached the fringes of the waters of the Cape on 12 May 1500, they saw a comet at nightfall. It was so large that its tail arched over most of the black sky, posing itself in the dark, unfamiliar space like some kind of cosmic question mark.

Comets in sixteenth-century Europe provoked much the same sort of foreboding as, according to Shakespeare's Julius Caesar, they did for the Romans. Whatever relief Cabral's fleet may have felt at the disappearance of the comet on the eighth night, as the ships lay becalmed off the Cape, went instantly when a sudden, sharp squall, as if aimed by the gods at them, struck the fleet. Four ships capsized and the rest were scattered, driven on by the storm and powerless to rescue the survivors of the four ships who were still clinging desperately to the keels of their upturned vessels and clamouring for help. The *Caravel*, commanded by Diaz, was one of the four, and he himself found his grave in the waters that he had been the first European to discover.

From that moment on there appears to have been an increasingly apprehensive stirring in the European spirit, and a climate of imagination that produced such disturbing examples of art and literature as the *Adamastor* of Camões, Coleridge's *Ancient Mariner* and the music we were discussing. Despite the diversity of location and form, all were concerned with transgression of natural law, as for instance the *Flying Dutchman* whose captain had made a pact with the devil and was seen by other sailors, themselves imperilled in some great storm off the Cape, sailing arrogantly into the wind with all its sails set.

It soon became a whole era's equivalent at sea of the Unidentified Flying Object of today, and, as I listened to Wagner's equinoxial music and matched it with Captain Coulthurst's talk, I would look deeper into the underlying causes of so strange a manifestation in the spirit of rational men. It seemed to me to have much to do with the fact that ships and their voyages from the beginning have been among the greatest of all symbols of the passage of man through time and space in search of meaning. That is why the voyage of Jason in search of the golden fleece, Odysseus's ordeal by sea, storm and wind on his long journey back to Ithaca and on to countless others, seemed to continue to attract the imagination as few other themes do. And as important as the ships and those who sailed in them, of course, was the role of the wind in all these re-orchestrations of this primordial imagery. For the wind in all times from the Stone Age to Greek and Roman has been the most evocative symbol of the spirit, urging man to a renewal and increase of the quality and nature of his being. That is why, in Coleridge's description in the *Ancient Mariner* of a ship 'becalmed on a sea so wide and still that God himself scarce seemed there to be', that total absence of wind, day after day, continues to strike such horror in those who read it now. Perhaps that is also why in our own day the conclusion of Paul Valéry's poem *The Graveyard of the Sailors*, "Le vent se lève, il faut tenter vivre", can still give us the shock of instant recognition; that this ancient image is still intact, alive, and as transforming as ever. The wind rises, the spirit moves, and one must try to live on into a new area of life. Looked at symbolically, it was as if this legend of a ship sailing against the wind, the mind and the will of man going against the movement of the natural spirit of life, was warning post-Renaissance Europe that this new traffic between Europe and the Far East had elements of dangerous excess in it.

The excess implied was all the greater when measured by the power of storms of which Captain Coulthurst spoke. They were unique and, in so far as the inexplicable was explicable, he stressed, it was due to the fact that there is, for thousands of miles, no land to impede the power of winds that come pounding down from what Englishmen under sail came to call the Roaring Forties. These winds out of the Antarctic drove Atlantic water before them in waves of an ascending order to disastrous heights. He had a word for those empty stretches of ocean that breed these storms, "fetches". The fetches of ocean between the Cape and the Antarctic are the greatest of all in the Seven Seas. He and all the many sailors I came to know who had done their time under sail, and therefore knew far more of the power of wind and water than the sailor of today who receives the whole of his ocean education under steam and in much larger vessels, had a special note of awe in their voices when talking about these things, and only when they came to the cause of their origin would they allow the word "fetch" to come from their lips. It was a note in the human voice which Christian fundamentalists reserve only for their rare references to Genesis and the Almighty.

As if this "fetch" between the Cape and the Antarctic were not long and lethal enough to give gales and hurricanes all possible latitude for their charge against the southern-most tips of Africa, they had also sinister allies in the submerged shelves of land which project far out into the sea along this coast between the Cape and Agulhas and from there to Port St John, together with the Benquella current which flows like an invisible Amazon through the Mozambique channel to meet them racing up from the south. The result is that when these long-maned gales hurl their phalanxes of waves over the barriers of submerged shelves of land and against the great stream of water going in the opposite direction, the head-on collisions throw up seas of cataclysmic proportions. For instance, a mere forty-knot wind from the heart of this immense fetch will produce waves averaging some forty to forty-five feet in height. I could not imagine, therefore, what monstrous things could happen when a gale of some eighty to eighty-five knots tears out of the tormented heart of this great fetch.

There was even another element, totally unpredictable, thrown in at the height of these storms which frightened the wits out of him and all sailors he knew who had experienced it. His voice became a whisper as he talked of it. At the height of the storm there seemed to be black holes opening up at the base of the master wave and threatening to suck ships in as forcefully as the weight of water and wind was driving them under. He was not a superstitious man, he said, but it was almost as if these were holes in reality itself, and even the best ship could pass down and through to vanish forever. If this were not so, how could the famous *Waratah*, new, soundly built in the best of German shipyards and only on her second voyage to and from Australia, have vanished in these waters some fourteen years before and, despite a search of several months, leave not a spar, lifebuoy, raft or any scrap of itself to mark the place and manner of her disappearance?

These descriptions became indelible in my memory as a result of a personal experience. I was invited by the officers of a naval sloop of the Flower class, *Protea*, to sail with them on one of their periodical surveys of the coast. Somewhere off Port St John we were caught with terrifying suddenness in an eighty-knot gale. For some thirty-six hours it was all that our ship and its experienced engineer and naval captain, Commander Wodehouse, could do to keep the *Protea*'s nose into the wind and ride out the seas that were thrown at us. Some of the waves were so high, long and wide that I thought so small a ship as ours could not possibly surmount them. The captain assured me that the smallness of our ship, far from adding to our danger, was an element working for our safety. The bigger ships, he calculated, would have been less able to ride those gigantic waves. Yet I remember well that, in spite of his reassuring voice, there came a moment on the small Bridge when this brave and distinguished officer, who had not long come through the First World War, appeared to be near despair and fearing the worst. It happened when a wave far bigger than any we had seen and obviously one of those master-freaks of storm and wind that sailors fear so much, suddenly appeared ahead in the fume and spume and spray, and drew itself up to such a height that the whole sky, screaming with the fury of the storm, vanished from view. As the wave drew itself up, it seemed to suck the water from all around it in order to achieve its full stature. The water in-between immediately seemed to vanish, and our small vessel looked to be plunging into a hole which had suddenly opened up in the sea, seconds before the water came crashing upon us. It was just as Captain Coulthurst had described it.

I have never been able to understand how the little *Protea* emerged. The weight of water that broke over us was so great that it nearly put out the ship's fires. But eventually the ship rose again to the surface and shook itself free of the streams of water. I was not surprised, when at last the gale passed by us and we could resume our way, that the Commander told me that it had taken the full power of the ship's engines just to keep us head on to the storm. In those thirty-six hours of gales we had made a bare seven miles.

Moreover, what we were suffering was shared by all the other ships in the vicinity. This was confirmed when the *Protea*'s Radio Officer came labouring up on the Bridge with an S.O.S. from a Greek cargo ship which his instruments had just plucked out of the crackling and groaning air. I do not, alas, remember the name of the ship but we had passed it early one morning and I still have a vivid picture of it in my mind. Like most Greek tramps of those days which made a habit of carrying as much cargo as they could to justify the cheaper rates they charged for their services, it was far too heavily loaded and accordingly too deep in the water for our captain's liking. As we passed by comparatively close, he turned to me and said in the kind of voice he reserved for members of the ship's company who had failed the high standards he set, "That fellow's not only asking for trouble but deserves and may well get it."

However he would have liked to have gone immediately to the ship's rescue but, as he explained, with so small a ship as his in so great a storm, he could not possibly turn about without almost certain risk of losing his own vessel. Luckily another and greater ship, a Union Castle liner called the *Edinburgh Castle*, was on hand to answer the S.O.S. call for help. The brave and instant manner with which it did so remains for me a precious memory. Like ourselves, the *Edinburgh Castle* was riding out the storm a bare mile away, and we had hardly taken in the full import of the Greek ship's plight when I heard the captain, who had just examined the

Edinburgh Castle through his glasses, exclaim, "Dear God, it can't be true
– she's going about."

Indeed, this ship of some thirteen thousand three hundred tons had
suddenly gathered way and, with engines at full speed ahead, moved
some cable lengths into the storm. Judging the interval between the
mountains of the waves to perfection, she came about as fast as possible
while we watched with our hearts in our mouths. Sometimes it appeared
to me that she was lying on her side almost level with the water and about
to capsize. At others she vanished in one of those ominous canyons
between the ranges of water in the sea while the wind whipped the black
water over the place of her disappearance into a great explosion of spray
before dropping what seemed like large funeral wreaths of white sea
anemones, to mark her grave. Several times I thought the lovely ship had
gone for ever, but always she reappeared in the dark area between us,
suddenly shining with heraldic reflections when the light of sulphur from
above found her glistening sides and she climbed, triumphant and
heroic, yet another summit of alpine water and at last accomplished the
last deft swing to put her stern on the storm. And when the captain, who
by now was watching it all through his glasses, let them fall down at last
on to his chest, he yelled more at the storm than at me, "By God, she's
done it. Dear heaven, what a ship!"

He sounded near to tears with relief and admiration. Knowing by then
that the *Edinburgh Castle* had safely accomplished her perilous about-turn
and was on her way as fast as she could go, we watched her disappear
into the murk of the storm, hoping and praying she would be in time. But
she never found the Greek ship. Not a lifebuoy or oar was observed, and
she became just one of the many ships since Vasco da Gama's day which
have vanished in those same waters.

No wonder Captain Coulthurst never lost an opportunity to stress that
one had not only to love the sea but to fear it in as great a measure. In a
strange way which he could only understand musically and not express
rationally, *The Flying Dutchman* sailing against such storms carried a
warning that none of us could afford to ignore. That warning was to stay
with me and to grow the more I heard of Wagner and read about his life
and experienced the retrogression of the classical European spirit in
Europe between World Wars and, above all, in Germany, but the full
meaning burst into conscious recognition was to come only much later
on one of the most frightening evenings of my life. This was at the famous
Walpurgisnacht March at Nuremberg when, in the midst of the
pounding of thousands of goose-stepping Nazi feet under the flame and
smoke of a lava stream of light, I would recall those moments back in
Captain Coulthurst's flat. I still do not know precisely how, but I believe
more than ever that the horrors which followed that evening in
Nuremberg had been foreshadowed during those Wagnerian days in the

captain's room and ocean-going cabin, where all that was potentially so wild and prophetic in the music was underscored and made more emphatic by the movement of the tug bringing the rhythm of the sea to swell on its own and those of the instruments recording a fatal sound. If that were not so, how could I at such a moment in Germany, so remote in space and time, recall the man, the music and the sea so vividly in that hurricane of sound in which not only Nuremberg but the whole of Europe seemed about to vanish as if through one of those holes in reality of which we had spoken then?

CHAPTER THREE

The Singing Whale

FOR all that I owed the frail little *Protea*'s introduction to the classical storm of the Cape, my real initiation in what it meant actually to make a living out of the sea came soon after, in a more archaic manner. I used to walk the docks from end to end not only in the course of my duties but often also when I had the leisure. Watching the long line of ships discharging their freight, loading new cargo, preparing themselves generally for the uncompleted segments of voyage to destinations in a world I did not know, I was acutely aware how personal experience still was necessary to round my understanding of the role of the sea in history and the imagination.

When released from work I would often find myself reluctant to go back to my rooms or visit my new friends but preferred to return to the docks to be among ships. Each ship seemed then to be also a rare and enticing volume and evoked the same excitement I always felt when walking into a bookshop. They contained another idiom of life and meaning that I longed to master and read for myself. The longing was always increased by the glimpse of some Blue Peter – the prettiest of signals in the harbour among all the rich and handsome display of flags designed to enable ships to converse with one another in their esperanto of the sea. Caught in a fragment of light, it would indicate that the ship would be making for the open sea at daybreak. So I would pass the last ocean-going ship in this mounting mood of longing until I came to the less sheltered water near the open sea in the outer basin where the ships of the whaling fleet would gather to rest their crews. They would lie there, usually linked to the quay, three abreast and, because of the greater stress of water and tide, somewhat loosely held to their berths so that they were always shamelessly bobbing up and down as if with excitement and eagerness to resume the hunting which was their vocation. They looked small almost to the point of insignificance against the ships which towered over them, noses high in the air, as if full of distaste for the smell of whales which always clung to whalers. I realized that there was perhaps a kind of justice in their lying at the head; the readiest, the most eager and nearest to the sea because they were the most contemporary expression in Port Natal of what was first and oldest in the story of man's relationship with the sea. They were the sea's equivalent of the hunter, the Esau, the first-born of the earth.

I was to think back on this moment and find confirmation of the validity of my reaction many years later when I became possibly the first European to explore the great Okovango Swamp in dugouts or Makorros as they were called by the Africans who evolved them. Knowing that these Makorro men had been driven some two centuries before from the dry land of Africa to live caught between the Kalahari and this improbable swamp, I wondered how they had learned to make these dugouts. Their leader's reply to my question came with a note of pity at my ignorance. "The First Spirit saw our need and gave us the vision, and we just made the Makorro accordingly." He paused, pointing to some weaver birds making their nests from the tips of the long, tasselled papyrus of the swamp. "See how well they build but, even so, you would not think that they went to a mission school to learn how."

There was a snigger at the mention of a mission school. He instantly silenced them by saying gravely, "Both they and us, we learned our craft in the same school."

Those whalers, although I could not put it into such words at the time, as our latest version of the aboriginal image of the ship, made me warm all over with an assurance of unbroken continuity between the first man and myself, standing there alone in the dark and watching the riding light at the tip of a mast moving up the channel from Umbilo and Congella like the point of a cardiographic pen, to chart on the black sheet of night the rising beat of yet another ship in a fever for the sea.

By morning the whaling fleet would be gone and the berths in the harbour occupied by the duty tug of the day. When the morning light had cleared itself of the brown murk of the night and I had a clear view over the sea, there was no sign of a whaler between me and the vast curve of the gleaming horizon. The authorities had long since decided that it was as absurd as it was a waste of expertise and time to impose pilots upon the whalers' coming and going, and left them to manage these things for themselves. There were the rare occasions when I did catch sight of a dark blot of smoke, indicating that, early as it was, they had already either harpooned their first whale or had marked one for the killing.

It was astonishing to me how abundant whales were in those waters during the winter months off Natal at so late a day when all over the world the alarm was being sounded that whales were a fast vanishing species. The warning was already spelt out clearly that, unless the tide of killing was reversed, the whale would go the way of the Dodo of Mauritius. What was so disheartening was that European man seemed to have learned nothing from the destruction of one of the rarest, most wonderful forms of life over the centuries in his own northern hemisphere. We should have taken to heart the warning implicit in the way we had killed the whale in the North Atlantic and Arctic waters. We would have seen to it that we did not repeat this amply discredited error and remove from our impoverished selves yet another source of wonder on which our own power of creation and renewal depended.

Perhaps some faint excuse for the killing could be found in the fact that the numbers of whales may have appeared inexhaustible and, the deeper south whales penetrated into these new seas, the more abundant became the supply. To men who had never seen whales before, the sight of these great mammals coming out of the waters at them produced feelings of alarm and aroused superstitions which had been suppressed, but not yet totally overcome, about the wisdom of this venture into what had been from time beyond record or memory, a forbidden world. For so superstitious a breed of people as sailors, the shock to the senses caused by animal life in the sea on a scale they had never imagined, and for which no conscious instruction could have prepared them properly, accounted for the flood of terrifying stories of encounters with sea serpents, dragons and monsters and, in rare moments of lyricism, of mermaids and mermen. These stories became a common ingredient of conversation of ordinary men, even the most land-locked among them, all over Europe and Britain; so much so that one historian declared in despair, "Nothing can either be painted or imagined more fearful than the terrible sea monsters the early navigators encountered."

Even we in the interior all knew and laughed at the story told in our histories of how the sailors in a Dutch East Indiaman one day in June of a year as late as 1629, say that a gigantic monster with seven heads and a mouth "so large that one could have thrown an ox into it", shot out of the sea with such force that "had it bumped against the ship, it would have been riddled with holes and everyone gravely imperilled".

There was Thomas Herbert, who was to stand beside the abandoned and most sensitive of Kings, Charles I, as he knelt at the scaffold. He wrote that in the ship which took him to the East they were escorted all the way by whales to their anchorage in the bay below Table Mountain. He called them the "Sea's Leviathans, who after their manner thundered our welcome into Aethiopia, fuzzing or spouting part of the Bringy ocean in Wontonnesse out of their oylie water hole bored by nature on top of their prodigious shoulders like so many floating islands concomitating him".

Even here superstition can be seen at work, presenting 'facts' that were sheer illusion or hallucination inspired by fear. We know today that whales may talk in the most tender and moving voices to one another deep down in the privacy of the sea, but they do not, even when harpooned, make any audible sound on the surface, not even in their flurry of death accompanied by the cruelest of pain. No, the validity of these records, covering nearly five centuries of coming and going between Africa and Europe, was something else and they left no doubt that our part of the South Atlantic Ocean was in the beginning full of a unique life of its own. It was stocked with whales, walruses, seals, fish, penguins, dolphins and porpoises, just as the air was alive and vibrating with the wings of the most amazing variety of birds, from Mother Carey's chickens and stormy petrels, which were our own Aeschylean portents of tragedy around our capes, to the albatross, gliding out of the remote Antarctic on the longest span of wings ever granted to a bird since the days of the vanished pterodactyl. Yet life in the interior had already taught me enough of the terrifying proportions that destruction of its unique plant, animal and bird life had achieved to make it a source of powerful emotion and subtle resolve in my being. But the natural life of the sea I had somehow assumed was still a different matter. However, my experience of whaling at Port Natal quickly made me realize how all-encompassing and frightening was the acceleration of destruction in the greatest dimension of all. It happened not long after my *Protea* experience.

Among the new friends I made in Port Natal were two Norwegian families, the Egelands and the Grindrods, who controlled most of the whaling industry of the port between them. I had only one day to confess my secret but strange longing, and they at once offered to make arrangements for me to sail as often and for as long as I wanted in their whalers. My only problems were personal ones. In addition to working at least an eight-hour day six days a week on my evening newspaper, I had other work to do as well. The five pounds a month my newspaper paid me was not enough, even in those days, to keep me in clothes, food and shelter in Port Natal. I found other work by translating the letters which passed between the wool-brokers and other rich merchants in Durban and farmers up-country, who knew as little English as those on the coast knew Afrikaans or High Dutch in which the letters were written. I seemed at the time to be the only person in Port Natal who spoke both the official languages of the country. As a result I was soon so much in demand that I was making more out of my translations than out of my newspaper work.

The newspaper proved no problem. Both the Grindrods and the Egelands represented great financial interests in the city and had considerable influence with the management of the paper. Besides, my own editor welcomed the idea, because he was as horrified as I was amazed when we discovered, from the files, that no-one had ever before thought of writing about whaling. My translation work proved a greater obstacle, but knowing how whalers were continually coming back with their catches to the factory, I made an arrangement with Mr Clark and Mr White to report the movements of the whalers in which I intended to sail to the Indian head waiter at my favourite hotel. He had undertaken to organize a shuttle service run by his unemployed relations between the ships and the hotel, with these letters for translation.

There must be something of a sailor in the heart of every male human being, because even to this day, one of the most pleasing aspects of the experience was the spirit of excitement with which everyone, from my editor to the smallest Indian messenger, joined in making the adventure possible. Even the reticent Mr Clark and Mr White appeared to approve of what I was going to do. Mr White, however, delivered himself of one reservation. He wished that it had been possible for me to have my introduction to life at sea not in one of those smelly whalers but one of His Majesty's smaller ships; a mine-sweeper or, at the very most, a sloop. However, in the reprehensible absence of any appropriate warship he thought a whaler, on balance, a more manly way of getting to know the sea than steaming up and down the cost in a liner.

So there came an evening in the early winter when I found myself walking down to the docks dressed in the warmest of casual clothes, carrying a well-worn duffle-bag, lent by Mr Clark, filled with things I would need at sea. Even the loan of this bag had an extra meaning for me. It was the equivalent of something which happened on an expedition into the bush up-country. A famous hunter of the day, both a friend of the family and the original of the Afrikaans hunter in John Buchan's novel, once handed his own rifle to me and asked me to shoot the buck we needed for food to last for several days to come. Young as I was, I knew it was the greatest compliment of which a hunter was capable.

For a Sunday the harbour was unusually quiet. For once, there appeared to be no ship in a hurry to leave, and no late-night coaling, as there generally was, of even one ship to break this round of vision and sound. It was only when I was passing the last of the line of ships, just before reaching the weekend berth of the whalers, that I began to catch up with isolated groups of men. Judging by the firefly light that came and went among them, they were smoking pipes in the course of a measured conversation as they moved in the same direction as I did towards the Point. They could only be members of the Norwegian crews of the whalers returning from the small Scandinavian church near the entrance to the docks. It was built, like so many of the little houses around it, of sheets of corrugated, galvanised iron, and looked like a shanty-town church. It was probably the poorest and least prepossessing of all such buildings in Port Natal, but, unlike its richer and more presumptuous fellows, it was always crowded on Sundays. These whaling men, I was to discover, were as religious as they were superstitious. It was at times difficult to tell where precisely the frontier between the two could be drawn.

They were dressed in their best suits; all the same homespun quality and cut, rough, enduring and obviously meant to last out a lifetime of weddings, baptisms, funerals, religious ceremonies and, above all, the great ritual for stepping ashore in foreign parts. This fact alone stamped them as a breed of seafarers different from all others because, whereas the crews of the predominantly English harbours and the inhabitants of Port Natal took their Sundays casually and dressed accordingly, for these men it was still the 'great and terrible day of Our Lord'.

For this reason they refused to kill on Sundays. At the very most I would occasionally see a whaler labouring back to harbour on a Sunday morning because of a late kill the night before. But usually by Saturday evenings these little ships had completed their week's business of feeding the hungry slipway that led to what I was already calling the whale crematorium at the foot of the Bluff.

I soon passed the sailors with what could have seemed indecent, if not blasphemous, haste, so concerned was I now not to be late for my appointment with the captain of the whaler in which I was to sail. With the tide out, the decks of the ships were below the level of the quay. I had only to walk down a plank on the first of these, cross the narrow deck between the foremast and bridge, climb over the low rails, board the second whaler and cross over to my own ship, the *Larsen II*, on the far side. Halfway across I could already see in the half-light someone smoking hard by the platform on deck underneath the bridge. The smell of dead whale which still clung to the ship, despite its weekend ablutions, was thick on the air, but not opaque enough to prevent what must have been the strongest tobacco ever put into a pipe penetrating it and smarting in my nose. This was followed by the smell of Schnapps.

I was about to climb over the railings when the pipe glow vanished and a deep voice with a rasp to it, obviously accustomed to being as direct as commanding, demanded if I were the young gentleman from the newspaper. My association with Thor Kaspersen had begun.

My first impression was that the association would be neither long nor agreeable. Without another word he beckoned, pipe in hand, and led me down a steep companionway into a small saloon deep in the ship. When I joined him at the bottom of the ladder, he was already standing there, feet wide apart. I was to find that this was his usual stance even on land, as if everything in him presupposed the necessity for perpetual balance against the movement of the sea. He was of medium height, solid, and looked shorter than he actually was because of his broad shoulders. I had always thought of Scandinavians as the fairest of Nordic peoples. But to my surprise, he was swarthy, and belonged to a dark strain in their race – darker even than the average Latin complexion. Although he looked to me no more than forty, the brown hair was already greying and contrasted oddly with a much younger look on a well-formed though much-weathered face. It was a face dominated by determination and self-confidence, with eyes large and wide above strangely Mongolian high cheek-bones. The eyes had a variety of expressions that I was never capable of unravelling. His look could be as young as it was old; as experienced as it was innocent; as aggressive as it was solicitous.

At that moment, however, I was aware only of the expression of command resulting from Kaspersen's unfailing sense of unlimited responsibility. I was immediately corrected and told that I was never again to come down the ladder with my back to the saloon. I must watch and he would show me. I was pushed aside so that he could go back up the steep companionway. At the top of the ladder he turned round. With his hands on the railings and feet splayed outwards he came down fast and lightly to join me in the saloon, explaining that that was the only way to make certain that the ship's movements, always sharp and unpredictable, did not catch one out. He also assumed that it was necessary for him to speak only once to be understood and obeyed. Without waiting to find out whether or not I had any questions, he straight away told me to follow. He parted a pair of mushroom-coloured curtains hanging over what I assumed was the entrance to a cabin, and disappeared from view.

I joined him in what seemed little more than an apex in the *Larsen*'s sides, with just two bunks in it. Underneath were a couple of mahogany drawers; there was hardly room for the two of us to stand side by side without touching the steel hull of the ship itself. This, I noticed, was so thin that the light lapping of the water against the sides even on this calm evening was audible from where we stood. It sounded like a brush of wind among the reeds and bulrushes of a marsh. Every now and then, as if to stress the point, the sea would move the little ship up and down sufficiently to extract a sound as of strain unfairly imposed on such thin metal.

"I sleep here," Kaspersen pointed to the bottom bunk, "because I come and always I go first." He put his hand on the top bunk and added, "You sleep here. You come and go as you like."

I had just enough time to see that both bunks were made up with rough blankets and a single pillow with no cover. Then he bent down, pulled out one drawer and said, "Your things go there . . . but first you come and drink with me."

The table at which we were to drink was almost triangular, and we sat on two broad, inbuilt coaches that ran parallel to it and were, I discovered later, used as beds by two members of the crew of seven. The engineer and the boatswain slept in a cabin on the port side of the *Larsen II* and had curtains of the same mushroom plush to mark the entrance and give them, too, an illusion of privacy. The rest of the crew slept in what I was to find was the smallest of fo'c's'les. The saloon table too was covered with a cloth of the same plush as the curtains. On it, in front of Kaspersen, a large Bible lay wide open. On either side of it stood two big earthenware bottles of the gin he called Schnapps. I found myself thinking that they were odd proxies for the candles others might have thought appropriate in the circumstances. Beside the gin were two tumblers of thick, smokey glass. To my amazement he filled one of these almost to the brim, pulled it towards himself and began filling the other, presumably for me. I was just in time to prevent him from filling only a third of my glass with a quick gesture, which seemed to disconcert him as he exclaimed, "You do not drink?"

His tone and my fear of having offended him made me uncertain, and I tried to redeem the situation by saying with a lightness of tone that I was far from feeling, "But I do, though mostly on and off – as the lobster said".

It was the first and last time that I ever attempted to joke with Thor Kaspersen because a look of extreme bewilderment was aggressively presented to me and he asked:

"You have brought a lobster with you?"

It took a long time to explain the allusion before we returned to a more or less settled conversation that consisted largely of a series of questions rattled off by Kaspersen and my somewhat desperate answers. For, in spite of my faith in the Egelands and the Grindrods, I was beginning to suspect the wisdom of their judgement in sending me to sea in a ship under the command of someone who, by the minute, looked more and more like an incorrigible alcoholic. Between his first question and halfway through my answer he had not only downed a glass of gin but refilled it, and tried to reprove me for barely having sipped mine.

"You," he said, pointing a strange long finger at me. "You, they say, come from the interior, no?"

"Yes, I do" seemed the only possible answer.

"You then, Boer? And Boers shoot much and shoot damn well." It was as much an assertion of fact as a question.

"Yes, some people say we do shoot rather well, but perhaps too much," I told him.

He ignored the reservation in my reply and rushed along the direction of what was clearly a predetermined and impetuous theme of his own: "Then you shoot and know elephant?"

The question was clear and direct enough and required the easiest answer. However, looking at Kaspersen, I hesitated because I knew that an enormous amount depended on the manner of my answer.

On any other occasion this sort of question from a stranger might have seemed natural enough. But his expression of profound longing, for which the hope of fulfilment had evidently been often deferred, could have had nothing whatsoever to do with me. Indeed, I felt that as a person at that moment I myself was as irrelevant and perhaps in danger of being discarded by some change in the mood of this strange and unpredictable man.

The thought came to me unbidden and could easily have been induced by a sudden movement of the *Larsen II* which, as I spoke, was lightly lifted and pushed sideways against the whaler moored to her side. It emitted a note of apprehension from its thin body of steel.

I don't know how long it took me to arrive at my answer – it came more by intuition than reason, perhaps because of a respect for the instinctive and a reverence for the imponderables of life that I had learned from the primitive people of Africa, who had been my companions from birth. I found myself saying slowly, staring into my barely touched gin, "I don't know that I *know* elephants. But I love them and respect them perhaps more than any other animal in Africa."

I hesitated, conscious that I was replying in terms of feelings that might be the last things sought by the matter-of-fact person I took Kaspersen to be. But, looking up at last, I saw that he was far from impatient and seemed eager for more.

Encouraged, I went on. "I don't know if anyone can really claim to know elephants. . . . My family have known a hunter who shot nine hundred and ninety-nine; his greatest ambition was to convert the number into a round thousand, but he was prevented from carrying it out. One day he realised that he had done all that killing without ever bothering to learn what elephants were really about. Once he realised that – and in fact he was a religious man – he was afraid he would be transgressing some law of God if he shot another. From then on he shot no more. Instead he has been spending the rest of his life trying to get to know them. Though the last time he stayed with us, he confessed that he knew less about them than ever –"

I could not go on, for I was interrupted by Kaspersen. In a voice suddenly humble he exclaimed more to himself than to me: "Ah – I think I know what that hunter must have felt!"

He paused. As if in the grip of the thirst of a man who had just crossed a great desert, he quickly filled his glass, tossed down the Schnapps again and commented, with this newly uncovered humility, "I, Thor Kaspersen, would give all my season's bonus and more to meet such a man. But, please to go on."

"There's really nothing to add," I answered more confidently. "Except to reply to the other half of your question and to say that I have had to shoot elephant myself."

This admission was the turning-point in our relationship. In the short time it took to utter so short a sentence, I had emerged no longer a stranger and intruder, but a member of the company that was élite to him. However new to the ways of the sea and whales, my status as a killer of elephants made me ever more welcome than any professional whaler. All this was confirmed by a new look in his eyes. Now he saw me clearly as some sort of equal, and suddenly his literal cross-examination had ceased to be important.

The rasping quality in his voice gone, he asked quietly, "And when you have shot your elephant, what is it then that you yourself feel? Good? Yes – no?"

I could answer that without hesitation. "I hate myself for it, not only at the time but for days afterwards. Even now I don't like to think of it."

"Hate?" He threw the word at me almost casually, as if he had expected the answer and merely wanted endorsement of my reply. "Hate? Why?"

Again I could answer immediately and without reserve. "Because, at one moment there is so much life through the sights of my gun; the next I pull the trigger and there's nothing at all. It's as if a great black hole had been made in the day – and everything around suddenly goes dark and empty."

I wanted to enlarge on this spare statement because by now it had become immensely important to me that this strange man of the sea should understand what I was trying to say. But I had tried explanations out on others when, as so often in the interior, men talked about hunting and shooting. Yet I had quickly abandoned my attempts because nothing that I had said had made sense to anyone. If those who knew the interior could not understand me, how then could this man who had never seen it all? How could such a man, for whom Africa must be a continent of sun, understand a fall of darkness at high noon? The darkness, of course, could have been explained metaphysically or metaphorically by the religious to satisfy the literal-minded. But as far as I was concerned that would have been less than honest. It had not just been a psychological event, but a darkening of the day. The impact had been all the greater because I had been totally unprepared for it. There was nothing in my surroundings at the moment of shooting even to hint at darkness. There, where I stood, the yellow sedges and bush swayed like some kind of seaweed in that burning swell of light breaking over the red earth between me and the elephant. Indeed, so much was there of fire and light that there was no room for shadow. There was just a paler form of sunlight under the trunks of the tallest trees. Even sound had become an equivalent of light, and imposed an element of hallucination on my ears by extracting from the land a hissing like the surf of the sea. It supported a chorus of millions of sun-beetles packed row upon row like insect cherubim and seraphim in the galleries of dark, green branches of the devout spires of the trees. There they were raising their platinum voices to heaven in an hysteria of ecstasy. How could there then have been any premonition of darkness in the mind of anyone standing where I stood in order to do, for the first time, what I had to do? Some such experience, even if transmitted vicariously, was necessary if the shock to come was to be properly understood. Still I hesitated, fearing that any account of so great a preamble of light to darkness would only make the event sound so paradoxical as to be judged 'far-fetched' by someone like the Captain who had never experienced it.

However, to my amazement, I was instantly exonerated. I found a rough harpooner's hand briefly put on mine, and a voice truly objective with pity exclaimed, "Ah, that hole . . . that darkness . . . that emptiness. I, Thor Kaspersen, I also know them well."

He broke off. I thought from his expression that it was because of a fear of not being understood. Had he misread the surprise on my face? But he hastened to explain: "You see, the whale is the elephant of the sea. I have killed more whales than the nine hundred and ninety-nine elephants of your hunter friend and –" his voice dropped into a minor key that was heart-rending, "and still there is no end to the killing. . . ."

The look of helplessness which accompanied this conclusion was like the expression I had once seen on the face of a man who, together with his horse, had been swept to his death in an African river in flood. I was not surprised when he refilled his glass before adding, "Tomorrow I shall show you some of that killing. For sure, you will be able to judge which makes the day darker, the killing of the whale or the killing of the elephant."

These introductory talks with Thor Kaspersen, as far as I can remember them, made such an impression on me that I am fairly certain that this is an almost accurate rendering. I have learned from the natural peoples of Africa that there is in fact a positive aspect to 'forgetting'. A memory cultivated out of honour for the quality rather than the quantity of human experience will retain all that is necessary to make it a compass for man in his search for truth on his zig-zag passage through life. Indeed, in the mere writing of what is only an introduction to an experience which has already inspired a long story I have called *The Hunter and the Whale*, I realize that my memory retains in detail much material of a totally different kind. For in the weeks that followed, when Thor Kaspersen and I were together, either on sea or land, our interchanges of thought contained more in shared experience, because so joined were we in mood and content that on a level of meaning outside space and time they merged into one. But this, our first meeting, came to an abrupt ending. Suddenly Kaspersen ordered me to bed. He told me we would be leaving harbour at four in the morning. He said that if I wanted to know what a whaler's life was really like I should start at the beginning and be on deck the moment he cast off.

I did as I was told readily enough. Now that the need for our conversation was removed, my mind was free to realize that the excitement of anticipation which accompanied me on my walk down to the sea had gone. The nature of what I was doing seemed to have undergone a sort of deep-sea change. I appeared no longer bound just on a personal adventure, but subtly contracted to something unforeseen in a way that I cannot explain. I felt bound as if on a pilgrimage with Thor Kaspersen as the itinerant High Priest charged with performing the requisite sacrificial rites. So I climbed into my appointed bunk and lay half-dressed on rough woollen blankets. All I had to do in the morning was to pull on my boots, slip on the thickest of my sweaters and be ready for the call.

But, alas, the weariness which affected me was caused more by a troubled spirit than physical fatigue. I lay there for a long while before I could sleep. Whenever I opened my eyes I found that the slit in those incongruous, plush curtains over the entrance revealed enough of Kaspersen to show him in the same position in which I had left him; except that now he was reading from his Bible, pausing every now and then to refill his glass. This impression of the man became a kind of portrait hanging in a long gallery of memories of evenings that were to end just like that first Sunday evening in *Larsen II*. These were all to become, as it were, just one unchanging moment held with singular clarity under the microscope of whatever there was of meaning in our first encounter. Consequently I feel compelled to do what I can to examine it in detail before I record what followed.

(It is true that between ultimately falling asleep and being woken by an alarm clock and the tramping of feet on the deck overhead, no time seemed to have passed at all. Half awake, I dressed and rushed through the curtains, barely conscious enough to note that the glasses and Schnapps had vanished from the saloon table, but that the Bible was still lying wide open. Kaspersen had been reading from the Book of Revelations as his preparation for the day.)

The sight of this man at the table was, as I have said, like a portrait. Only his face was lit; all the rest of him was in shadow, as in a Rembrandt self-portrait, dominated by an empire of darkness as yet not transformed. There was no artifice in his 'pose' since the subject, at that moment, could not have been more vulnerable to exposure. He sat there as if stripped naked of protective cover and, of course, without any knowledge of being observed.

Far-fetched as the parallel may seem, I suspected that this life of a deep-sea hunter had in it something of the meaning for Thor Kaspersen which painting had for Rembrandt. Whaling for Kaspersen was not just a means of earning a living, or of bringing solace to a spirit so unquiet from birth that the excitements of the drama which were the inevitable accompaniments of hunting on so vast a scale, were his only tranquillizers. It was, instead, the only way he knew of realizing in himself the greater self-knowledge that he needed to rescue himself from the feelings of individual meaninglessness in which he was trapped.

In this connection I thought of Baudelaire who himself once sailed the same seas where Kaspersen was doing his killing and who was, as he sailed, perhaps even more deeply perplexed and troubled than my captain. In one of the most moving appraisals of painting ever written, he labels painters 'lighthouses' and puts Rembrandt high in his roll-call of the grand masters of this ancient order.

As I lay there watching Kaspersen, my mind turned again to the last self-portrait by Rembrandt. To me it remains an almost unbearably moving testament, wherein the painter bequeathes the totality of himself impartially to all who have eyes to see. And the emphasis must be on the totality, because gone at last are all the special pleadings, evasions and excuses that men use to blind themselves to the whole truth of themselves, discovering in the process their portion of the estate of aboriginal darkness to which they are the natural heirs and successors.

It is true that the effects of constant drinking, probably as ardent as that of Kaspersen himself, made the face of the painter much more flushed than in previous portraits. The veins inflamed by spirits are more pronounced. The wrinkles on the forehead, the creases at the eyes, the lines on the swollen features are deeper than mere erosion of time warranted. The warts are multiplied and not a defect is spared in this final re-examination. Yet, no matter how much greater the defects revealed, there is at last, unblurred in those blood-streaked old eyes, a look of a certainty of pardon, and an intimation that through total surrender to the truth of himself he has been emancipated from error and discovered something greater than even his art to carry him on beyond the advancing moment when painting would end. How far then would Thor Kaspersen have to sail, and how much more killing would he have to do, before he found his equivalent translation?

So began my season on the *Larsen II*, and yet after two seasons whaling I was still to be reminded of the unanswered question one afternoon when I was alone on the little bridge with the captain. I thought the moment singularly blessed with beauty. The sky was as blue as the sea, and the sea as illuminated as the sky with the westering light. One could easily have fostered an illusion of our ship as a bird hovering in space rather than a vessel afloat on the Indian Ocean. The horizon was clear and precise and round with a glint of the first yellow of evening and so without flaw. I felt at one with it all and centred, as I had only done before on privileged occasions in the bush, until I was startled by a loud outburst by Kaspersen.

"You like those horizons, my young David, don't you?" he growled, calling me by the name he had given me ever since I told him of my first encounters with elephants. (These had instantly become all one in his imagination as Goliath, just as all whales had long since been the one and only Leviathan.)

I was about to tell him how singularly beautiful I found the horizon about us, but he gave me no time for he rushed on to say, "Well I, Thor Kaspersen, I hate them. Always they follow me around. I travel and travel and travel, and I sail and sail – and you would think they would all be finished – all used up. But always there are more. I look into the sea; I look at the sky. No horizon, only a point where you don't see – where the eye fails you. All right . . . fair enough. But the horizon, it is not a place where my eyes cannot see. Your eyes especially, David, your eyes, the best eyes in this damned ship, can see God-damn further, but then comes this horizon and says, 'No, so far, but no further'; it's like being in a travelling prison. No matter where, how far, how often, always you in the middle of this prison. No? And I tell you I hate it."

At the time I thought the outburst merely a result of the tensions which stretched him to well-nigh unbearable extremes as the time between one expected killing and another had lengthened. But I soon realized that he was unknowingly speaking in the code of his own inner unrest. He had reached the point when the human being realizes that no amount of knowing diminishes the amount of the unknown. Knowledge moves and searches for meaning, just as our little ship was moving through the sea and looking for whales with the inexorable horizon insisting on keeping us always at the same distance. It was as if the unknown infinite made a mockery of the known infinitesimal. One needed something more than the knowledge which has brought one there to carry one beyond. But what was that something?

Clearly that was the dilemma which confronted Thor Kaspersen more and more acutely whenever he had neither killing nor drink to divert him from the matter. Just as clearly, the dilemma raised issues of profound import, and I was to come to my own tentative inklings of how one made one's peace with what 'horizons' meant in the alphabet of the imagination. But at that moment I could only feel as startled as I was sorry to find my captain so at odds, so locked out of a composition of the sea and sky and light contained in a pure circle of gold that brought me personally an ecstasy of peace beyond articulation.

I was even presumptuous enough to be somewhat critical and, with the arrogance of youth, to feel that I could think of nothing more awful, symbolically, than running out of horizons; of yet more things to know and more places to discover. But, fortunately or unfortunately at that moment, although I was so much lower than the look-out in the crow's nest, I saw a flash of mystical silver vapour cloud a patch of luminous blue. Hating myself for it, I had to point it out to the captain. Within the hour, another of my 'black holes' had been made in the day to aid and abet the fall of darkness which later hastened to join it.

There are, I know, human beings either doomed or blessed to a way of life which deprives them of any of the normal indulgences and privileges of their communities. It is almost as if they are charged with a role that sets them apart and makes them incapable of joining in the society of men. They are, as it were, victims of a kind of Ancient Mariner compulsion of the spirit, which makes them incapable of participation in the round of life which is symbolized in its essentials by the church of collective humanity. It is not that such men do not feel human in themselves and seek to shun the company of their own kind. If once known and understood, as I was to come to know Thor Kaspersen who was also of this uncomfortable breed, it is clear that the longing within them to be part of the ordinary community of man is greater than that of any of those who daily take for granted such belonging. They are rather like those lone animals I have so often observed in the heart of Africa who can neither join nor leave the herd, but are forever moving around the far perimeters of their fellow creatures to lead a form of existence wherein they have to learn to make their peace with the fact that the most they can achieve is a satellite companionship. I can remember times without number when I have seen these satellite souls rebelling against isolation and the fate which compelled them to face alone the daily dangers of life in desert, veld and bush; dangers in which even the numbers of the herd were no certain protection.

It occurred to me in time that this kind of separation, even in the animal, was necessary to create a greater awareness which it was impossible to acquire in the context of sympathetic numbers of their own kind. In the years I had already spent in devout observation of the creatures of Africa, it was most striking how these lone phenomena developed senses so keen that the beasts who preyed on them and their kind would leave them alone, because they realized they were no match for the qualities of vigilance produced by loneliness and isolation. It was, in fact, far easier to prey on animals who assumed that there was safety in numbers. If this were true and necessary for the increase and renewal of animal awareness, I often wondered how much more necessary it was for the human being. Unlike the animal, the human had no sheer, blind obedience to the will of nature which is instinctive. On the contrary, he had an inspired kind of disobedience to the laws of nature which led to a recommitment of life in a more demanding law of individuality designed for the growth of consciousness. This growth set the implacable pre-condition that any new awareness had to be lived out in isolation before it could be understood and known, and made accessible to society. I believed that Thor Kaspersen was just such a spirit.

Unlike most of a score or more of whaling captains that I came to know, he was the only one who had started his life by going to sea as an apprentice in a passenger ship and persevering until he had qualified for a Master's certificate. But with that done, he could not wait to leave what was to him nothing more than a 'floating hotel'. Working in hotels of any kind was not what he had been born for. As for cargo ships, they were just the 'trucks of the sea', and although his whole imagination was devoted to the sea he had not yet found the special ship on which to sail. But one season on a whaling fleet in the Antarctic was enough to convince him that no other form of life at sea would suit him. He felt, in a way that he could not define, that this was part of the purpose of his existence. He was uplifted by the feeling that he was at last on his chosen way. When the Antarctic season ended he had no desire to return to Norway, and only regretted that the hunting season in which he had participated was over. Then he discovered that as many ships as could find crews and captains were going to follow the whales north into the warmer waters for as long as the long Antarctic winter froze their seas and made life there impossible even for the whale. Again he had no difficulty in obtaining command of one of the few chosen whalers and a crew of like-minded hunters. And that, he said, was how he came to Port Natal some thirty years previously. In all those thirty years he had not once been back to Norway.

I cannot pretend that this outline is representative of the whole of this man. It does not convey any of the imagination and sensitivity he put into the task of educating me in the school of the sea, and particularly in its course of hunters of the sea. This concept was all the more striking because his relationship with the small crew under his command was almost perfunctory. Provided they did their duties well in his little ship, his behaviour towards them was little more than correct.

I know this may sound perhaps more reprehensible than it was in fact. None of the Norwegian crew ever seemed to be much in need of command or direction. They seemed to be sufficiently in command of themselves and to know what to do, and so make unnecessary the sort of discipline that one became accustomed to later in the Merchant Marine, and even in its tramps. They were by nature farmers, and indeed most of them, once their round of whaling off the coast of Zululand was ended, returned to farm their marginal land in Norway. Their relationship, not only in Kaspersen's ship but in all the other whalers, tended to be more that of a farmer with his labourers who were almost a part of his family, than that of the captain of his ship with his crew. It was enough for them that Thor Kaspersen handled his ship better than anybody else in the large whaling fleet. He was by far the best gunner; brought in more whales to their base by the Bluff, and so earned them the highest bonus in the fleet.

However, with me it was a totally different story. Starting from a point where I was truly bewildered, dismayed and even mistrustful of the man, I soon discovered a far-ranging kind of fellow-feeling for him which uncovered unexpected depths of compassion and a sense of the importance of an understanding in the human spirit that is beyond censure and judgement. Also he taught me things about whales and the sea during the three seasons of spasmodic winter whaling that no-one else, I believe, could have done.

What did he teach me? He taught me, of course, all that there was to know in those days about the life of a hunter at sea. He provided my imagination with the aboriginal point of departure in the evolution of man's relationship with the sea which is necessary for the sense of continuity with our primordial beginning. He passed on to me all that he knew about whales and their hunting – in greater depth than anyone else could have done because of his own compulsive interest in what he did. This interest of his ensured that in our discussions there was always two-way traffic between us. The elephant in the bush of South Africa was to him then what the whale was to the sea. But I was gradually to find out that there was much more to it than that.

This happened on another of those Sunday evenings when I found him alone in the midst of one of his ritual sessions with a glass of Schnapps by his left hand; a dark brown earthenware bottle of Schnapps on his right and, in between, on the table, his big Bible, open again at the Book of Revelations. There was an exceedingly heavy swell breaking over the breakwater at the entrance to the harbour so pressing that it ran on through to the dock where we lay at the head of the line of whalers. The range of this swell lifted, pushed, pulled and tugged at our little ship so persistently that the thin steel plates of its flanks groaned and squeaked like a mouse caught between the paws of a gigantic black cat just playing with it before the inevitable kill. However different the vocation of whaling may have been in the other vessels of the large fleet, there was no doubt in my mind that whales and the sea where overwhelming and unknown symbols in Kaspersen's imagination which played with him as a cat with a mouse, and could one day, I feared, lead to his death.

On that evening, as soon as he saw me he remarked in his gruff voice, "Ah, young David, there you are. Damn late! But better late than never. Please to sit down at once."

There was something so urgent about his command that I instantly dropped my duffle-bag on the floor beside the bench opposite him and did as I was told. One look at his face told me that he had reached the climax of his drinking, when he became paradoxically most sober and clear-sighted.

I had barely settled in my seat when he asked, as peremptorily as ever: "You dream, young David, of elephants?"

The 'young' was always a reassuring sign, because it was the greatest token of verbal endearment of which Kaspersen was capable

"Not for a very long time," I answered, now certain that something unusual was stirring in his spirit.

For a moment he looked alarmed and asked even more hurriedly than before, "But you do dream, don't you?"

"Yes, of course! Lots. . . ."

He interrupted me before I could complete what I was going to say.

"But not of elephants!"

As I nodded in agreement, it occurred to me that I had seen and known so many elephants that, in a sense, there was no need for me to dream of them.

I said so without realizing how superficial such an explanation would be to my captain's mind. "Well! I, Thor Kaspersen, I dream of elephants and whales and whales and elephants all the time, and I know nobody see more whales than me."

He paused, glared fiercely, not so much at me but at the obdurate and obscure face of whatever it was that obsessed his thinking, before going on more gently. "But perhaps something in what you say, young David: I dream more about one whale I have never seen and an elephant which I also have never seen. Ach! If only I could see and catch them at it . . . I . . ." He did not finish but let a sigh of some unutterable longing break from him.

Only then, after a pause and more Schnapps, did it come. Ever since he was a young boy and had first started reading adventure books, he had always had a recurring dream. In it he was watching a great black river wind out of a dark, dark land covered with very thick bush, broadening as it made for the sea from which he appeared to be observing it on a ship of some kind. A red dawn was just breaking; so quickly and with such violence that it seemed more like an explosion. It stained the river so that its waters looked like a stream of blood. Swimming up the centre of the stream he saw what he now recognized to be a sperm whale of gigantic size. He had hardly seen this when out of the dark forest there stepped a black elephant as great and majestic among its own kind as the sperm whale now swimming towards it. The elephant walked to the edge of the river, stood still on the edge of the dawn-stained water, and stared for a moment at its oddly quixotic reflection. Then it threw up its trunk and trumpeted a sort of réveille for the day. As it did so, the whale at once changed course and made again for the elephant.

The elephant saw the whale and waded into the water to meet it. The solicitude with which the elephant went towards the whale had something Biblical about it . . . perhaps like the story of the father going to welcome the return of the prodigal son. Did I know, Kaspersen interjected, that the whale, far back, had been a land animal that took to the sea? Then quickly going back to the dream, it seemed that when the two met, the elephant put down its trunk into the water beside the dark head of the whale and they conversed together. Kaspersen explained that he could not breathe at that moment and came near to crying because of the emotions aroused in him.

As the phrase 'conversed together' fell from his lips, my captain suddenly thumped the table so that some of the Schnapps spilt on to it and he shouted, "Yes, talked to it, God damn, David! Talked to it!" Then he went on to say that when they had 'talked', the elephant turned round and walked rather sadly back up the bank and vanished into the bush, leaving only a shimmer of gold on the trembling olive green leaves to mark the place where he had re-entered his own element.

"But the whale," Kaspersen said with a strange look of exultation, "did a slow, dignified circle like a king leaving a palace. When back in the middle of the water he . . . can you guess what he did?" He paused and then whispered, "It began to sing and, still singing, went on out to sea."

He had such an intense look that it made me acutely uncomfortable and I wanted to turn away. But he resumed in time to prevent me and said in a far-away, outer-space tone, "I tell you, young David, that singing was really something! Never did I think of it but I long to sleep and dream of it again. I go back to my bunk saying – tonight I must dream of it – but no, it does not come when called. When I do not expect, then suddenly it comes. And then the music clearer than ever, and I wake up to more trouble with myself than before. "Now, young David, what the hell do you make of that?"

Happily he did not wait for an answer, declaring that all he knew was that if some great person could interpret the language of the whales for him, then he would know the secret of life. For he, Thor Kaspersen, was certain that life and its purpose were precisely what the talk between whale and elephant had been about.

I was to think of the 'singing whale' many years later when I heard the first recording of the sound of whales talking to each other in the deeps of the sea; I was almost unmanned by such delicate, vulnerable and tender music. I realized only then how Thor Kaspersen's dream had also been prophetic, because at that time whales were thought to be incapable of sound. We were then united in the conviction that, if these great animals were capable of sound, the horror and pain of their killing, the shattering impact of a harpoon fired deep into their tender flesh and the explosion of the grenade enclosed in the harpoon, would force some sound out of them – a sound so overwhelming that whaling would have stopped there and then.

However, at the time, all I could tell Kaspersen was that, although I was as dumbfounded as he was about the nature of the communion between the elephant and the whale in his dream, the actual physical meeting was not as improbable as it sounded. I told him how the Zulus had told me that in the past many whales had come swimming up one of the greatest rivers on their coast, the Umkomaas, and there, away from the swing and unrest of the sea, had suckled their young.

Naturally this anecdote immediately excited Kaspersen and convinced him that Providence had brought him to this part of Africa precisely so that he could pursue the meaning of whatever it was in this recurring dream which haunted him so much. The immediate consequence was that instead of making his intuitive sweep for whales, we steered, in the early hours of the following morning, for some forty miles from Port Natal to where the Umkomaas, then still covered with thick indigenous bush, broke out of the great continent of Africa, rising steeply with the dawn light behind it to reveal the deep, raw gash it made in the land on its breakout to the sea. We went dangerously close to its mouth, almost to the edge of the surging swell, so great was my captain's urge to observe it thoroughly.

He studied it long and hard. But when sunrise revealed the first rash of red-roofed houses on the coast close to both banks of the river he winced. Shaking his head, he exclaimed with dismay, "Yes, sometime, maybe, that was the river where whale and elephant met, but not today, nor tomorrow, nor evermore."

With that he began spinning the wheel as fast and easily in his gunner's hands as if he were a dark croupier of fate measuring its law of chance in some game of roulette. Then he began to turn our ship about towards the open sea and rapidly rising sun. He was muttering as if desperate to convince both himself and me that, nonetheless, all was not lost. "I don't know about elephant, David – but somewhere out there I am certain there is still the singing whale swimming around . . . and sometime I, Thor Kaspersen, will find him."

He paused, and I noticed that he was looking for once as if the horizon, which always bothered him, had ceased to exist. For once perhaps, he was able to see something beyond it coming towards us from the margin of the day of light that might have been present at the beginning, but was no longer in the world in which he and I were enclosed. Then, like somebody coming out of a trance, he tried to dismiss the matter with one of those brave jokes which men turn to when a sense of pre-ordained tragedy that would be too much for flesh and blood to endure, threatens to invade them.

"When that day comes, young David," he said, doing his best to smile, "I'll send you telegram and quit whaling. Then perhaps you and me go looking for talking elephant." He spun the wheel to steady the rudder amidships and we made for the open sea to resume our routine of killing.

From the beginning to the end of my three seasons with this man I got more of the feel of the sea, and more experience of its role in the life and spirit of man, than I came to have in the many ships which I subsequently sailed in across all the oceans of the world. It is something for which I remain intensely grateful.

One of the greatest of those moments, by some great good fortune, came on my very first day out at sea with Thor Kaspersen. He was still suffering from the physical after-effects of a long weekend of intense drinking and the subsequent shock of the sudden and total abstention that he never failed to observe at sea, and I was obviously and understandably an encumbrance to him on that first day. I was not surprised therefore that when the look-out appeared characteristically without any command, at the foot of the rope ladder to take up his position in the tub at the foremast, Kaspersen dismissed me from his side with a curt though impersonal suggestion: "You want see whales? You go with him and look damn good!"

With that he called out something in Norwegian to the burly farmer-sailor who already had a foot on the first rung of rope. Consequently the man waited and, after telling me to do exactly as he did – in serviceable English – led me up the ladder of the foremast. Soon we were up together in the tub at the crow's nest looking out over the dark blue water under the morning sky. It was a day without a single cloud and, as always, the sea heaved with the long, deep swell which is the everlasting thrust of life coming straight from the steady beat of the heart of that great ocean of history. There was not a ripple to mar the reflection of the day in its water, nor blur our vision of it.

For a moment just after sunrise, far too brief for my liking, the light of morning was so charged with gold that it became curiously substantial and gave one the feeling, high up there at the masthead, that we were not moving through air so much as through honey. Those mornings, and for that matter the evenings too, had the longest and most level shafts of sunlight that I have ever seen. For me they were always the true glory of those miraculous Zulu-winter days at sea. As I adjusted myself in the narrow space of the tub, I could see the dark land lying like a bangle of pewter on the edge of my vision. Just below it, a strand of pure gold which the sun had made of those dunes of sand at last reached the sea. I watched it fall away behind us until the strand of gold merged into pewter and, ultimately, the pewter itself was drowned in the water. I realized there was something about the scene that, beautiful as it was, puzzled me greatly.

Instinctively I turned to my companion who was already scanning the sea, not only straight ahead but also to port and starboard. He had told me that when he was on watch he spoke to no-one, so that nothing could spoil his concentration on the sea and he could be certain that if a whale did blow out there he would be the first to see it. He would prefer it, therefore, if I too kept quiet. As a result, the question which had arisen out of my puzzled feeling was stillborn, and its answer postponed to the end of what turned out to be an eventful day.

The question was why, on all the immense stretch of ocean, there was not a single ship or whaler in sight. Considering the long line of ships of the fleet of which we were part, it seemed strange to me that we should have so huge a stretch of water to ourselves. But as I was to discover that evening, this was deliberate. Kaspersen not only disliked hunting in a pack but also avoided them for practical reasons. He went more by intuition than by statistics or information fed into the ears of his fellow captains by one another. As a result, throughout the time I knew him, I could count on the fingers of my two hands the times when we went out hunting for whales in convoy.

This first morning, therefore, was as good an introduction as I could possibly have had into Thor Kaspersen's whaling ways. Moreover, the courage of his own intuitions, which was so marked a feature of his character, was amply justified by its consequences, so that his crew had implicit faith in his judgement. They staunchly defended him from the criticism to which he was constantly subjected by their fellow whaling men, who regarded this behaviour as contravening all the laws of reason and common sense. And when they saw how, season after season, Thor Kaspersen's men obtained larger bonuses than any other crew, they developed a steadily increasing resentment against the captain and, to a lesser extent, against his crew. This resentment mainly had to do with the feelings aroused in all men when they are confronted with the unpredictable and disconcerting workings of the laws of chance. Out of their own passion of protest at this lack of regard for law-abiding human beings and their concepts of justice, they condemned us with phrases like 'the devil's luck', as if men like Kaspersen had his inordinate success because of some secret, diabolical pact with life. Indeed, I was to be present at many discussions between the men of our ship and those of others over glasses of beer in one of those smelly bars so abundant in the harbour area. At such times similar phrases were freely used, and I heard Kaspersen frequently accused of having made a bargain with the devil.

However, unaware of all this on my first translucent morning at sea, I did the best I could to deal with the situation in which I found myself. I turned instinctively to something that I had learned from the Bushmen and Hottentot trackers who used to accompany us into the bush. It was, perhaps, the most important law of hunting in the bush: one person always had to be responsible for the track ahead and another for that behind; otherwise one could not only miss one's quarry, but could also be overtaken by disasters creeping up on one in the rear. I therefore turned myself with some difficulty in the tub so that my companion and I stood back to back. By leaning on the sharp rim I could, with a slight turn of my head, observe the sea in an arc from due east on to the south and west.

I had no experience of what we were looking for except, of course, the vicarious one that I had gained through reading books on whaling. But again, thanks largely to what I had learnt from indigenous trackers, I did not impose any wilful expectations during my watch over those immense, heaving but vacant waters. I remembered above all one of the basic tenets of their teaching: one kept one's eyes, as well as one's heart, free of any preconceived notions, so that whatever was to happen would find neither heart not vision shut against it, and consequently the event could fall unimpeded into one's senses. My reward came after about an hour of watching the ocean. By then I had forced myself to take no further interest in the way our ship was constantly swinging like a pendulum from what seemed like one perilous pitch to another. At an acute angle to port of the tiny deck below us, one would be staring straight down into the sea and one could often then see the fin of a shark or the shadow of a great fish pass with infinite ease like that of a cloud over a field of grass in the deep, green-blue water below. Then we would swing back to another equally acute and dangerous angle over to starboard. But fortunately my eyes and mind were open to receive one of those happenings for which precisely such a suspension of the senses had been prescribed.

The surprise was so great that I could not, at first, quite believe in the event. Midway between *Larsen II* and the horizon, and east-south-east of our port bow, it appeared as if the sea had suddenly come alive. It first breathed in deeply and then, from behind a high amber swell, it breathed out so powerfully that it sent a fountain of steam soaring upwards from the summit of one of its longest ranges of water. Briefly, the fountain stood there like a palm of silver in some mirage before the eyes of a traveller lost in a great wasteland. My lack of recognition, however, did not last a second and quickly transformed itself into a sense of being witness to something miraculous; so much so that I remember that the hair at the back of my head was tingling. Some reflex of my whole being made me grasp the sleeve of my companion's jacket violently and call out, hoarse with emotion, "My God, something is blowing there!"

He came about so fast that he nearly squeezed the breath out of me as he pushed me against the side of the tub. He was just in time to see the same samite-white fountain of life rise out of the trough between two great waves. Loud as my own call had sounded in my ears, it was as nothing compared to the voice of a Minotaur which broke from him. It sounded as if it had come straight out of the pit and drum of his stomach. As his shout in Norwegian of the first 'blast' of the day fell on the deck below, our ship came to life. I saw Thor Kaspersen on the bridge look up with great speed at the crow's nest in order to take a bearing on the direction in which my companion was pointing and, at the same time, spin the wheel to point the ship in the right direction, setting the bells in the engine-room below to ring out the most immediate message: 'Full speed ahead!'.

I found myself marvelling at the way in which the propeller bit into the water and sent the *Larsen II* jumping forward like a hunter whose rider has just seen his hounds flush a fox. As for the rest of this action that followed with equal speed and despatch, no detachment of the best brigade of Guards trained for a Trooping of the Colour could have carried out words of command as faultlessly as did the men of this ship.

The bo'sun was already at his station by the donkey-engine, winch and drum, on which a cable was wound and which led across the decks up to another yellow coil on the gunner's platform where the end was securely tied to a harpoon in the gun set and ready for firing. There, too, one of the crew was already checking that both harpoon gun and coils of the rope were arranged in perfect sequence and firing order. All this had sprung from an apparently indifferent, almost somnambulant moment in the progression of the ship towards the first blow of the whale. For the two of us in the crow's nest watching this blow, it was a transfiguration of a spout of silver into a white mist that drifted slowly sideways over the waters. It was, for me, the most magical blow of all the three that followed. But to my companion, who had watched everything through German field-glasses which he always wore round his neck, it was obviously just another business affair. He began to speak more to himself than to me, something to the effect that it was the blow of a Blue Whale, a big one too, and that it had 'sounded' and would now remain submerged for at least twenty-five minutes if not more; he paused and reiterated with passion that it was indeed an unusually large whale, since it had left behind it so large a design of smoothed-out water where it had dived back deeply into the heart of the sea. Then, when he was quite satisfied that the ship was making exactly for the target on which his glasses were focused, he proceeded to shout down his deductions to the Captain on the bridge below.

All this combined to make me acutely aware of how every person in the ship, and even the ship itself, had been joined together in an overwhelming singleness of purpose. The *Larsen II* already was travelling so fast that I was amazed by the facility with which she gathered way and sliced the heaving water. I was amazed also at how the quickening bite of her powerful engines and the turning of the screw had been transformed into a kind of wild music which was mounting to a crescendo which even produced a reaction from me. My own pulse quickened with an excitement greater than I had ever felt hunting before. I had become so caught up in this new language of communication that I felt as if I had known the man in the nest beside me all my life. At that moment the man started thinking aloud in a staccatto English for my benefit, as we went racing to the place where the Blue Whale had vanished.

"Big, very big . . . but patch of water vanishing fast . . . must hurry . . . if we to arrive before it goes. Why ship so slow today? Don't they know patches don't last long, not even biggest ones? God in heaven! Thor Kaspersen can do better . . . and to hurry quicker . . . quicker. . . ."

He paused and appeared to be on the verge of shouting at the Captain to this effect. But a quick glance was enough for him to show how tense everyone was down there below, and ready at their stations to do whatever was necessary.

After about twenty minutes from the time we had seen the last blow of the whale, my companion, whose eyes had been darting from the patch to his wrist watch, shouted something down to Kaspersen and pointed to his watch. Kaspersen acknowledged the shout with a wave. Immediately the bells in the engine-room rang out, loud, serious and clear. The thrashing of the screw ceased at once. The engines silent, the ship began to lose way immediately, until it drifted, hushed into an almost tangible drama of expectation. It no longer galloped over the water but moved gently, with lessening speed, conforming more and more to the abiding rhythm of the sea itself. Soon it was so quiet that I could almost hear the movement of the light air passing through the rigging. Then the unused head of steam, which had been building up in the boilers for that spent burst of speed, took over and began to emerge, first hissing beside the funnel and then finally turning into a high-pitched hysteria of impatience. Obviously the lookout and Captain had judged that they were so close to the place where the great whale mined the deep waters below that they could not risk any danger of sound coming from the ship which would send it hurrying away out of danger. Even up there in the crow's nest my companion talked to me in almost inaudible whispers. Down below, it seemed to me that the men were communicating not so much by voice as in a pantomime of gestures.

I noticed, for instance, how two of the crew, each armed with long steel cutting-lances whose function was as yet a mystery to me, were waved into position one on the port and the other on the starboard bow. Another was also summoned by hand to take over from the bo'sun who was still by the donkey-engine. A mist of steam had already appeared at the side of the engine and obviously satisfied the bo'sun that it only had need of one pull on the long lever attached to the winch, to set it unwinding the cable leading to the harpoon. For the first time, I noticed that the harpoon cable coiled by the gun led back to the eye of a great steel spring, itself firmly fixed to a stout derrick that was clasped tightly in bands of steel to the base of the foremast, but again, I had no inkling why. No sooner had the new man taken up his station by the winch than the bo'sun, without a word, turned his back on him and made his way forward, went fast up the ladder and joined Thor Kaspersen on the bridge. That, my companion whispered portentously, was a sign that Kaspersen was convinced he would get that whale, for he was preparing to hand over the wheel to the bo'sun when the whale reappeared, giving him the freedom to take up his position on the gunner's platform.

But what most struck me at that moment, as we were drifting along, hushed and swaying eurythmically, was the appearance on deck, through the engine-room, of the most un-Norwegian looking member of the crew for whom no act of my imagination could have prepared me. I watched with utter amazement as a tall, magnificently made Zulu, his black skin like silk with the sweat of his labours below deck, suddenly appeared. Obviously the amount of unused steam screaming beside the funnel was convincing proof that the fires he had been tending below needed no feeding for the moment. For, judging by the amount of coal dust on his naked shoulders and his rough seaman's trousers, he could only have been *Larsen II*'s stoker. He walked to the side of the ship with a balance as if he were a native of the seas as much as the earth of Africa. He drew in the fresh air of morning rapidly and with obvious satisfaction, stared out to sea for a moment and then suddenly, almost as if a voice had called him to do so, he looked up at the crow's nest.

He saw me at once and appeared to be as taken aback as I had been by his appearance. For a moment we stared hard at each other, he amazingly steady and balanced on deck; I and my companion swaying more than ever through a slow, wide arc as the ship, losing way, became more and more vulnerable to the movement of the sea. Some sort of communication passed between us, great as the distance was, informing him too that we were natives of the same land. A wide, vivid smile fell from his black, distinguished face. He lifted his hand high above his shoulder and greeted me as one of a people whose sense of manners is, perhaps, the most fastidious in Africa.

It was my first introduction to 'Mlangeni, who had already been serving on the *Larsen II* for seven years and who, I was to discover, though with no English himself, was closer in being and in feeling to Thor Kaspersen than anyone else in the ship. For the moment, however, my astonishment was uppermost and too great for other speculation.

I had barely returned his greeting in kind when our Captain signalled to the Zulu to return to his fires below. He crossed the deck with the long stride and heavy grace which is so marked an accomplishment of his people and vanished into the engine-room. Almost at once my ears were sent ringing by the loudest 'Blast!' yet from my companion. I was just in time to see another and even greater ectoplasmic manifestation of warm-blooded being in a winter sea disperse on the air. I also caught a glimpse of the whale itself and judged it to be about half a mile away, some fifteen degrees to the north of our port bow. The last of the silver mist created by this blow was just dispersing when on the summit of a wave there appeared the dark back of the whale, shining like polished marble and carved like an arc of triumph over the trough between two waves. Then it plunged, like a great submarine, into the next wave. All this, observed merely through my own eyes, was impressive and exciting enough for me. But for my companion, who had been watching it through his glasses, it was almost too much.

Although this time the response to his shout of 'blast!' was even more instantaneous and impressive than to the first, the quality and power of what he had seen, as well as his fear of losing it, made him find fault with everything. There must be something wrong with the engineer . . . he did not know why Kaspersen was being so slow today . . . that black stoker should never have left his fires at so critical a moment . . . there must be something wrong with the engines – if we couldn't go any faster we would not catch up with the whale in time to harpoon it . . . and if it vanished this time, alarmed by all the noise, we would never see it again.

Inexperienced as I was, even I would have thought that he was taking things too far, if I had not realised that it was his way of dealing with the tension which this critical phase of the drama had made almost unbearable.

Soon there was another, though somewhat lesser 'blow' than the first, and we had another glimpse of that great back, announcing that we were doing far better than my companion's criticism had implied, and that we were closing in fast on our catch. But even this was not enough for him. A glance at the bridge had showed him that Kaspersen was still at the wheel. Again he wanted to know why he was being 'so slow' . . . but Kaspersen was clearly a better judge of distance than my companion because on the third blow, after another view of the whale's back, my companion's excitement left him. In a whisper full of disappointment and foreboding he announced lugubriously, "Not near enough; too late. We lose him now."

And indeed, at the end of the series of declining blows, the whale vanished and, judging by the expanse of strange, smooth, almost ironed surface of water the whale had gone deeply under again. So once more the engines of the *Larsen II* were silenced. We drifted slowly but unerringly towards the patch of smoothed-out water with the ship more hushed and silent than I had know it. There was one difference, however; no-one on this occasion changed their positions in the ship. Even the Zulu stoker stayed below by his fires, and the only new element was that Kaspersen had handed over the wheel to his bo'sun. Then slowly, almost stealthily, without a word to anybody, he crossed the deck looking more like a cat burglar, climbed up the little ladder and, with his feet well apart, stood at last behind his gun. Although it seemed much longer, we drifted in this manner for a further twenty minutes without a word being spoken in the ship. After one hoarse whisper of approval of the Captain's action and a sombre indication that it all might have been in vain, even my companion felt the need to observe this great armistice of silence on the sea. At the end of this time Kaspersen, for about the twentieth time, aimed the gun at some imaginary whale on the surface of the water and did a practice swing from extreme left to right, and then back again. He then let go, turned about satisfied with the smooth action of his piece, and looked all over his ship as if to make certain that no man was out of position. When ultimately he turned his back to grasp the butt of his gun and put his hand on the trigger, my companion said, "Nearly twenty-five minutes: if he doesn't surface soon he will have gone."

The time passed and still no whale nor any sign of one appeared.

"He vanished – or he is extremely big and gone very deep to stay under for so long!"

My companion's whispers were full of doubt. Judging by the way I seemed to detect a lessening of concentration among the men on the deck below, it seemed to me that he represented the feeling of everyone except, of course, Kaspersen. Kaspersen was now crouching over his gun and expecting to shoot at any moment.

Just on half an hour, not a furlong away and straight ahead of us, the sea suddenly came alive and there arose the greatest blow of all. I saw it rise some thirty to forty feet into the air. Now that we were near, I realised I had never seen anything more beautiful and moving – a beauty indeed great enough to give the phenomenon of a kind of sanctity and make it inviolate and total in my imagination. As this, the highest of all the blows, established itself like a fountain in the air, it was followed by the reappearance of the whale, slowly and majestically arching itself over the crest of the swell. I had become so totally absorbed in what had now become for me an act of almost Biblical revelation, that the desperate reactions on the ship were of secondary importance. It seemed to me that we must be almost within firing range. Indeed, when the second blow came and that broad back reappeared so close, the crystal silence was suddenly shattered by a burst of gunshot, and in my line of vision appeared a cable with something heavy at the end of it, wriggling like the fastest of snakes towards the whale. It hit its target in the centre of the back just as it achieved the greatest arc over the sea. I winced in participation of the shock and pain of the harpoon's entry into the warm flesh and blood. At once a spurt of mist appeared where the harpoon had entered deep into the flesh of the whale, and a slighter and lesser thud from within the inner tabernacle of its body reached my ears.

A great sigh of relief and satifaction broke from my companion. As if he had never doubted Thor Kaspersen at all, he explained, "Got him, harpoon, grenade and all! No gunner like him now or ever before!"

Then I had no time to ask him what he meant by 'grenade'. It was only later that I learnt that each harpoon carried a grenade between its flukes set to explode within seconds of hitting its target so that for both humanitarian and practical reasons it killed the whale as quickly as possible. But even so the sound of the gun and the explosion made me assume that there had been an instant killing. I soon realised I couldn't have been more wrong.

The whale, on being hit, sounded at once and dived with such power and speed that the coil of cable on the gunner's platform unwound so fast that the slack between it and the drum on the winch was spent almost at once. I watched the spring on the derrick through which the cable passed suddenly forced to take all the strain of the whale's steep dive. It stretched so violently that the gap in the coils widened alarmingly. The steel moaned, thinned and yawned so much that I feared it would snap. But the man beside the donkey-engine had begun to release his brake on the winch by pulling on the lever just in time to allow the cable to unwind and give neither too fast nor too slowly, but just enough to hold it taut enough to act as a brake on the whale's precipitous descent. I reacted with revulsion at what seemed the unfairness of it, for both winch and spring were the modern technological equivalents of 'playing' one of the greatest of the sea creatures as if it were no more than a trout which had been deceived by a fly on some English river, and so had impaled itself on a hook. Since then I have never taken kindly to the thought of either when done for so-called sport. The moment, indeed, was the tip of the iceberg of a paradox remaining potentially below the surface of my imagination ever since the shooting of the elephant.

On one hand I could not deny the excitement and the acceleration into a consummation of archaic joy which the process of stalking and hunting, even at sea, had evoked in me, although I was present now only as an observer. On the other hand, hard on these emotions came an equal and opposite revulsion which nearly overwhelmed me when the hunt, as now, was successful and one was faced with the acceptance of the fact that one had aided and abetted in an act of murder of such an unique manifestation of creation. The only dispensation of the paradox ever granted to me in the past, unaware as I had been of the immensity of it until revealed to me in this moment at sea, was that in hunting out of necessity, all revulsions were redeemed by the satisfaction one felt in bringing food home to the hungry. That such satisfaction was not an illusion, nor a form of special pleading in the court of natural conscience, was proved for me by the profound feeling of gratitude one invariably felt for the animal who had died in order to allow others to live. But this form of killing and this battle for survival going on down in the sea, far deeper than any Shakespearian Prospero could ever have sounded, what could this possibly have to do with the necessities which were essential for the redemption of the act of killing? Once, in the days when the Arctic and Northern seas were witnessing a similar elimination of their whales, it is probable that some of these essential preconditions of redemption of this sort of killing might have existed. But, in this increasingly technological moment of my youth, when control of life was passing more and more from nature to man, and when there were already available all sorts of artifical substitutes for the essential oils of which animals like the whale had once been the only source of supply, what, I asked myself bitterly, could justify such killing except the greed of man for money, and money, moreover, acquired in the easiest and cheapest way without regard to the consequences? Worse still, I was certain that our imperviousness to the consternation caused by such killing in the heart of nature could be the beginning of an enmity between man and the life which had brought him forth, that would imperil his future on earth itself.

Kaspersen himself, from what he had told me, was in part aware of this, and related it to one of those 'black holes' which appeared in the day when one had shot elephant (as I described the phenomenon to him). But physically, there was no black hole – nothing had been taken from the day because only the prelude to the first of this kind of killing I had witnessed took place on the surface of the sea and only partially in the light of day. The greatest and fatal acts were being continued in an area of the darkness of deep water which I could not even begin to imagine. The parallel held only to the extent that something of that same darkness was already creeping into my spirit.

For some twenty minutes I watched the surface consequences of this struggle between the doomed whale and the master rod which the *Larsen II* had become in the hands of the supreme angler Thor Kaspersen. I watched with a mixture of equal extremes of excitement and distaste. As always, I was moved by the way in which an animal, without exception, never accepts defeat as long as a flicker of life vibrates within its being. Equally, it was now a point of honour for this great whale, doomed as it was, to go on fighting until not it itself but the life within it decided that primordial honour had been satisfied. As for myself, who could influence the affair as little as I could control the longing to do so, I thought that the whale had already fought long and hard enough to be free to be dismissed into the peace and company of all that had fought likewise before it.

The ship had long lost all the way given to it by its last burst of speed and was now totally at rest in the sea. She was swaying so widely from port to starboard and back again that I thought my companion and I would fall out of our tub, so I hung on as hard as I could to its edges. All I could see below were those shadows of great sharks and other beings of the sea passing now like the shadows not of cloud but of an unbelievable dream moving through the half light that the sun had made of the opaque water. All there below looked indeed as if the sea were carrying on business as usual. But not so *Larsen II*. As it lost the last of its way, it began to heel over towards the angle of the whale's desperate plunge.

The silence, every half minute or so, was broken by the sound of the winch being released to give the whale more line. Even the bows of the ship, with Kaspersen still standing at the gun with his back firmly averted, began to swing in the direction of the plunge. So it took all the bo'sun's skill at the wheel and the experienced judgement of the man at the winch to keep on playing the whale at an angle which would not get the harpoon cable entangled with the poop and gun platform of the ship. This Olympian tug-of-war between whale and ship went on for perhaps some twenty minutes. Then a deep "Ah yes, he has seen the light at last!" was uttered by my companion. The expression, even at the time, seemed inappropriate enough to be almost blasphemous. An adaptation of Mark Antony's 'And I am bound for the dark', would have been a more precise and dignified comment on its condition. But it was apt enough to the extent that the pull between the whale and the spring through which the cable passed to the ship had suddenly slackened and the winch itself was sent spinning rapidly to take in the excess.

The whale surfaced – still alive but with its flank covered in blood. It was fighting not so much the cable and harpoon but life within itself which was leaving it for good. In the process, it was thrashing and beating the waves with that beautifully carved tail; perhaps one of the finest and most precious ornaments and decorations worn in the brave order of the animal kingdom. It was my first experience of what whalers call, so aptly, 'the flurry of death'. Indeed, soon the thrashing ceased and the whale floated inert on the surface. It was as if a black grave-stone had been raised on the surface of the sea as sign of the guilt of the everlasting Cain in man.

The whale was immediately hauled to the side of the ship. The two crew members with the long steel lances joined forces to cut deep openings into the side of the whale. At the same time they produced a long rubber tube with a nozzle at the end, inserted the nozzle into the holes they had made and set the donkey-engine going again to pump it full of air, so that once the nozzles were withdrawn and the holes securely plugged, it would be able to float by itself like a buoy on the water. It was only then that I realised how huge an animal it was. My companion informed me that at a guess it could be between a hundred and a hundred and twenty tons in weight.

We watched it being lashed firmly to the side of the ship by the head, the middle and the tail. That tail, which had been so alive and violent a moment before, was now still and cold as moonlit marble. The dead whale had parallel grooves in its skin which ran from chin to tail and were there, according to my companion, to allow for expansion to fill it with the breath it needed for diving so deeply and for so long.

Once all this was complete, Thor Kaspersen signalled to me whether I would not now like to come and join him on the bridge. But somehow, for the moment, I had taken against him and all the others below. Despite my own part in the affair, I preferred to stay up there in the tub where I could have myself to myself. This I made plain with an abrupt, almost offensive, signal. Even so, Kaspersen still had thought for our welfare, because he sent the cook up the rope-ladder to hand to us a huge flask of hot, sweet tea, mugs and sandwiches made of thick slices of Norwegian cheese between even thicker slices of brown rye bread.

As the day was still young we continued our hunting and, before sunset, we had killed two more whales in a similar manner. In the first fall of darkness we began heading back to harbour, with enough light to show me two things that have stayed with me ever since. First, there were suddenly great white sharks of the Indian Ocean to follow us brazenly on our way, every now and then turning upside down so that the strange, phantom-white of their bellies, and that angle of death implicit in the line of their jaws set underneath their faces, showed up quite clearly like a grin on a skeleton. Upside down they pulled huge chunks of flesh out of the whales we were taking back to port, with an expertise that no pack of hunting dogs, whom I thought supreme masters of the art, could have equalled. Most of all I remember a baby whale, wary and nervous as it was of the sharks, following us at some distance, compelled because its mother was one of the whales tied to the ship's impervious flanks. It kept us company until, forlorn and desperate, it too was lost in the final fall of night.

This description of my first day's hunting in the *Larsen II* is as good a model as any of the basic experience I was to have at sea. Always there was the inexplicable and indefatigable paradox of the 'pull-baker' of the excitement of the hunter in man and the 'pull-devil' of revulsion against the cardinal provision of life that made such things plausibly essential.

Although the wind and weather and location at sea obviously varied enormously over the years, the function of *Larsen II* itself and the conduct of the men in her never changed. Nor did my ambivalent reaction to it vanish. Similarly I was certain that no matter how the expression of the enigma of it all in the spirit of Thor Kaspersen may have varied, the problem remained unsolved and therefore daily more obdurate. Yet it would be misleading, if not false, to leave out of the picture the many compensations for me that came of my fitful participation in these seasons. There *were* moments of beauty and happenings of meaning that, both at the time and in retrospect, seem almost miraculous.

In so small a ship, even looked at from the tub of the masthead, I was far closer to the sea than people ever are who sail the oceans of the world today in their great tankers, container ships or hotel liners designed for self-indulgent cruises. I saw things that nowadays even professional sailors, from their positions on high in mammoth vessels, can never see. These are things that are not only more difficult to see because of the distance between the observer and their natural element, but also because with the galloping exploitation of nature that is the main trend of our time, even the abundance of life in the sea is alarmingly diminished since the period of my youth in which I learnt many a new idiom of beauty in the life of the sea. Yet if asked beforehand, I would have declared a conviction that I could only be repelled by these manifestations of being – that ultimately revealed so strange and cinderella-wise a beauty to me.

There were, for instance, those white sharks I have mentioned, among the three hundred or more varieties of shark to be found in those particular seas. Physically they were certainly the largest; one of more than thirty-six feet was caught in the heart of our whaling grounds. This suggested enormities that none of us had ever imagined, and in my own indoctrination – through all I had read about them – I could not have believed for one minute that they would not look as monstrous as their habits were supposed to be.

Since this ocean, Thor Kaspersen's hunting ground, was so full of these sharks of all kinds, we would often return in the evening to release our kill at the slipway of the whaling factory underneath the Bluff to see scores of men at the end of the breakwater on the Point fishing for sharks; using empty four-gallon petrol tins as floats. According to my newspaper, they were regularly catching sharks well over two thousand pounds in weight and would, in doing so, often be in battle with a hooked shark for half a day before it ceased to struggle and they could haul it ashore. The record was already said to be well over a ton.

None of this surprised me. In this business of whaling I thought I saw even greater ones at sea. At first the sight of their dorsal fins cruising round our ship would make me feel what all the adventure books of my childhood had conditioned me to feel: fear, horror and loathing, as if these sharks were things of unmasked evil. But these reactions soon passed, to be replaced by an unexpected wonder at the beauty of their form and their perfection of design for their function. It was almost a revelation of their own right to be there, and also of the dignity of being that they shared with all living things. If there was a question of *being* at sea it was relevant not to the sharks but to the men in the *Larsen II*, and all those who conceived and supported the vocation of whaling. As the element of lawlesness in the calling of the *Larsen II* and its kind became more and more painfully apparent to me, the clearer appeared the need of these sharks to be part of the totality of life at sea. They helped to keep the sea obedient to the laws of proportion which increasingly appeared to be the great overall commandment of life. This necessity of their presence was soon demonstrated for me with the accuracy of a mathematical proposition when I encountered the first manifestation of what everyone in Port Natal, even those not in the least interested in the sea or fishing, called the 'arrival of the sardines'.

This was a phenomenon which occurred regularly once every year. There would suddenly appear from somewhere out of the South Atlantic, in the direction of the Roaring Forties, a great stream of small fish which flowed like an Amazon up and along the south-eastern coast of Africa. The news that this great, charged river of little fish was flooding steadily north and nearing the coast of Zululand, caused immense outbreaks of excitement, not only among humans but also in nature.

In the human beings it was understandable, because of the daily news reports. Nature did not have newspapers but that did not mean that it was without a system of intelligence of its own. How it worked then, and still operates to this day, remains a mystery to me. But aboard the *Larsen II* we would have recognised it from the change in the life of the sea and its birds. Both seemed to know of the coming of the sardines before we did. It was my companion on the lookout who first drew my attention to the change by asking me if I had noticed how the numbers of sharks, porpoises and dolphins had increased in recent days. And, had I noticed how many more gulls and gannets were flying past us in such numbers that the sun blazing down on the white feathers made them look like a blizzard sweeping down to the Cape? It could only be, he explained, because the sardines were coming.

When we brought our next kill into Port Natal and picked up the ship's newspapers we read, under banner headlines large enough for a declaration of war, that the river of sardines was flowing deep, wide and purple in our direction. What was so singularly awesome about this was that the small, vulnerable sardines, exposed to attack by fish-eating birds and larger fish, had only one ultimate weapon against their many enemies, their numbers. Their role in nature had been conceived on so abundant a scale that no matter how they were attacked, nothing could really eliminate their species. The sardines could, at the end of their strange pilgrimage into those waters of the Indian Ocean, be reduced to what one might call their lawful proportions in the life of the sea. But they could never be annihilated. And it was one day when the *Larsen II* found itself sailing into the first flood waters of this Amazon of sardines, that this truth became clear to me.

Everywhere I looked, the sea had darkened. It was shaking and quivering with new life which at moments, when its swollen streams of fish were forced to surface, made the water look like a dark wine of Aquitaine. All round us the killers of the sea were already in action against the sardines. The birds were attacking them without cease from the air; there was hardly a patch of sky ahead that was not taken up by squadrons of gannets, who never ceased to amaze me by the way they would rise sheer from the sea, then suddenly fold their wings and dive, a streak of white, back into the sea, strike and devour yet another sardine, and then rise once more to repeat the almost endless process. On land, wherever I looked through the field-glasses, I saw dark lines of fishermen along yellow sandy beaches in close ranks like infantrymen ready to repel an invading force from the sea. But as I studied the line I noticed that, like the sharks and the gannets, they too were busy taking part in this carnival of killing that was going on all around us.

The fishermen had nets and lines and hooks cast in such profusion that one experienced a kind of illusion as if we were an ocean-going Alice who had just gone through the looking-glass to discover that what had passed for reality before had been but its reflection. For it was not the sharks, the gulls, the gannets, the fishermen and the anglers who were 'playing' the sardines; it was the insignificant sardines in their inexhaustible numbers who were 'playing' the people on land and compelling them to conform to a pattern of some overall design of life wherein they had precisely the role of provoking this compulsive reprisal from their large fellow-creatures. Before the end of the day, for the first time in my life, I was astonished to see all the killers on land, sea and in the air, turn their backs on the sardines.

They had fished and killed until they were over-full and disgusted. Yet the next morning the scene would repeat itself as before! So I could now see how, if it were not for the sharks and the carnivorous creatures of sea and air, some form of life could exceed its natural proportions and by sheer force of numbers take possession for itself of what was the lawful allowance and ration for others. The sardines, indeed, would be like those cells in the human body which are necessary for the daily well-being and renewal of the whole. But through something which we do not yet understand, they can suddenly break the laws of being to exceed their own role until, totally out of control, they proliferate to kill off the body they were designed to renew, through the phenomenon known as cancer. These sharks which I had so abhorred in my first reading, therefore, were one of nature's remedies against the cancer of numbers at sea.

I learnt from them, as I had already learnt to a certain extent from the canivorous birds, animals, reptiles and insects of life in the interior, how they often had a perfection of design and colour and movement which other forms of life, beautiful as they were, lacked. Once I acquired this basic idiom I was free to admire and enjoy the extraordinary elegance, and the effortless ease of speed and movement they daily manifested in the water around us. Indeed, this law of the aesthetics of the authentic destroyer and its necessities of being was demonstrated superbly in my first encounter with a pack of killer-whales.

We came across them on a morning almost as perfect as my first morning on the ship. I was once more in the tub (ever since the official 'lookout' had told Kaspersen of my sighting of our first whale of the day at the beginning, they insisted I should always keep the lookout company; they thought I brought them luck). I was enjoying this pure, still day of crystal beauty at sea. It was brought more alive and sparkling by a school of dolphins passing hard by us, leaping out of the water where the sun wrapped haloes of gold round their graceful bodies. Others dived out of sheer delight from one side of the ship, under the hull, to the surface on the other side, or else merely raced alongside, clearly visible just below the surface. When they had had enough of our company they put on a burst of speed so great that it made the ship look pedestrian and awkward. When the dolphins finally left us I found the sea, for the moment, deprived of magic and looking singularly empty and forlorn.

The dolphins had made music in my senses and seemed a clear indication, as well as justification, of the role of their progenitors in historical legends. Classical man had found a kinship with them which after this morning no longer seemed strange to me, and which I myself, even after so brief an introduction, was beginning to feel. I had no difficulty, henceforth, in understanding how and why they should have come to be thought of as the 'friend of man', and why they were held to have been devoted to music. Even in the kingdom of Neptune they were addicted to the art of Apollo. When Kaspersen told me that they were in fact of the same species as the whale, it seemed to bear witness to the validity of the whale which, in Kaspersen's recurring dream, went singing out to sea.

On this particular day the school of dolphins had hardly ceased to play with us when suddenly the porpoises took over with an act peculiarly their own. They moved through the water with a grace possibly even greater than that of the dolphins. Only rarely did they ever leap clear of the water, but preferred arching their backs briefly over the waves, as the first great whale of my experience had done. But they had one special delight and unique accomplishment of their own which never ceased to amaze me. They seemed to love picking out the highest wave or swell on the ocean, whether in storm or calm; swim within it as it moved along and, before it reached its summit, the porpoises would turn and move upward to the crest of the wave and then gracefully loop the loop several times before descending into the trough behind the swell. They appeared the most accomplished acrobats in a circus high in a circle of light, as if they knew that they too had a great audience observing anxiously the beauty and danger of their movement. The drama of it was that, near as they came to the edges of the water, no awkward flip of their tails or other error of calculation at speed should break the cover and mar the shape of the wave. In consequence one saw this whole movement as it were in a comb of golden honey.

I had hardly finished with the porpoises when there was a great shout from below. Startled, both my companion and I look downward thinking we had missed a vital 'blow' at sea. It was both an imperative and strangely angry shout. Kaspersen was pointing to something I should have noticed if I had not been so obsessed with dolphins and porpoises. A pack of killer whales, their superb dorsal fins gracing the picture of a morning sea like some Picasso abstract, was going by with the speed and elegance which far exceeded that of any shark I had seen. They were going due east, straight into the morning sun. So still was the air that we could clearly hear Thor Kaspersen's repeated words, "Watch them, you two up there . . . watch them! They are not going east for nothing. They know more about whales than you and Thor Kaspersen put together."

I had barely need of this exhortation to watch this strange new and beautiful phenomenon. I marvelled at the ease and speed with which they were increasing the distance between themselves and the *Larsen II*, leaving the sea between us all a-tremble and the water shaken out into flake upon flake of silver in the morning sun. I was watching them dutifully, as commanded by our Captain, when I became aware that our ship was coming round into a bearing in line with this broad highway of silver that the killer whales were leaving behind them. Although neither of us in the tub had sighted any whales Kaspersen, as always, was practising what his experience and intuition preached. The bells in the engine-room rang out loud and clear, inducing an element of excitement which this call for further speed never failed to arouse in us up in the fore-top. If the Captain's decision to follow the killer whales was to have any tangible result, obviously it would not do to cruise nonchalantly in their wake. Already they were barely visible to the naked eye and their wake declining by the minute. Once our ship was steady and in exact line with the head of the pack, I noticed with gratitude that our bearing was sufficiently north of the sun to enable us to observe the whales with no interference from the dazzle of light. Moreover, every moment as the ship gathered way, we gradually started closing in on the pack. But long before we had overtaken them, this feeling in my senses was immeasurably heightened by the sudden eruption of another school of whales directly ahead of us.

They announced themselves not by their customary 'blow' but in a manner that was far more dramatic. Two bodies, whom my companion judged to be more than ninety tons in weight, suddenly leaped like salmon out of the sea and with unbelievable grace rose high into the air before coming down with flattened horizontal bodies on to the sea and hitting the surface with such powerful 'blows' of their perfectly fluked tails that a great fountain of broken water shot up high into the air. It hung there briefly like a chandelier and then fell back and shattered on the platinum water. This they did with the enemy pack perhaps no more than one hundred fathoms away and closing in rapidly for what must have been thought to be an easy kill, because they were not sperm whales, who are armed with powerful tails and real teeth of their own. They were in fact the harmless and exceedingly vulnerable blue whales who, although they were the largest whales to be found in the seas, had no special armament to protect them from enemies such as the killer whales except, of course, their tails and the courage which they now proceeded to display.

Unexpected and new as all this was to me, it was obviously not so to Thor Kaspersen. It was precisely something of the kind that he had expected because he hailed me loudly from the bridge with a proud warmth in his voice that I had not heard before, "Closely look, David, closely look. Those men . . . real men. Fighting for their women and children."

The sound of his loud, rasping voice, more than normally smoothed out with emotion, had hardly died away when not two, but three males emerged from the sea and repeated the manoeuvre I have just described, though now from diverging points of the compass. Obviously the pack of killer whales, on the first response of the blue whales, had moved to out-flank them and were now trying to attack them from all sides and angles. It was only then that I saw, moving south as fast as they could, four females and four little ones streaking through the twilight a fathom or so below the calm sea like those multitudinous shadows of one of my *Coral*

Island dreams. From the masthead above I could observe these haunted, simple beings clearly. Kaspersen reduced the speed of our ship so that it merely idled along to keep us, as it were, in the stalls of the shifting scenes of this fateful drama that was being acted out.

I could not say then or even now, how long it continued with this pack of foremost killers of the sea continually probing the defences of the blue whales and as consistently being forced to hold back from those gigantic 'blows' of body and tail that the blue males were delivering straight at their aggressors. But I do remember that it lasted long enough to produce an incredible agony of fear in me lest the blue whales should lose out and the killers should win – an unendurable result because already it seemed to me that the blue whales had enemies enough in fleets of whalers of which the crew of the *Larsen II* was only one. I was protesting in fact that in releasing such a pack of dedicated and professional destroyers against them, nature was abandoning the law of impartiality to which it was subject, and was tipping the scales of all natural justice against those whales and their young.

I longed for Kaspersen to call for more speed to take our ship in the midst of the battle, to separate the killers from the blue whales. I had somehow assumed from his warmth of tone in directing my attention to the conflict that he would also wish to redress the balance of fate by joining in the battle with the whales. But as time went on the males kept coming out of the water and the battle became more and more desperate. I realised that, with all the energy it took to hurl such gigantic bodies so high in the air, they must soon tire against such a Tartar enemy, who could apparently manoeuvre with such agility, grace and lack of effort, with no noticeable loss of speed, accuracy or energy.

Perhaps it was as well that I did not succumb to the temptation of going down to Kaspersen to intervene, for I would have missed one of the most moving demonstrations of how courage, which truly endures against fear, can be vindicated in life even when all odds are against it. Well before noon the brave blue whales leaping out in advance of their females and young were reduced to only two; soon after there was only one, and then suddenly there were none at all. The torment of the dark blue water around us mysteriously subsided, the pattern of conflict on its surface was erased, and one long rhythmical swell after the other took over to show that the Indian Ocean was once more breathing calmly and steadily, as if nothing abnormal had disturbed that perfect winter's day.

Then another shout from Kaspersen, this time pointing to the north, showed the pack of killers in full retreat.

It was only in the seasons to follow that I was to realise fully how unusual the outcome of that battle had been. I was to see many demonstrations of how the cunning of the killer whales more than matched their speed and power. I was to see them hunt, for instance, just as a great hungry pride of lions would hunt impala and springbok. They would fan out in a crescent moon formation and drive porpoises and dolphins towards a corner of the ocean where the most experienced killers in the pack would be waiting with the impervious calm of the born executioner, and then they would launch themselves at the fugitives to kill them in numbers until the translucent swells over that sea were no longer a single illuminated colour, but were like some patchwork quilt in which scarlet was predominant.

For the moment, however, I was immediately obsessed by another fear that Kaspersen, now the battle was over, would set about his own business of picking off an exhausted group of blue whales one by one, as he could so easily have done. But to my delight and amazement, our ship began to swing round and take a new bearing on the pack of killers and follow them at our usual cruising speed to the north. This in due course led us at the foretop to observe a whole new series of 'blows' and guide the *Larsen II* to what seemed a legitimate killing.

Before dark we were able to head back for port. And that night in the saloon while we were revictualling at our usual berth in the harbour, I asked Kaspersen why he had not intervened. He looked at me as if I were just entering some kindergarten of life. Then, like an Old Testament prophet, he rebuked me for what he obviously regarded as a blasphemous suggestion implicit in my question. He told me that no man, however great or important, had the right to tip the balance of fate either way in these conflicts of one species and another set up by nature both on land and sea. If man did so, he was certain that some terrible retribution of fate would follow. At the same time he believed that no man ever had the right to take advantage for himself of the consequences of those conflicts. Had I not noticed that he had turned away from an easy killing and gone to look for his own whales?

But why then, I asked, had he bothered to follow the pack of killer whales at all, knowing that he was neither going to join in the battle, nor profit from its result? He said with unusual humility that he had asked himself this question many times and did not fully know the answer. It was, in part, because he could not help himself. What we had seen that day was, to him, one of the most beautiful, pure and helpful experiences that had ever come to him in his long life of hunting whales.

If ever there was any sign that a repeat performance of what we had seen that day was about to happen again, he would be compelled to follow, to observe and then humbly accept, whatever was the verdict of fate delivered in the conflict. Strangely uplifted, he looked away as if embarrassed by so intimate a confession. Then he added that he felt some little hope in the hopelessness of the contemporary scene, namely that courage and faith in the life entrusted to animal, fish or bird, defended and pursued with patience, courage and endurance, was the only honourable option for us all. What else did I think kept him on course and gave him something to steer by?

Even had I known, I could not have asked how and why he did not interrupt the acting out of so great a parable as we had seen that day? He was compelled to tear himself away to pursue his own calling in such a way that what followed was purely a matter unaided and fairly pitched between the whales and himself. I might have said something about this introduction of a man-made concept of fairness into the discussion, and drawn attention, perhaps, to what seemed one overall flaw: the kind of whaling in which he was engaged, by the day, made the contest between the hunter and the whale more and more unfair because the headlong rush of technology was interfering on an increasing scale with the provisions of fate in the organisation of life at sea, increasing the power and means of the hunter to kill, while leaving the whale with no more than it had possessed at a beginning when the whale might, perhaps, have been more than a match for the hunter? I did not do so for two reasons. First, I always remember a proverb which had been a favourite of my mother's and handed down from her Huguenot ancestors: 'Never to let the best be an enemy of the good.' Second, Kaspersen's answer obviously was that, in the context of his time, he conducted the battle as fairly and as well within the terms of the original deed and act of life as he could. Yet I knew him well enough already to realise that underneath all he said there were not only these, but many other questions far deeper, which were constantly at work in the depth of his own imagination; and also that he was still as far from answering them as he had been in his own beginning.

But soon I discovered that, as far as I was concerned, the hunting and the 'doing' at sea decreased in importance. It was just the feeling of the sea around me, and a sense that something of tremendous importance which could change the course of my life would emerge from it at any moment, which really mattered most to me. There was far more food for my imagination and spirit in this aspect of my whaling seasons than any injections of excitement from hunting down the whales. I always wished that all activity could have been confined to observation. Just watching the sea and waiting for what it might reveal was for me a full-time and almost inordinately exciting occupation. I found increasingly that I could hardly wait for the dawn to break, so that I could claw my way up the ladder of the mast and take up my position in the crow's nest. There I would have the most resolved, peaceful and beautiful moments that I had ever experienced. Not a day was without a character and beauty all of its own.

It was odd, nevertheless, how in the midst of so much water I would often think of metallurgical metaphors drawn from the alchemists rather than from a poet's or modern journalist's vocabulary. For instance, there were those incredible days of peace of the Zulu winter which reached out far from the Dragon Mountains of the interior to beyond the horizons that might have been drawn with the precision of a blue pencil fixed in a mathematician's callipers, so flawlessly and symmetrically were they coiled around us. However much Thor Kaspersen might rage against such days, for me they were frontiers of worlds of shining time and space traversed by endless hours of delight and wonder, and heightened by constant expectation that at any moment some whale might come to meet us as we compelled them to recede and regroup themselves again and again on our shimmering thrust towards the eastern quarters of the Indian Ocean. This clarity of an almost analytical kind was perhaps the greatest quality of those winters. The light was purest in the early morning and was so bright and magnifying that up there at the foretop I would feel as if I were enclosed in the heart of a diamond. But the evenings, just before the sun set, those spectacular colours bade a profound, mythological farewell to a day that it had warmed and illuminated with such tenderness and precision. Then suddenly the sky and sea would be transformed into a magic lantern slide of all possible colours and nuances. The diamond quality vanished and I, the ship, the sea and all in them seemed suspended in the heart of an opal.

There were other days, too, particularly towards the end of the season, when the sun began to acquire some of the aggression necessary to force the earth out of the slumber of winter and impregnate it with the growth of summer. On such days the dawn would be as full of vivid colour as the sunset. Soon after sunrise one could think of all the alchemical metals and mediaeval visions transforming one into the other, until they reached their fulfilment in ultimate gold. I still seem to have a calendar in my mind wherein they are labelled not in months and weeks, but as days of steel, brass, bronze, cast-iron, platinum, silver and up and on into gold. It is still impossible to define what these days did to our spirit. All I know is that, young as I was, they led me to the discovery that sailing this great Indian Ocean in the hunter's way was taking me to places in my own mind where I had never been before.

Yet there was one species of day, connected with the metal which filled me with an awe almost akin to fear, and also to a discovery within myself of a startling kind. These were days of lead, which the alchemist regarded as the basest of metals. They were usually the sort of day that would send most of the whaling fleet scurrying back to port. Yet, not surprisingly in so paradoxical a character as Kaspersen, those days of lead would cause him the greatest of all excitements, and send him heading as fast as the ship's engines could go in the direction where the atmosphere seemed to be at its most leaden. They were the days which preceded the most violent storms.

Our whaling grounds were close to the fringes of an area of the Indian Ocean wherein monstrous cyclones were born. There, from the tropical waters around Mauritius, Reunion and Madagascar, they tyrannized land and sea for hundreds of miles at times, even penetrating the world of Mozambique. Although those cyclones never broke through into the waters of Port Natal during my time there, they came close enough to raise a gigantic swell on the ocean, and ultimately filled the sky with such tumult that the mornings would break into days of a sullen, almost negro-brown colour. Such a day was at its brightest at noon, yet barely lit with a glow of sulphur. Then towards evening it could break into an awesome brown before turning in the blackest of starless nights. One of the strangest aspects of those 'days of lead' and our venture into the 'kingdom of lead' was the fact that, although not a breath of wind itself reached us and the air itself was still with fear, yet the presence of a wind of almost unbelievable power somewhere over the horizon to the east was made manifest by a strange kind of moaning and groaning. At times it was a heart-rending whimpering which overcame the normal sounds of our little ship ploughing through the sea with, as it were, its crew of farmers culling their flocks of whales in that uneasy pasture.

On one occasion we had to come close to the tormented frontier of one of the greatest of all cyclones. It was bearing down through the Mozambique Channel over brown waters whirling, twisting and hurling all its energies at a horizon that seemed to be a-flutter in warring, demonic, unworldly and transparent shapes of substantial shadow, as if the irresistible storm were about to make dangerously ponderable matter of itself.

We did all this because Thor Kaspersen had a belief that sperm whales, whom he admired above all others, loved to live and feed along the margins of storms and in great schools of both old and young. He would tell me that those storms which cast unrest so deeply into the atmosphere and so widely over the sea, reached down also to the depths of the ocean and there disturbed the giant squid which he claimed was both mortal enemy and favourite food of the sperm whale. As always, not only the storm but also the sperm whales and the squid were basic images in his naturally symbolic spirit. The squid for him was a personification of evil pervasive in all forms of life; the sperm whale was its heroic opponent, and the storm was the element where evil challenged good. The confirmation of the cosmic imbalance of the ensuing of battle between good and evil was self-evident for him in the fact that the sperm whales in their victory, wherein the giant squid was devoured, transformed its ingredients of evil into the good of ambergris. In those days, ambergris was still one of the most rare and valuable of products of the sea and a fundamental element in the making of the finest perfumes of Paris.

All this had an Old Testament parallel for him in the encounter of Samson and the lion and the honey that was made in the skull of the lion, and indicated how the lawful exercise of strength against destruction always brought forth the greatest sweetness. Kaspersen, his enigmatic face glowing with vision whenever he spoke of this encounter between the sperm whale and the squid, would illustrate his theme with plausible stories of giant squid which were worthy of Jules Verne. He would recount how he had seen an octopus dead on shore with tentacles forty feet long but also, using it as his ruler, he had seen a sperm whale surface on the edge of a cyclone with an octopus three times as big in its mouth with its own head ententacled like a monument entwined in giant ivy. He told me about a schooner, I think called *Pearl*, of some one hundred and sixty tons, becalmed in the Bay of Bengal in the 1870s. Suddenly on the calmest of evenings, the schooner was embraced by a squid of unbelievable proportions until its tentacles had the mast so firmly in its grip that it could haul itself on board and slowly, with its great weight, capsize the ship. Both the survivors of the *Pearl* and crew of another steamer nearby who rescued them, confirmed that it was no seaman's fantasy. If that, he asked me with a childlike look of wonder, really happened in calm seas, I could imagine what the bed of the sea produced when paralysed by storm and successive upheaval. Fortunately, the sperm whale needed no storms to tear such evil from its deep sea bed. It dived deeper than any other whale and forced them to battle. It was not surprising that he called the sperm 'Emperor' or 'Kaiser' of the sea.

All this produced the most hazardous form of hunting whales that I was ever to experience. There were the difficulties and dangers of harpooning whales and securing them in a sea with those immense silent waves bearing down on us and a cyclone keeping the dark state of the sky deprived of the faintest scribbles of light. Through the constant vigil we kept in the tub we were directly affected by the violent motion of the ship swinging not only up and down but increasingly from side to side. I was compelled to grasp the sides of the tub constantly with such force that at the end of each day I clambered down to the deck and into my little bunk utterly exhausted and aching with stiffness.

It was astonishing to me how over my three seasons with the fleet, Kaspersen's belief in this by-product of storm at sea was never proved to be in vain. He was violently criticised for it by his fellow captains, and was repeatedly warned by the directors of the company that he was putting his ship and crew into undue danger and that, in reality, he was acting out of superstition. The fact that he brought in more sperm whales, the most valuable of all – in my last season one sperm whale was found to have in its stomach some seven hundred pounds of priceless ambergris – was passed off as being sheer luck. But nothing would make him mend his ways. Fortunately, his crew by this time had been converted to his own faith and were almost as fanatical about observing it as he was himself.

Kaspersen had no poet's command of words, but by nature he was a poet in action and an incorrigible romantic at heart. Some of the most memorable moments I had with him were during those evenings in the disturbed antechambers of the storm, while outside the ship was taking the giant waves in the darkness like some steeplechaser takes the hedges of an unknown field. And it was precisely during those times that he would talk to me endlessly about sperm whales, squid and what he thought of the invisible life contained by the ocean.

There was only one other experience during those seasons that I enjoyed even more and that was the Sunday evenings spent in harbour, with the crew away either in church or enjoying beer in their favourite pub. Kaspersen, 'Mlangeni and I would then take the air on deck on the side away from the quay. We hardly spoke on those occasions, except for an odd remark which I would translate for Kaspersen or 'Mlangeni. But we were usually silent, with Kaspersen giving 'Mlangeni one of his cheroots to smoke and the three of us pacing the narrow deck and its gangway with a kind of companionable fire-glow from the burning tobacco lighting up the whiteness and darkness of their two strangely resolved faces.

It was extraordinary how close 'Mlangeni and Kaspersen were to each other, in spite of lack of any means of verbal communication. It was my first intimation that human beings communicate in a way far more profoundly with one another through the nature of what they are, than through their words. They were two of a pair, because in a sense they were equally de-tribalised. 'Mlangeni was, perhaps, less so because he still returned annually for some months to his kraal on the edges of the great blue-black hills of Zululand where the Zulus say 'the shadows gather, and the spirits of the ancestors meet'. But he was already deprived of all the tribal supports of his people, because he had broken through one of the greatest ritualistic inhibitions of his nation. His people always had a profound aversion to the sea and what came out of it, not without reason as their history at the hands of the Europeans might prove. This aversion went so far that no Zulu man would then eat fish, because, as a potential warrior of a military people, they were sure that eating fish would 'turn their hearts to water'. Yet, in spite of this, 'Mlangeni, a Zulu of aristocratic origin, had taken to the sea without anger or regret.

Often I would go down, on a calm day, for a brief moment to fetch refreshment for my companion and myself in the crow's nest. I would pause at the entrance to the engine-room and hear 'Mlangeni singing his own poems to music of his own composition. He would sing in a deep voice, audible above the beat of the engines below, for the men with whom he was sailing and, above all, to his captain. But the greatest song was dedicated to fire and proclaimed to his people that fire could be made on water; and that fire and water combined could lead them to greater things. It was perhaps not for nothing that he was called 'Mlangeni. It meant 'man of the sun'. And how right he was in having the courage of his convictions which took him to the sea, for the Zulus have since become the greatest ocean-going nation in Africa. There is hardly a port in the world that I have visited where I have not seen Zulus manning ships of the commercial navies of southern Africa, and doing so with an efficiency, discipline and ardour that has no equal on the seven seas today.

All in all, on those Sunday evenings in the shadow of a Bluff which was still under its primeval African cover, and a sky so full of stars that it produced another reflected sky in the smooth harbour waters stretching away from the side of our ship into the darkness towards Umbilo and Congella, the three of us seemed to be enclosed in a timeless moment wherein the beginning, the now and the future, like ourselves, were utterly at one.

To this day I think back to those moments and the orchestration of whaling activities out of which they came, as an indispensable part of the road to exploring my own self. The only sadness connected with it was that I had to break with it all without a chance of saying goodbye to either 'Mlangeni or to Kaspersen. I was suddenly called upon to undertake a long journey, and on the particular Monday when I accepted the summons, I hastened to the harbour to tell Kaspersen. But his ship had already gone. I returned several times in the midst of the many preparations I had to make, but he had not come back to his normal berth. I had to leave with only the briefest of farewell notes which he did not answer. I never saw him again.

This in a sense should not have surprised me for in all the time I had known him I never saw him write a personal letter or even postcard to anyone. Nor did he ever seem to receive what looked like a letter from a member of family or a friend. The only letters he received looked official, ones which he left for days unopened on the small triangular table in that tiny saloon of the ship. Over the long months between one whaling season and another, when he was in the Antarctic, he never wrote to me once. But what did surprise me, even hurt me until I understood, was that he virtually vanished without leaving any record or explanation behind him. For some five years, whenever I returned to Port Natal, I sought out my friends in whaling circles for news of Kaspersen. For about four years they always confirmed that he had been back as usual for another round of winter whaling. I was assured that my letter explaining my departure had been delivered to him and that he had taken it away without comment. But on his subsequent visits to his company headquarters he left no messages, nor did he enquire about my whereabouts.

This seemed to me at first an unnecessarily hurtful and insensitive thing to have done. But in time I had only to think of all the considerations he had shown me, of the honour he had done me in revealing his innermost self and the way he had tried to pass on all that he knew about whales and the sea. I had to realise that there was no system of human communication whereby he could have sent an appropriate message. He was one of those rare people who could never have proxy of man or pen to stand in for him. I knew that if we ever encountered each other again, our relationship would prove undiminished. In fact, the conviction of the fallibility of communication through human beings and the system which he had already found woefully inadequate would have been the ultimate heresy to him. Perhaps in observing this silence, he was paying me the compliment of realising that it said more than any letter or verbally transmitted message could ever have done.

I think it was sometime during the sixth year after our last parting that I heard that he had not come back for another season. No one could tell me why. And although I continued to try and find out what had happened to him whenever I came across someone who might have known, I really never culled anything more definite than a rumour that he had died tragically in the Antarctic. The rumours, vague and circumstantial, varied from person to person, but all seemed to have one element in common: his death had been the result of an encounter with some unusually great whale and in circumstances that provoked a fatal accident.

All I remembered of the man instantly rejected that feckless explanation of 'accident' at the time, as it still does today. I already had a suspicion, which has since become a belief, that there are rare individuals in life in whose evolution and so-called 'end' the phenomenon of chance knows far more about what such individuals need than they do themselves. It is as if what they are and have within themselves constitute a subtle power which, out of all the manifestations of circumstances and external events that surround them, draws to them the circumstance they need. I personally needed no more proof than that Thor Kaspersen was just such a person. What human beings in their approximate way call an accident was perhaps nothing so trivial, but rather an essential element necessary to complete the end for which that human being had been born.

All this happened more than half a century ago. Yet I still feel it all so keenly that I cannot altogether suppress the deep longing I had then for details of his end, if only to confirm what I believe. But, at that immediate time, having no inkling whatsoever that I would never see him again, I was filled with excitement and anticipation for this great new voyage which had suddenly become possible to me, which was to have greater consequences even than my friendship with that enigmatic and foredoomed hunter of the sea.

CHAPTER FOUR

The Ship and the Captain

IT was ironical that in spite of a predisposition to an awareness of the role of chance in life, I had no hunch even then, when I had failed in my effort to say farewell to Thor Kaspersen, how this had already infiltrated the pattern of my future.

I have always loved games, both for the physical well-being they generate and as a form of relief from an unrelenting preoccupation with things of the world within. The exacting nature of my work in Port Natal had forced me to abandon most of the games that I had enjoyed as a boy. All I could do was concentrate on Sundays – usually my day of leisure. I had found that Sunday hockey was an increasingly popular game all over southern Africa. I had captained the team at school for some three years and played for my province, so found no difficulty in getting into the First XI of a club called the Nomads which, as the name implies, drew its members from those who were, in a sense, exiles in Port Natal. In my first year, the Nomads won the Inter-club Championships of the province, and I became one of five players from the team to play for our new province. In my third season with the Nomads, which was also my third season with Thor Kaspersen I was, to my surprise, chosen to captain the provincial team for a larger inter-provincial tournament which was to be held in the Transvaal. We competed with teams from the Cape, Eastern Province, Griqualand, the Orange Free State, Transvaal, Southern and Northern Rhodesia, and finally managed to win the final against the Cape. By the end of the final I had had enough of hockey and its players, and hurried out of the dressing-rooms to be on my own.

It was a very cold, early winter's afternoon with a thin, piercing little wind blowing from the Drakensberg and the Mountains of the Night in the then Basutoland. I hurried to my favourite coffee house where I knew I could consume the kind of national delicacies which had been our special delight as children in the interior. I had hardly started my coffee, together with a generous helping of waffles, honey and cinnamon, and was at last beginning to feel warm, when suddenly I heard a shrill feminine voice crying out in Afrikaans, "I won't have niggers in this place! Get out!"

I looked up, disturbed more by the note of hysteria in the voice than its loudness and offensiveness. The woman behind the long counter at the entrance, pale with emotion, was glaring down at two comparatively small men in belted macintoshes. They were standing, hats in hand, puzzled and surprised at so rough a reception. They looked at each other for enlightenment and, finding none, back to the woman again.

Their hesitation raised the pitch of her hysteria and she shouted again even more loudly, "I told you I can't have you in here! So get out, will you?"

Automatically I rose from my table as fast as I could, went up to the woman and asked her, in her own language, what was the matter? The sound of her native tongue had a soothing effect and she explained more mildly, but still with an underlying passion, "Well, it's obvious, isn't it? I can't have any coloured people in my place."

"Why not?", I asked her. The question seemed to agitate her so much that I thought it would provoke another outburst. But controlling herself with difficulty, she said, "Well, if I do I'll lose my customers."

"If you don't', I told her firmly, "you'll lose mine."

Before she could recover, I turned to the two men and said in slow, deliberate English, "Would you gentlemen join me and do me the honour of having some coffee with me?"

The café, fortunately, was not crowded or I think the woman may well have appealed to her other customers for support. But I had acted so quickly that she really had no time for reflection. As she saw myself and the two strangers in full possession of my discreet table, she had no option but to serve them.

I had from the start recognised the two men as Japanese and now found that they were, like me, newspaper men. One, Mr Shirakawa, was from the *Osaka-Ashi* and the other, Mr Hisatomi, from the *Osaka-Mainichi* which were then perhaps the two greatest newspapers in Japan. They were not merely representing their own newspapers. They had been chosen by the Japanese as part of a far-reaching design to develop trade between Africa and Japan and to explore the possibilities of advancing the new policy of Empire on which the country was about to embark. It was a policy that included visions of the creation of Japanese spheres of influence in Africa, with an ultimate dream wherein Abyssinia, the last of the non-colonial countries in Africa, would be turned into a colony of Japan. To give some idea of the professional stature of the two men, I should add that Mr Shirakawa was destined to become one of the most distinguished foreign correspondents in the history of journalism in his country, and was to be chosen to serve his newspaper in Washington during the critical years leading up to the war. Mr Hisatomi's development ultimately took a different line because, as a distinguished sportsman himself, he became a leading sports executive and controlling influence in the making of his country into the great sporting nation it is today. But of course at that particular moment they were only concerned with the new task of making their own people more aware of Africa and its economic and political potential for Japanese ambitions.

I took to both men immediately. It was astonishing how I seemed to have no difficulty in understanding not just the literal meaning of what they said to me in their very careful, calculated English, but also appreciating the idiom of meaning which predetermined their choice of words.

Consequently I gave them my address in Port Natal and for about a fortnight I entertained them and helped them all I could in their work. Yet, when they ultimately came to say goodbye and sailed for Japan (which to me seemed physically just about as remote as any country could be), I did not think I would ever see them again. However, some months later, a colleague, Gerhard Pauli, who had taken over my duties as shipping correspondent, came to me with the news that the Japanese had started a new monthly service between Port Natal and Osaka, and that the captain of the first ship to pioneer this service had arrived in harbour that morning. He had asked if he knew me and whether a meeting could be arranged?

If it had been anyone but Pauli who had asked me at that moment, I believe I would have said no. I had almost more work to do than I could manage. In addition to more exacting work as 'specials' writer I was doing for the newspaper, I had joined the poet Roy Campbell and writer William Plomer as one of the editors of the literary magazine *Voorslag* which Campbell had founded. My association with these two and the work on the magazine, which was the first major literary venture of its kind in my native country, had already led to my abandoning the weekend games which had meant so much to me. I could only do the magazine work by spending the weekends with Campbell, his wife Mary, and Plomer, at a little place called Sezela on the south coast of Natal. I would arrive there late on Saturday night; the most exhausting day of an exhausting week. I would have to walk four miles from the station along the railway track through the black bush to reach the little house just short of midnight, and start work straight away.

Stimulating as I found both the company and the work, the magazine, which from the beginning had provoked a hysterically hostile reaction from both public and newspapers, had developed internal problems of personalities and management of a disastrous potential. So I had not been surprised, the weekend before Pauli brought me his invitation, when Campbell resigned from the magazine and Plomer and I, of course, had followed his example.

The resignations had merely shifted my problems from one dimension to another. Neither Campbell nor Plomer had any income. Moreover, they now had to leave their shanty bungalow at Sezela and for the moment had nowhere else to go. Mary Campbell, who had been through this sort of thing repeatedly with Roy and was a person of unusual fortitude, was nonetheless singularly dismayed by a homeless prospect as she now had two little girls to care for as well. I was the only one of our little community to have a reliable source of income, and so my invitation that they should come and live with me for the time being was accepted. In fact I had already made arrangements with my landlord to take a larger flat and was engaged in trying to find ways and means of increasing my salary to meet the formidable responsibilities I had now incurred.

I had a strong instinct, therefore, to say no to Pauli. But he pleaded with me to go that very evening to dine in the ship with its captain, as he was so sure that the three of us should meet. Pauli had become a special friend, so I took all this seriously. He too came from indigenous Africa; his father had been a much respected missionary in Zululand. He spoke far better Zulu than I could ever hope to, and was an enormous help with our Zulu contacts. We shared a love of literature, music, things of the mind and, of course, an interest in the sea. We spent much time in the harbour discussing and arguing over books we had read, from Heraclitus to Bergson; from Homer to Nietzsche and Dostoyevsky. We would explore the origins of the passionate colour and racial prejudices of our society which we abhorred as much as we were mystified by them. He was, and still remains, one of the most selfless and lovable human beings that I was ever to meet. I could no more have gone against his instinct in this matter than against my own. And so, that evening I accompanied him to the docks and to the berth of a ship called the *Canada Maru*.

By that time, since it was still winter, it was too dark to have an adequate grasp of the vast, mixed concourse of shipping of which the *Canada Maru* was part. The ship, therefore, was in the best possible situation to make its own impact upon one's senses without interference from comparisons, which was probably just as well as she was in fact wedged between the Royal Mail ship *Walmer Castle* and the British India passenger liner *Khandala*. Though the stern of the *Walmer Castle* and the bow of the *Khandala* towered above the *Canada Maru*, the bulks of those imposing vessels were vague and indistinct, and they made a kind of frame within which this first vision of the *Canada Maru* has hung as a swift impressionist painting in my memory.

There seemed at first nothing at all remarkable about her. She was merely a cargo ship of commonplace design; funnel, bridge and officers' quarters amidships; two slender and tall masts in the centre of the deck between the bridge and the raised poop over the crew's quarters in the fo'c's'les, with the after-deck similarly raised over the berths for the rest of the ship's company in the stern. Both masts at the base were equipped with derricks to serve hatches on either side, which made it clear that the

Canada Maru was accustomed to calling at ports where there were no cranes to load and unload. Those derricks at that moment were neatly folded back against the masts. Even her tonnage, as far as the usual run of cargo ships at Port Natal went, was unremarkable, if not insignificant. She registered, Pauli told me, six thousand and sixty-four tons gross. Moreover, within those limits, she was so designed that she looked even smaller, and one tended to overlook the fact that for a mere cargo ship she had rather an elegant line, which provided enough of an arch to her structure to strengthen her decks against the onslaught of storm and wind in her own typhoon waters.

The hull was painted a ceremonial black, but, just between the hull and upper decks, two clear bands of white conformed to the swerve of the line and ran the whole length of the ship. Her upper decks, even in that dim light, were a refreshed, stainless white. The 'lookout' post some two-thirds of the way above the deck caught just enough of the reflected light to remind me of the foretop in the *Larsen II* and her captain, who was never far away from my thoughts. The funnel, upright like the masts, was tall as befitting a ship which still burned coal and trailed a long pennant of black smoke all over the blue of its beat around the world. It, too, was black except for two bands of white just underneath the rim. The davits and the lifeboats tucked into them were white likewise, as were the ventilators. The mouths through which they breathed, however, were a startling pillar-box red.

It is only by concentrating on physical details such as these that I can prevent my initial impression from being overwhelmed by the ship of legend that the *Canada Maru* was to become in the evolution of the story. It is insignificant that I was later to travel in warships, sloops, frigates, destroyers and luxury liners, some of the latter more like floating baroque palaces conceived in a hubris of Rheingold fantasy. Yet none of these gave me as much as did Captain Coulthurst's tug, *Harry Escombe*, Thor Kaspersen's *Larsen II* or the *Canada Maru*.

However, as Pauli and I went up the gangway in the flickering electric light, I had no inkling of what was to come. I only remember that I was struck by the neat and orderly appearance of the deck. This order and circumspection seemed personified by a young sailor who appeared at the head of the gangway, not dressed casually as were his counterparts in cargo ships but in a smart uniform and a sailor's hat with a pair of black satin ribbons on one side. He saluted and bowed politely. Then, as he came out of the bow, he revealed an expressive young face of regular features, with an uncharacteristic expression of fun, if not mischief, in his eyes. It was my first meeting with a young quartermaster whom I was to know well by his first name of Gengo. He led us across the deck and, although I did not know it, across a far frontier in my own mind.

Apart from my brief meeting with Messrs Shirakawa and Hisatomi, I had never had any individual exchanges with the Japanese. Such encounters as I had had in the past were confined to boarding ships of the Nippon Yusen Kaisha when they called in at Port Natal crowded with emigrants for Brazil. Then, somewhat painfully, I tried to extract what I could of interest for my readers, from scrupulously polite but monumentally solemn and dignified first mates and captains who spoke hardly any English. However, there was still a difference about those ships and their inhabitants from which I derived some savour of an excitement. Japan itself I knew only in terms of the broad outlines of European history; also in what, retrospectively, appear to be scandalously sloganised forms of half-truths which passed for profound conclusions, as exemplified in H. G. Wells's *History of the World* – "Japan has made up the leeway of centuries in two generations." Japan, however, was not just as an element in history observed from the Mont Blanc of an amateur historian of genius, but had a story in its own right and a record, therefore, of some two thousand years in the life of a people and culture evolving their own unique forms of the living spirit. My own obsession with the meaning of history had tempted me to have a slightly closer look at Japan than my companions. I had been stimulated by a typically French flirtation with Japan in Pierre Loti's *Madame Chrysanthème* which, brief and self-indulgent as it was, had some importance as a cross-pollination agent for the European spirit out of all proportion to its substance. I bought Professor Chamberlain's *ABC of Things Japanese* but so far had fingered it sufficiently to marvel at how two incredible brothers dedicated their lives to serving foreign nations: one, the great Basil Hall Chamberlain of Japan, the other, Houston who, in the typical way of the converted, became more fanatically German than the Germans.

It must be added that in a country like my own, so firmly in the grip of racial and colour prejudices of all kinds, warnings like those of the Kaiser about the 'yellow peril' were still echoing from the walls of our little world which was so much closer to the heart of the matter. They rendered more plausible the protest of commerce and industry which saw another form of danger from the new Japan. They held that Japanese businessmen and manufacturers were not to be trusted; that they stole their industrial designs from the West, pointing to the common charge that Japanese pencils, which were apparently driving their own out of the market, had lead only at the ends and were hollow in-between. It did not seem to strike anyone that all this could have been honest errors due to the haste in making up the 'leeway of the centuries'. Only a few years previously the Japanese had been arrested in a feudal system so involved that consequently only three per cent of the population were engaged in business, and the makers of money were almost a caste of untouchables. Considerations of money, indeed, had been matters of contempt to Japanese gentlemen and their followers for so long that a whole system of rituals of almost metaphysical complexity had to be evolved for bringing financial reality to their notice by these despised orders who were engaged in so indispensable a traffic. Although I could add to the detail, this summary is in essence representative of the flimsy preparation I had for this new experience and for my evening's encounter.

I do not know precisely what I had expected, but the surprise which accompanied my first impressions was considerable. Instead of the usual bare, rather cramped wardroom in which officers fed in most of the cargo ships that I had known, I saw before me a ship's saloon worthy of the highest mail-boat standards. The single long table was covered with a dazzling white cloth and was laid with silver, cut-glass tumblers and crystal wine goblets. In front of each pointed, neatly folded white napkin was an elegant, silver menu-holder on which was printed what seemed like an incredible number of courses.

I saw the table first because it was closest to the entrance but almost immediately my attention was drawn to a group of Japanese officers beyond it. There were six of them, standing and talking in the liveliest manner, with an obvious relish and gift for conversation for which both Pauli and I were unprepared. As we entered, one officer separated himself from the group and approached us. I assumed he must be the purser but as I noticed on his sleeve four rings of gold braid, I realised that he had to be the captain. He was in fact Commander Katsue Mori.

He was barely thirty-six years of age. He was obviously young for the command even of a single ship, let alone for heading what I suspected was no ordinary commercial mission. The directors of the Osaka Shosen Kaisha, as well as the unseen makers of the long-distance policies of Japan, must have been fully satisfed that he possessed qualities which outweighed any claims of seniority. In a quick appraisal he appeared to establish himself as a man of unusual quality and authority. Confidence and dignity appeared to sit easily upon him and to fit his spirit as if tailor-made for it, like the immaculate, navy-blue uniform that he wore. He was of more than average height for a Japanese of that period. I say 'period' because the Japanese today, thanks to their improved diet, are taller than they were when I was young. In addition, he was of a well-proportioned and almost athletic build. He had a face of regular Japanese features except perhaps that the forehead was broader and the eyes wider apart than usual. His eyes were full of spirit, intelligence and a suggestion of defiance. His hair was closely cropped in the new American style referred to as a 'Brutus' haircut. However my intuition told me that I was confronted with an unusually complex and paradoxical personality, resolute, experienced and stern, and yet underneath it there was perhaps a childlike innocence. With all this went a gift of laughter belied by the air of entrenched authority and pre-determined dignity which presided over our welcome. This sense of paradox seemed to me confirmed at once by what may appear trivial today, but which I found conclusive then.

As he shook my hand I realised that his comparatively stern face had, growing on its determined chin, a straggling beard, looking as out of place as did a top hat I once saw on a Hottentot. It had not intruded on my first impressions of him as it was so thin, but it was typical of this observant man that he seemed to know what was in my mind and made it a point of honour to tell me all about his beard as he stroked it with an immaculate hand. It was, he stressed, unusual on the face of a Japanese for they were not usually a 'hairy' people but, please, I had to understand that it was there neither by accident nor lack of forethought. So what did I think of it? His question was accompanied by a laugh, which was to be his habitual way of saving the other person from the embarrassment of having to think of something polite to say. His officers, he said, had just been teasing him, urging him to shave off his beard in the interests of better relationships with this new world which it was their mission to cultivate. No-one, his officers had said, could possibly be impressed by a growth like that, and he had better be off to his bathroom at once before he put his guests off their dinner. Even the managing director of the Osaka Shosen Kaisha, a man to whom Mori was devoted, had hinted in a subtle, gentle way that he might serve his country better without that caricature of a beard to fly in the face of strangers. He had replied to this director that, no matter how ridiculous it looked to others, he had decided to ignore all conventions and social prescriptions. Not only people of his own nation but foreigners as well, he asserted with confidence, would only have to look at his beard to know he was not just a machine carrying out the orders of others, but a man in his own right.

There followed one of the best and most unusual dinners I had yet had in a ship, let alone a cargo ship. The food was good; a provocative blend of European and Japanese, a sort of culinary demonstration of this halfway-house sense of history that seemed always present in me. I chose to drink saké, the rice wine of the East, warmed the Japanese way, and drunk out of a small and beautiful porcelain cup. The conversation too, although most of it was in the unfamiliar sounds of a language I did not know, seemed, as the evening progressed, to become a musical recitative improvised for the occasion. It felt more like a banquet, celebrating a meeting of nations. In fact there were, besides Pauli and myself, only six others at the table – Commander Mori at the head, with Pauli and myself on either side of him, then the first engineer and purser facing each other; the first mate opposite the chief radio operator, and finally the doctor at the far end of the table.

If anything more were needed to confirm the fact that the *Canada Maru* was bound on no ordinary mission, it was the menu and the presence of the doctor. No cargo ships I had ever seen had carried a doctor. And although Commander Mori told me then that he had accommodation for twelve first-class passengers and fifty more in steerage, he had carried no passengers on the voyage out and the prospect of any passenger traffic between Japan and Africa was very unlikely at the time. Yet it was typical of the spirit of thoroughness with which he and his masters had thought out their purpose, that they were not going to be caught out in any way. Consequently he had persuaded a newly retired doctor to accompany them, more for the fun than the duty.

With regard to the dinner itself, when I had complimented the captain on the food, he told me he had taken great pains to select one of the best Osaka Shosen chefs. He had a feeling that good food served in his ship might be of very great importance to the success of the mission. Stroking his beard even more fondly and assiduously, he explained that his feeling had already been justified. At the port of Mombasa he had suddenly found himself approached by a member of the staff of Sir Edward Grigg, who was Governor of the Colony of Kenya, and had been asked whether he could give the Governor passage to Dar-es-Salaam in the then Tanganyika where he was needed for an urgent conference with one of Britain's great colonial governors, Sir Donald Cameron. That he should be asked so early to be of service to the highest authority in one of the key countries within the orbit of his mission far exceeded the limits of legitimate expectation. Had they been without a doctor, and had they carried only a 'galley-hand' that passed for a cook in most cargo ships, he would have had regretfully to refuse. But as it was, the voyage had proved successful beyond his dreams.

During the dinner Mori told me about his mission, and the difficulties involved. Foremost among those were the racial prejudices he had encountered along the coast of Africa, not just against the black people of Africa but against any people of 'colour', including the Japanese. He was certain that he could try to make the Japanese an exception to any rule of prejudice, however profound. He honoured me also by talking about himself. As a boy he had rejected the chance of going to the appropriate Academy to join a navy that had obsessed the imaginations of most boys of his acquaintance. He had decided, after much anguish of self-examination, to join the Merchant Marine because the way of peace through legitimate commerce ultimately was more in keeping with his own seeking than was the way of war.

From this point he came naturally to what had apparently been the real reason for my invitation to dinner. Messrs Shirakawa and Hisatomi had told him all about me and, in particular, of my rejection of the racial and colour prejudices of my own country. Mori and his company, therefore, would like to join Messrs Shirakawa and Hisatomi in inviting me to come to Japan to see it for myself and so help to show my people how ridiculous were their prejudices; also how great were the opportunities they had for trade with a rapidly expanding and dynamic Japanese economy.

Even at this distance in time I remember the excitement produced by this invitation. All I wanted to say was "When do you sail?" and accept gratefully. But then I recollected bleakly that I had just taken on the temporary responsibility of the Campbells and William Plomer. Sadly I explained that in principle there was nothing I would like to do more, but that just then I was pledged to urgent responsibilities which made it impossible for me to accept at once. But could we not do it later? I stressed and restressed both the wish and conviction that I could do it, given time, because of fear that much postponement might be the end of the matter. I was somewhat reassured by the philosphic way in which the Captain took my answer. That, too, was characteristic. He appeared incapable of being frustrated, and seemed to regard 'setbacks' in life as an experienced climber recognises false summits on the way to the top of a mountain. He had not expected, he said, that I could be ready for an invitation at such short notice. Perhaps, when he came back in five months' time it would be possible? Would I wish to, and could I arrange my affairs accordingly? My acceptance then was unqualified if purely intuitive, because I had no guarantee that my newly acquired responsibilities would end in so short a time. However, after promising that I would dine with him in a week's time, on the night before his ship sailed back to Japan, we parted on that outwardly optimistic note.

It was, I remember, a Wednesday night towards the end of August. In that tranquil clarity of sky and the cut-glass sparkle of stars so characteristic of the Zulu sky just before the cobweb haze of spring is spun across it, Pauli and I walked back home. So quiet was I with the effort needed to deal with the disappointment of my first opportunity to move out of this halfway state of spirit in which history seemed to have placed me and my country, that Pauli became quite concerned. He feared I had not enjoyed the evening? I had to reassure him that of course I had; it had been all and more than he had promised, and I could not thank him enough for such an opportunity. I left it at that and said nothing about my invitation, because I knew that he would have jumped at such an invitation himself, and might have been hurt by the fact that, as the first person to meet Mori, it had not been extended to him.

I persuaded my news editor to release me from duty on the coming Saturday which was the busiest of all our days on the paper, so great a burden was this whole issue. I took the first train to southern Zululand, and by eleven in the morning I was with the Campbells and Plomer who were still at Sezela. At the outset of our meeting I blurted out an incoherent description of my meeting with Captain Mori and his invitation to go to Japan. The moment I let out the news I felt thoroughly ashamed of such insensitivity. It would surely have made it now more difficult for them all to come and live with me in Port Natal. But to my amazement, far from being troubled by my impulsive confession, they turned on me and said, "But surely you weren't so great a fool as to say no?" Campbell even hastened to add, "My God, Laurens, watch it. You don't have enough imagination!"

Just for a second I thought rather bitterly, "It's you who lack imagination. Can't you see I did it for you and your needs?"

Happily, however, the realities of the moment were so pressing that my gaffe, as I supposed it to be, sank almost immediately and without trace. Soon after my arrival, a telegram came from Campbell's eldest brother in Port Natal, asking him to telephone him from wherever he could. Within an hour Campbell was back, transformed and jubilant. His family, on hearing the news of the *Voorslag* disaster, had been summoned to his widowed mother's home and, after consultation, decided to lend him a furnished cottage and make him an allowance of twenty-five pounds a month.

The sense of release produced in me was as if a dam which had arrested the main stream of myself had burst. A flood of ideas for a way out immediately presented themselves. I turned to Plomer and asked, "William, if I can persuade the Japanese captain to take you as well, would you come?"

"Can a duck swim?" Plomer replied.

By the afternoon I was back in Port Natal, calling on a surprised Mori. He had just taken his evening bath and was relaxing on the small deck between his night cabin and the lifeboats of the *Canada Maru*. He was dressed in an elegant kimono carrying the heraldic badge of his family on its sleeves. He looked at one with himself, relaxed; in another world and area of mind and spirit.

In some strange way my visit seemed not altogether unexpected. At the earliest opportunity I told him I would like to go with him to Japan as soon as could be arranged. I felt sure that it would do nothing but good for his maritime venture and also that my editor would let me go. There was only one difficulty. I told him about Plomer, and about *Turbott Wolfe*, a truly remarkable work of art, and relevant because it was by far the most devastating exposure of the evils of racial prejudice in southern Africa that had yet come from an English pen. I also told him about *Voorslag* and how it had been dedicated to the same ends as *Turbott Wolfe*. I told him that whereas I had only a local reputation, Plomer already had an international reputation and could add a dimension to what we were trying to do which I could not hope to achieve.

He listened to all this with growing interest and emotion which came from a feeling, I suspected, that once more his sense of direction was being confirmed by the relevant coincidence. At the end of a somewhat inadequate but over enthusiastic outpouring from me, he asked only one question, "Could you get Plomer also to represent a reputable newspaper?"

That would be no problem, I assured him, having already foreseen this and worked out a solution on my way back from Sezela in the train. I could let him have appropriate letters of accreditation by the Monday. If that were so, he said with firm and immediate decision, he would take Plomer and myself to Japan on one condition: he would have to ask Plomer to pay him thirty-five pounds which was the cost of a steerage passage from Port Natal to Kobe. For that he would, of course, carry both of us in the first class as personal guests. Moreover, he could undertake that once in Kobe the passage money would be returned by his company and we would from there be treated as honoured guests of his employers and country. In fact, if it had not been for the difficulties of making these arrangements by cable, as he had authority to take me but not for Plomer as well, he would not have charged anything at all.

I returned to my office and telephoned Desmond Young, the new editor of the *Natal Witness*. This remarkable man was to make the *Natal Witness*, one of the oldest papers in the country, a singularly alive instrument of reappraisal of our political, social and intellectual life. He had been a staunch defender of *Voorslag* and wrote in his autobiography, *All The Best Years*: "Alone in Natal and almost alone in Southern Africa I wrote in praise of *Voorslag*, the magazine that Roy Campbell, William Plomer, Laurens van der Post and Lewis Reynolds had started." One day I hope to do him the justice I have no place for here.

I had only to tell him what it was I had in mind for him to decide. From birth Desmond Young had always been an urgent spirit; his pace was naturally that of the gallant rifleman and officer he had been in the first World War and not the steady measure of the Guardsman favoured by military establishments. "You will," he announced firmly, "have a letter from me first thing on Monday, appointing Plomer our special correspondent in Japan and the Far East."

On that Sunday evening the Campbells and Plomer arrived in the ample home of Roy's mother in Musgrave Road on the Berea where we had a reunion full of excitement, emotion and hope of a new future. Until late in the night we discussed the journey to Japan. As for the Campbells, we made certain that we would all come together again either in southern Africa or Britain. The immediate problems were mostly concerned with finding the money for Plomer's provisional payment for his passage, also finding clothes for him to wear on an official journey. His informal, sub-tropical clothes would hardly be appropriate for such an adventure in the outside world.

We solved the problem by dividing my own wardrobe in half, giving him two suits, a sports coat and two pairs of 'Oxford bags'. The result was not altogether satisfactory because, although the coats themselves fitted well enough across the shoulders, William was taller than I and longer in the arms and legs, and the slightly incongruous effect of our makeshift alterations is evident in some of the photographs taken subsequently. We managed to borrow his passage money from a friend I had made in the course of my translation work, a remarkable Scot called Kepple Harvey.

The first meeting between Plomer and Mori was an even greater success than I had anticipated. My editor, as I had believed all along, could not have been more helpful. He knew that, whatever his own appraisal of my capabilities, I lacked experience, and that letting so young and new a recruit go on so prestigious an assignment might cause resentment among his staff, but did not flinch from the prospect. It says much for his magnanimous soul that although he apparently disapproved of Plomer's work and feared the effects on me of my personal association with him and Campbell, he did not allow the fact that now I would inevitably be in even closer company with Plomer to deny me so unique an opportunity.

On the Wednesday night we had a farewell dinner with the clan of Campbells happily reunited in Roy's mother's home, after years of feuding over his behaviour. The following morning I arrived on board the *Canada Maru* to find Plomer already installed in the best and most luxurious cabin, on the boat deck just underneath the bridge and connected by a bathroom to Mori's own cabin on the starboard side. I was late as I wanted to call at the whaling office to leave my message for Thor Kaspersen. I had also hoped to have had a last glimpse of Pauli but when we cast off at noon he had not appeared; I was not to see him for many years. Not long after our departure he himself apparently signed on as a deckhand in a rusty old 'tramp' and went on a beat of his own around the world.

The pilot assigned to the ship was an old acquaintance. When he boarded the ship he slapped me instantly on the shoulders, saying "So, you've done it at last! You'd better come on the bridge with me – I'm sure you know as much by now about taking a ship out to sea as any of us!"

However Mori, who had not been consulted, immediately showed how alive his sense of command was at all times. He countermanded the pilot's suggestion with the excuse that the bridge was not big enough for so many people. The result was that William and I, unusually silent, watched our passage out to sea leaning on a rail between two lifeboats. A strong wind blew across the harbour mouth and added to the height and power of the abiding swell of the Indian Ocean. Far from being winter clear, the sky was hazy with moisture from a certain quickening of the heat of the sun and turmoil of air. This movement of the wind, I remember even now, seemed deeply involved with the rhythm of my own spirit. It revived my own excitement and led my spirits on to the open sea at last and the East beyond. . . .

The Indian Ocean was already beginning to make itself felt long before we crossed the bar. Its swell, impelled by the wind, sent the little *Canada Maru* to stagger upward higher and higher and to dig down deeper into the sea until, just before we emerged finally from the harbour mouth, its finely cut bow went down so far that it plucked its first feather of foam from a wave and stuck it jauntily in its cap. Soon the town was blurred and indistinct behind us. Little more than a line of white surf remained bleached and yellowing in the hot August sun to mark the end of the land of Africa. I watched the widening distance complete this erasure of my native earth with mixed feelings, until suddenly the escorting tug came up from behind and pushed its nose firmly into the side of the *Canada Maru* and held it there almost directly beneath where Plomer and I were standing. I looked down and at once recognised the *Harry Escombe*. On the bridge, waving his hat wildly to attract my attention was Captain Coulthurst, his reddish, round face gleaming, and smiling from ear to ear. That set a seal on inevitable feelings of fracture caused by the step I had taken, which qualified the exhilaration of the moment. For somehow I knew in an inexplicable way that I had better let this impression of this musical man of the sea become fixed in my memory, for perhaps I might never see him again, nor hear his expert gramophones release their Wagnerian tides of music as they had done so often in the past for so many of us.

The tug had hardly got her nose firmly in position when the pilot, accompanied by Mori, appeared on the deck beside us. We just had time to shake hands quickly before the pilot was over the side and going fast down the rope-ladder. Both the *Canada Maru* and the *Harry Escombe* were now heaving up and down in an alarming manner. No sooner had the pilot been caught and steadied on to the deck of the *Harry Escombe* by two sailors than he doffed his cap and waved it in the gesture that I was to see repeated by pilots all over the world. As he donned his cap again at a 'Beatty' angle, a series of blasts on the tug's whistle soared steeply in the air and silenced the nostalgic voices of the attendant gulls. The *Canada Maru*'s siren responded in a deep, long, sustained note of irrevocable farewell.

Almost at once we were joined by the indefatigable Captain Mori to pose for the first of an endless series of snapshots which were, alas, all to perish in one of the fires caused by bombs in London in the last war. I now became aware that Plomer was becoming more and more disturbed by the movement of the ship which became more violent as it gathered way and pushed with increasing speed into the rising sea. However, always a person of great self-control, he survived not only the session of photography but the lunch that followed. Then, hastening to his bunk he confessed that he'd always been a bad sailor. Although he could endure the movement stretched out on his bunk or in a deck-chair, even the slightest motion when on his feet tended to unsettle him. As a result he spent most of his days horizontal, although incredibly he did not miss a single meal. When he did have to be on his feet he managed to convey nothing to Mori and his staff, and of course I said nothing.

I myself would probably have felt as bad had it not been for my experience in whalers and the rigorous training under Thor Kaspersen of whom, even then, I was being constantly reminded. For two or three hours, wherever I looked, I saw a whaler either on the prowl or galloping wildly after a whale over the white, uneven tundra that the wind had made of the sea. But soon they too vanished and we were left alone, all Africa lost below the wind-dimmed horizon and the last of the gulls reciting a deep-sea dirge before turning about, to drift home like crumpled tissue-paper cast away on the heaving air.

From time to time I would break from my self-imposed watch to look in on Plomer. He was usually sleeping and, in spite of months of sun on the coast of southern Zululand, he looked very pale. I suddenly realised that, although I knew a great deal about the externals of his life, the fever and dramatic circumstances of our *Voorslag* connection had not told me much of what kind of a person he really was. For the first time I wondered how we would get on in so confined a context for so many months. I had no doubts myself. Although I knew that he was exactly three years and three days older than I was, differences in age have never really mattered to me. I knew that I liked him as much as I admired and respected him. But I was aware of the fact that we had different natures and that there was not yet between us the same immediacy and rapport that I had enjoyed with Campbell. Perhaps there was a built-in impediment in our natures that might prevent our relationship from developing into real friendship. This is not the place to examine in depth an association which was to last some fifty years. It is enough to say that differences and inextricable impediments of being had nothing to do with race or upbringing, but were rooted solely in the fact that we had been born totally different psychological types.

I cannot pretend that these facts did not make a difference to our relationship but they were, mercifully, never strong enough to prevent a friendship which was ended only by Plomer's death. The differences can be summed up best by saying that I tended to live more by instinct, intuition and feeling than by reason, which probably explains why, although I had always longed to write, I knew that part of me would never be satisfied by just that. Whatever it was that caused this longing and need to write more and more would also need to be lived. Plomer, however, tended to live far more by reason and an intellect of unusual power, which he continued to cultivate for the rest of his days and moulded into a worthy instrument of his great poetic gift. Although by nature I believe he was more complicated than I, his purpose was perhaps simpler and his approach more single-minded. As a result he frequently sacrificed qualities that to me were just as important as the values which he served and upheld with such distinction. This was clear from the way that he had come to writing. He said that as a boy he realised he possessed two gifts of almost equal power: one for visual arts, particularly painting, the other for writing – above all poetry. It seemed to him that no man could serve two gifts simultaneously and so he had to choose. At the age of eight he consciously chose to become a writer and never again looked back on his choice. Yet the gift for visual expression continued to lead an underground existence in his spirit and tended to rebel whenever he relaxed his conscious watch on himself and his will to direct his gift to a preconceived end. It would surface then in a rash of sharp, bitter and brilliant caricatures, hinting at almost volcanic resentments which the urbane, civilised, articulate and pre-possessing external appearance utterly belied. Particularly significant was the fact that the most savage and frequent of his caricatures for many years before the War were those about royalty, suggesting that this profound conflict of talents was at war also with what was royal in his own values.

It was not surprising therefore that, a few nights before sailing, he had felt compelled to declare: "You know, Laurens, I think we had better agree on a division of labour for ourselves on this voyage. I shall take charge of letters and the arts; you had better look after sport and music."

I had not given the matter a thought and found the suggestion at the time unnecessary, if not also somewhat preposterous, and I mention it now only as an illustration of how much of a planner he was. It was the first and almost the last time that we ever talked about music. Plomer was basically unmusical and made little attempt to learn about it until many years later when I brought him and Benjamin Britten together in Aldeburgh. Britten, who thought highly of Plomer's work but had not yet met him, was convinced that he would be the perfect person to write the libretto for an opera based on the stories of Beatrix Potter.

It was another and totally unexpected form of music which restored Plomer to himself on our first afternoon in the *Canada Maru*. Our steward, who spoke nothing but a Japanese patois of his own and whose disproportion of features was redeemed by an unusual sweet-ness of disposition, had an unfailing desire to serve. One day he had unobtrusively entered our cabin and had barely presented us with a tray of delicious Japanese tea and biscuits when a sound almost like a roar of pain from a trapped lion came from the bathroom next door. Plomer looked at me with some consternation and exclaimed: "My God, what's happened? Murder in the first degree, I presume!"

But this outburst of sound then produced a pattern of tone and rhythm which, strange as it was to us, made it clear that Captain Mori was singing in the bath. This singing went on for nearly forty-five minutes. Soon after it ended Mori, changed into his most formal of uniforms, appeared at our cabin entrance to invite us to have a cocktail with him on deck and watch the sun go down. And, he went on to ask, how had we liked his singing?

For a moment I was at a loss for an appropriate adjective. Fortunately Plomer, whose natural wit never abandoned him, answered without hesitation: "We found it most impressive."

The word could not have been bettered. It not only pleased Mori's disposition but also fitted the purpose of his singing. He squared his shoulders, expanded his chest and said, not without a certain air of self-approval: "I was singing classical Japanese music. I do it every afternoon for half an hour at least because I find it such good exercise."

Exercise it must have certainly been to produce those volumes and enigmatic variations of sound to which we had been treated and were going to be treated precisely at that hour, no matter what the condition of sky or sea, until we arrived in Japan. By that time, as far as I was concerned, the music ceased to be strange and came to interest me greatly, but for Plomer it was another matter. In due course he produced a series of doodles of our captain in the bath going through convolutions and contortions of muscle and limb like a Hindu Guru, compelling the movements of his body to express the metaphysical complexities of his spirit, as he delivered himself of these arias. No off-stage accompaniment would have better suited this world of the *Canada Maru* to my mind and, had it been absent, I now feel it would have impoverished the atmosphere of our voyage.

The 'world' of the *Canada Maru* was exactly what it was: a microcosm of the macrocosm of Japan, a sort of Bonsai tree of the spirit transplanted into this miniature pot of its culture afloat on a foreign sea. International and contemporary as the ship was, everything in it was totally Japanese. Apart from Plomer and myself, there were no foreign ingredients to subvert an essentially Japanese version of the modern world. Few of the officers and none of the crew spoke even the most elementary English. They were insulated from any distortion of their own national pattern which contact with the wider world might otherwise have caused.

Far from being disturbed by this, I found myself reassured by a discovery which came to me on our very first night at sea. Plomer had taken to his bunk immediately after dinner, at which he had given one of his bravest and most convincing performances of someone at ease in a ship pitching steeply in a lively sea. But I was too stimulated to put an end to so remarkable a day. Since Mori had made it clear that we had the freedom of the ship, I set out to explore the *Canada Maru* in the dark. There was unfortunately too much spray coming over the bows, so that I had to go aft. I climbed down on to the well-deck, made my way over the hatches and up a short ladder to take up position with my back to the rails over the stern.

I stood there for a long time alone, rejoicing in the dark, undiluted by any artificial light except a firefly effect of the navigation lamps high up the masts. The wind was falling fast, and the cool of the night had cleared the air. The sky was a transparent black and sparkling crystal, and the stars as precise and near as in my own high native air in the Interior. The result was that I had an unimpeded view of the movement of the *Canada Maru*, which was perhaps fifty times bigger than the *Larsen II*. The increase in size imparted a certain majesty to the progress of the ship over the swollen sea which I had not experienced before. I could never have believed that any mast, even one so tall and slender as the *Canada Maru*'s, could reach so high that it stirred the stars like a finger feeling jewels in an Aladdin's cave of the night. Yet this is what it went on doing over and over again. With the explosion of water against her bows, the ship would rise shuddering and black with fern-coal skin all ashiver as it shook itself free from the wet, and lifted itself higher than ever to reach for the remote stars once again. But as the wind-raised sea declined, the regular swell took over and imposed its own symmetry on the movement of the ship. The vibrations of engine and propellers in the deck under my feet became more and more synchronised to the everlasting rhythm of the Indian Ocean. They were transformed accordingly from the mere mechanics of their origin not just into the heartbeat of ship but to a pulse of a universe and its dancing stars, which made me feel centred with an intensity that I had not felt for years. It was almost as if until that moment I had not fully believed in what was happening to me, and unconsciously feared that something could still come reaching out from the great dark continent below the horizon, to haul me back into the kind of prison that those last months in Port Natal, if not all the random years before them, had made of my life.

With this realisation, confident that I could take my new direction for granted, I turned about and looked over the rails into the wake of the ship. There the screws were releasing fragments of phosphorus out of the sea to vanish like glow-worms into the hedge of black which bound my vision. At that moment, from an open porthole in the crew's quarters under my feet, a strange music rose fountain-wise into the sky. It soared with singular lucidity and a noble purity as if inviolate and sheer from the same source as this personal sense of release into the freedom of movement of the universe that I had just experienced. It was, in fact, my first experience of the Shakŭhachi, the bamboo flute which participates even more profoundly in the symbolism that informs the spirit of one of the few peoples left who still lead a symbolic life, than it does in the almost countless practical needs of their existence. Since the bamboo itself rose out of the earth, as the music to which I was listening soared out of the silence, there was total reciprocity between the fashioning of the flute and the fountain of sound that came to me. That sound was spare and devout in its obedience to its own law of expression which commanded that it should convey all that was possible with clarity and simplicity.

I was to discover that the music itself was about some seabirds combing a secluded beach of yellow sand by the Inland Sea which, like a great lagoon locked out of the swing of the storms of the ocean, holds a vision of calm on the far frontier of a volcanic people's tumultuous history. As a result it was charged with nostalgia; a nostalgia just as much mine as it was Japanese. At once I was glad I had come so unprepared to this new experience. Books would have come between my natural reactions and Japan. For the first time I was unconditioned to let what had to happen come to me unimpeded and be received in my own natural way.

I went to bed as if warmed with wine, yet I woke early and went at once, just before an impatient dawn was at the rails outside our cabin, to look out over the sea. I had done it as if it were a most immediate task. From then on I continued to do so daily on board all the ships I have ever travelled in. It was as if I had to begin my day with the reassurance that the sea had not vanished in the night. Perhaps it was a relic of some ancient rite left over on the scorched scene of our contemporary spirit, like one of those broken columns on the way from the great plain of Troy to Jerusalem, a shaft of sunlight splintered upon it and deprived of the temple that it once supported. Primitive people often act or dance their prayers to creation, and this might well have been something similar. All I can say for certain is that never yet, after this silent contemplation of the day breaking over the sea, has the reassurance failed to carry me through the day as it did on this first morning on the *Canada Maru*.

I had hardly climbed back into my bunk when a low voice, using the highest of the many degrees of politeness used in Japanese, addressed me. Startled, I sat up. Our steward had, as the Victorian novelists had it, 'materialised' with our morning tea. Plomer was still asleep but seemed happy to wake to find that the wind had dropped and the sea had been reduced to the long, slow, rounded swell which passes for calm on the Indian Ocean. It was becoming warmer and so later, dressed in our coolest and most casual clothes, we had, without Mori, a breakfast which would have done honour to a Dickensian middle-class establishment on a Sunday morning. In high spirits, we established ourselves in the shade of the little deck outside our cabin and talked happily together. Plomer was one of the most natural and lively talkers; he was witty and equipped with deep insights of his own. Occasionally he would pounce on a word like a cat would on an unwary mouse. He had once remonstrated with me, "Come off it, Laurens, everything is a thing."

When the bells of the watch on the bridge overhead signalled that it was ten, Captain Mori joined us. He was already dressed in a dazzling white tropical uniform, the trousers ironed into lines fore and aft that would not have disgraced the clear-cut bows of the ship. It was then, as far as my own sensitivity was concerned, that the problem began.

No sooner had we all said good morning than Mori and Plomer began talking to each other as if I were not there. I sat there in silence for ages, expecting to have a chance to join in as indeed I tried to do once or twice, but with such a lack of welcome that I gave up. Discouraged, I recalled that I had noticed something of the tendency the day before at dinner but had assumed it was only of passing concern. However, the longer I sat there listening to the two of them absorbed in each other, I realised that a new and potentially formidable development was taking place. Finally when the ship's siren at noon signalled that the officer on duty on the bridge had just shot the sun, and eight bells rang out to mark the change of watch, I was amazed how swiftly even two hours on my own had passed. I hurried back to our cabin not without a sense of guilt. I need not have worried. As soon as Mori had endorsed the report of the ship's position and the readjustment that it demanded, he waved to me to join them. At the same time our steward appeared with ice-cold cocktails in which Mori toasted us and then began a general conversation. Neither seemed to have noticed my absence except that it was implied in Plomer's announcement: "You will be glad to know, Laurens, that Captain Mori and I have decided to spend each morning translating *Turbott Wolfe* into Japanese."

Remembering how wounded Plomer had been by the national outcry *Turbott Wolfe* had raised, I was indeed glad at this act of recognition from the ruler of our little world and knew how healing it would be. It no longer mattered that it would draw him and Mori into an association from which I would be excluded. These two hours on my own had been long enough for me to make my peace with a diminished role in what I had regarded as a partnership. It seemed, after all, right that Mori should be more interested in Plomer than me. Plomer was older, more experienced, more travelled, more articulate, interesting and amusing than I must have been. He had more to give Mori than I had of everything, except admiration and respect. Plomer was always far more critical of the Captain than I was and, though fond of him too, never ceased to weigh his heart and mind against each other in the scale of his personal commitments. This he did continually with all his friends. He maintained what he and others would have called a proper balance in their relationship. I still believe that his measure left some great imponderables of spirit out of the human assessment. But to the end of his days he continued this form of accountancy.

Only envy or jealousy could have clouded my understanding of why and how these things between the three of us had to be as they were now becoming. Fortunately, it was as if fate had given me fool-proof armour against such an assault on my sense of proportion. As a mere boy I had once experienced the full power and infamy of jealousy. In the remote interior, where I grew up, one of the most eventful occurrences always was the appearance of a stranger in our normal round of existence. There was magic about the unknown which a stranger trailed as a cloud of glory along with him until familiarity in the end restored him to his fallible and comparable human shape. A new boy, to me the most ordinary of boys since I already knew him at the public school to which I had been sent a year before, came with his parents to settle in our isolated little community. The effect on all my friends was like that of the Pied Piper and his music on the children of Hamelin. They abandoned me, and fled our vast garden where we had happily played for so long, in order to swarm to the stranger. Their injustice and ingratitude, which were to be my excuse, caused an eruption of jealousy that overwhelmed me. I struck back with an instinctive cunning, a mastery of intrigue and a capacity for the invention of calumny, that astounded and shamed me. Indeed, I behaved generally so badly that the recollection, even now, fills me with unbelief. In the end the whole village was involved in a scandal of my invention and ended only because, from somewhere, I summoned a last remnant of integrity to confess publicly what I had myself brought about. The hurt to pride and self-respect was extreme. But the horror over my conduct gave me, I believe, an immunity for good and the weapon of powerful emotions against jealousy. With me, if there is a priority among pitfalls in life, this deep, black pit comes first. I have only to think of how I once came to fall into it, to be strengthened in my determination never to fall into it again. Moreover, I had seen how great a role unrecognized envy and jealousy play in the lives of men, not only in the streets, institutions and temples of peace, but even on the battlefields of the world, with the foreseeable consequences in the death of thousands of lives. My resistances would have been in vain if they had merely been based on a resolve of mind and will, and had not been allied to the powerful emotions which this boyhood experience had permanently put at my disposal.

As a result, no temptation of envy, let alone jealousy, arose to complicate an inevitably difficult situation, because I saw the responsibility for containing it as singularly my own. Neither Mori nor Plomer ever realised that this sudden translation of myself from the centre to the circumference of our relationship could be the cause of pain for me. As the voyage progressed they became more and more absorbed in each other, not only during their work of translation but whenever the three of us were together. This, of course, meant a great deal to me at the time: but they did it so naturally and unselfconsciously that it became a point of honour with me to wish it to remain so. I succeeded in this well enough to be free to start a most rewarding ship life of my own.

The first intangible, but decisive, incentive was associated with that moment when, alone in the dark, I heard those lucid notes of music and discovered feelings of emancipation and nostalgia, a nostalgia greater than the one we all experience for what merely is past and cannot be relived. I have always had a great love of languages, and this love and the new kind of 'homecoming' to which this evocation of nostalgia seemed to be pointing, demanded, I believed, that I should learn the appropriate language for the occasion. So I started at once by asking the purser if he would teach me how to read and write Japanese in exchange for some lessons in English.

Outwardly the purser was the gentlest of men, unusually good-looking and with eyes wide apart, rounder and bigger than normal. They were the eyes of a thoughtful, introverted and contemplative man with deep and complex feelings. But often I noticed a shadow of something melancholy like a cloud coming between the sun and him. It was particularly so when he was engaged by our Chief Engineer, who sat opposite him at table and was incapable of leaving him alone.

The Chief Engineer was a totally different character. He was older, alert, mentally adroit and wise in the ways of the world, content, if not arrogant, with what applied science had given him as a way of life and spirit. He was the intellectual of our world: all the more formidable because he was not without wit. By training and nature his mind was interested only in the established and the known. He was scornful of the undiscovered and the feelings that it evoked. He was accordingly extremely well adapted to his own emerging world; anxious to impress his audience, to gather public applause, and happiest when he raised a laugh at our table against the purser over what was, of course, paraded as a joke but was knowingly or unknowingly of a deep, subversive intent; all attitudes and codes of behaviour which I felt were utterly alien to the purser. I very soon summed up their differences by calling them 'Science' and 'Religion'. And it was to 'Religion' that I turned for instruction.

I was immediately given the warmest of welcomes, and soon found how good a choice fate had led me to make. On the very next day, as Plomer and Mori started their translation of *Turbott Wolfe*, I began my studies of Japanese with the purser. The first character he taught me was that of a tree, perhaps feeling prompted to establish that what was about to happen between us was to be not an act of will and mind so much as a growth from roots deep in the dark and mysterious earth. The very first word I learnt from him was 'sensei': master or teacher. And it is as both master and teacher, in the indivisible connotations they had for the ancient Chinese and architects of the Japanese Renaissance, that he remains in my grateful memory. That classical concept had innumerable implications of great significance. It presupposed that a person could be a master to others only through continued seeking of the truth and humility of his own teaching. Someone could only teach others through what he had taught and mastered in himself. So ultimately, of course, he taught even more by living example than by words. It was a concept, therefore, concerned with the act and meaning of knowledge; and the heightening of responsibility of the human being in proportion to the increase in his knowledge. It was never just an uncommitted acquisition of knowledge for knowledge's sake, which passed as 'education' in my own world. So, though without dogma or doctrine, it had profound religious undertones. Also, through the nature of both Chinese and Japanese writing, the fact that it was a vast system of a condensed and highly evolved picture-writing, not related to sound and word but designed directly to express whole ideas, situations and complexes of feelings, one could not study it without being involved throughout with the growth of human consciousness.

As a result Japanese, as I now began to learn it, appeared firmly connected with its aboriginal beginnings and tended, to my delight, to express reality more in terms of feeling than in ideas and intellectual abstractions, at which the Chinese excel. For instance, in writing 'tree', divested of any phonetic obligations, one drew, in fact, a simplified picture of a tree, and in the process the imagination was enriched with all the associations it had with trees, in a way that is not possible by just saying the word. The 'sky', as something higher than the trees, was represented by another simplified picture of a tree and a line above it; 'heaven', as something beyond the sky, was yet another line above the line representing the sky. Tree, sky and heaven, therefore, were joined in a vision of organic unity from the earth wherein it was rooted, to the heaven at which all that grew from it was aimed. The East, for which we were bound, was not just a cardinal on a compass but was shown like an outline of one of those ancient stone lanterns that light the way to some shrine in Japan: as the lamp of the rising sun shining behind a tree from the direction along which light and life were renewed out of darkness and death. And so the process went on and on, to be orchestrated into a great symphony of more and more complex relations of forms as, for instance, in characters like that for 'rest', which is a picture of a man underneath a tree; or 'anxiety', which was an immediate favourite, i.e. a heart at an open window. Even the teaching in its preliminaries as I was experiencing it, to my delight, was a singularly dynamic image of a great stream, not of water but of words. Studying the writing and learning the words I was, therefore, automatically participating in the progression of the human spirit and its values, and groping after meaning without ever losing connection with man's first vision of the world about him. I could not fail to understand, therefore, why captains and kings, statesmen and artists or even priests and theologians did not represent what was highest in the aspirations of China and Japan at their best. In some such interpretation of what was happening to me, I was content and excited by the prospect that education of myself now held up before me. It could not compare with the prison fare that my last years at a conventional school had inflicted on me in the name of 'education'.

Accordingly I ended my first session feeling uplifted and full of enthusiasm, and hurried back to the deck outside our cabin to join Mori and Plomer. They, too, were just ending a good morning, both pleased and excited by what they had achieved. Indeed, Mori put down his pen with a deep sigh of satisfaction, obviously reluctant to revert to his duties as master of a ship again. Neither of them, however, appeared interested in the way I had spent my morning. Beyond teasing me about my 'return to kindergarten', the reckoning of the day as far as they were concerned was closed, and I made no effort to reopen it.

But in those strangest of hours at sea, between lunch and a late tea, when ghosts walk the oceans as they do the bush, and the human spirit is at its lowest and in need of exorcism, it was a different matter. Plomer and I were left to ourselves and he would give freely of his mind, imagination and fantasy and show that he could be as much a person as he was full of satirical edges and social graces. To pass the time, we used to invent games of the mind, such as sessions of outrageous punning, and limerick competitions. The fact that he always won hands down did nothing to spoil my enjoyment. I sometimes wished I had made a record of Plomer's limericks but perhaps if I had interrupted our play to record the material, the essential spontaneous element of the game would have been destroyed, as his limericks came so fast and there were so many of them.

The afternoon hours were also for reading the vast quantities of books we had brought on board. However poor in clothes and money, we were certainly rich in reading material. Plomer began the journey by reading Walter Raleigh's classic on Shakespeare, I by reading Lawrence's *The Plumed Serpent*, not long out at the time and reviewed in one of our editions of *Voorslag*. As it was my first Lawrence, he went deeply into my imagination, as deeply as any contemporary had ever done, except Conrad. For his part, Plomer discovered a burning involvement with English history which took him as much by surprise as it should have alerted me. He at once plunged into a deep-sea reading of the

Shakespeare plays, and all this burning English potential in him like wood laid on the hearth of some Tudor fireplace leaped into flame from a spark of a family legend that his family were somehow connected in blood with Shakespeare's own Ardens. I realised many years later that at that precise moment his first long poem ever written in England, 'The Family Tree', was born. After that, Plomer began to work his way back with conscious determination and against the stream of natural instincts which had inflamed *Turbott Wolfe*, *Ula Masondo* and *I Speak of Africa*. He returned to the England of his parents which he had known briefly at Rugby School. He began to push himself further and further away from the Africa in which he had been born. Indeed there would come a day when he could laugh off all his formative African years with the jest, "If a kitten is born in an oven, it is not necessarily a biscuit". It was the beginning of a process which was to be England's gain and Africa's loss; though I still have a feeling that in terms of the task of the individual to live a life so that it fulfills its own highest meaning, it was also Plomer's loss.

Towards the end of our second afternoon on the *Canada Maru*, Mori decided that the time had come for play as well as culture, and he invited us to join him and the crew in their daily round of games. Plomer declined with some agitation and urged me forward with a portentous and for once totally earnest, "Remember, you are minister of sport and recreation". And so, forth I went like a scapegoat to sacrifice on a strange altar.

The games were staged on the after well-deck where the hatch covers had been removed and the hatches themselves been opened wide to the palest of blue afternoons, still and warm. On the deck, thick rice straw mats, beautifully woven and of a warm, glowing yellow, had been laid to cover the steel surface. We looked down on this stage, I dressed in tennis clothes, Plomer in brown corduroys, open-necked shirt and his large, dark, horn-rimmed glasses, possibly as much a shield against the eyes of an alien crowd, I used to think, as the light. Everyone in the ship who was not on watch was collecting in the alleyways connected to the deck, galley-hands, stokers, quartermasters, stewards, sailors and officers. Too many of them for my liking were dressed in those coarse, off-white three-quarter length trousers and short kimonos tied round the waist with a girdle. I knew the sport from my reading at the time as jiu-jitsu, and had relished the mystique made of it in the *Boy's Own* paper to the point of acquiring a certain fearful respect for the extent to which the physically small and vulnerable, equipped with its skills, was able to master the uninitiated giants of muscle and brawn pitted against it. I was physically bigger than any potential opponents and considered myself in good condition. But there was a self-assurance in the eyes of the smaller men dressed for the occasion, and in the dismissive appraisal of me that they seemed to make depressingly quickly, which confirmed all my premonitions.

Plomer picked up some of the unease as well. He decided to stay where he was, out of reach of any more pressing invitations to participate. He watched what was to follow with the looks of a Christian compelled to bear witness to the martyrdom of a member of the faith in order to make a holiday for a Roman crowd massed in their Colosseum.

I went on my own to meet Mori down below. He was in high spirits and already dressed suitably; totally uninhibited by the temporary loss of uniform and badges of rank. He was on the easiest of terms with all the crew. It was indeed one of the nicest things about the moment, and a revelation of the underlying feeling that the Japanese seemed to have of all being united and equal in spirit through obedience to an order which seemed, on the surface, to be designed to divide them in as many social layers as a Neapolitan ice. I could have been walking into a family or clan gathering rather than into the assembly of a highly disciplined and differentiated crew.

One glance from Mori and my best tennis outfit was condemned. I had to strip off shift, trousers, shoes, socks and was made to put on the largest spare kimono uniform on the ship. In this I sat with the kind of inadequacy of a Victorian orphan most incongruously upon me; for something of amusement and a disagreeable relish of anticipation seemed to shine in the dark eyes of the massed *Canada Maru* clansmen, even though not a muscle or nerve twitched in their composed and courteous faces. By the time Mori had led me barefoot to those heavenly, soft mats, the rest of the gathering were squatting, legs crossed before them, upright and still, as if in imitation of Bhudda, all round the open square of the hatchway. Once in the middle, Mori bowed to the watchers. I followed his example, awkwardly, with all the 'new boy' feelings of initiation which I thought I had left behind for good in my first months at school. He then addressed his audience in simple, forthright, unfaltering and appropriate words, all uttered in the voice of one born to command. He then turned and spoke to me in the same authoritative tones. He was, he declared, first of all going to teach me how to fall without doing myself injury in the process. He then said that, as in life, one had first to learn how to fall before one could learn how to get up and rise; first master the how of losing properly before one could be worthy of winning. So I was to be engaged in a sporting contest that was also the acting-out of a Japanese parable, and the well-deck was as much a place of religion as of fun and games.

After the exercises of correct falling and tumbling, Mori proceeded to throw me about with disdainful ease and an enjoyment which, at the climax, made me suspect that he had ceased to be the captain that either he or I knew. I was no longer the welcome guest. He briefly became all of

Japan and I became all that had wounded and was still thwarting Japan in the west. But even then, magnanimous as he was at heart, he brought himself and me back into focus, putting an end to the throwing.

My turn ended, I joined the spectators, sitting among them trying hard also to be a credible version of Bhudda, although for some while a western Bhudda shaken, hard of breathing and not in any way interested in his navel. Mori, however, continued to have a few bouts with some of his sailors and demonstrated how he practised what he preached by falling with as great a grace as the ease with which he had thrown me. Later, over our evening cocktails he told me, obviously for my comfort, that the great Emperor Meiji, who had so successfully led Japan out of its feudal state into the modern age, even as a boy had insisted that his tutor in judo and the kendo (a form of fencing which I was still to learn) should be not only the best but one who would show him the respect of extreme disrespect. There it was again, the instinctive assertion of the national awareness of the play of opposites in the totality of truth which I was daily noticing more clearly – throwing the Emperor about and trouncing him as if he were the lowest of his subjects. Endearingly, too, he told a story against himself even more to the point.

It happened on the voyage south from Mombasa to Dar-es-Salaam with the Governor-General of Kenya, his Lady and staff on board. He had staged a display of judo and kendo for their amusement. At judo he had engaged a particular sailor who was in fact the best on board and had allowed himself to be thrown about in a way that could just, he thought, humiliate his image in the eyes of his distinguished guests. So the very first time he managed eventually to throw the sailor, before the response which would demolish Mori could come, he whispered urgently in the sailor's ear, "Use your brains, man!" From then on, Mori was allowed to win and impress Lady Griggs so much that she exclaimed, when he bowed at last to all as the victor, "Oh, you are wonderful, Captain!"

As for Plomer, the lamentable and inept performance of his 'minister of sport', and anxiety over what he himself might have suffered, increased his dislike of sport in general and of this new form of it in particular. In fact he did just grab me by the arm after my first session and in a voice of outraged concern exclaimed, "What a how-d'you-do; quite, quite unbelievable. Are you sure you are all right?"

Mori never appeared on the judo scene again. But he did regularly at kendo and took great pains to teach me this ancient, equally allegorical sport of fencing with long, two-handed bamboo swords as substitutes for the steel used in feudal times. For this I was made to wear a helmet and padded suit, as Mori did. I soon found out how essential this protection was. Once Mori had nursed me into an elementary capacity for self-defence, he began to belabour and prod me without mercy. Without helmet and breast-plate I would certainly have been hurt. As it happened, the physical discomfort was trivial and the hurt more to my confidence in my own ability at games, making me understand why my Japanese teacher found 'Kyuba-no-Michi' (the way of the bow and the horse) older and more complex by far than the 'way of the warrior' and why the simplistic Bushido, of which kendo was a part, was increasingly popular. I thought there was a note of regret in his comment and a dismissive tone in his delivery of the word Bushido, which was absent in the exalted way our Captain spoke about it. I noticed, at my lessons, that in writing the ideogram for bow in that free, highly individualistic style of calligraphy called So-sho, the simple, asymmetrical image was one of the most evocative and dynamic with all the implicit tensions of nerve and muscle, and effort of gathering together of mind, eye and hand of the archer when the bow was stretched to the utmost and the arrow about to be released.

Henceforth, all my days at sea were to pass in a blend of elements such as these. From the beginning to the end of the adventure, I could not imagine even the best of European ships offering a fraction of such food for the spirit and imagination of anyone so young and impressionable as I was. Nor for that matter did I subsequently ever come across another remotely like it. Much as I was to love travelling in European ships, they were poor vessels in comparison with the *Canada Maru*.

In this fashion, after many days, almost precisely at noon (wherein, my teacher had just informed me, midnight is born), we had our first glimpse since we had left Port Natal, of the coast of East Africa. It did not cause much excitement in Plomer or Mori, for whom it was little more than a landmark proving how faultless his navigation had been. Seen from the shade of our little deck on a day white with heat and a haze almost as dense as fog, it was indeed not a dramatic sight. It appeared just a dark line drawn firmly on the infirm air; now and then surges of white of a ghostly kind against it marked the breakers of the Indian Ocean swell. But through the glasses that I borrowed and when Mori told me it was the port of Mozambique, it was another matter. I could then just see the flamingo pink coral below the line of the land, and the fort or castle resting still intact on the foundations Vasco da Gama had provided for it (so legend had it), in stones carried as ballast from Lisbon, four hundred and twenty-eight years before. I knew both legend and history well. The Portuguese had had to fight desperately against disease, malnutrition and stubborn rivals like the Dutch, in order to maintain their hold on that strategic base on their route to the Far East. Immediately that mere outline in the light of a day so fierce that it hissed like a serpent in my ears, became an ideogram of the kind that I had been studying that morning with direct access to all the emotions of endeavour and seeking that had brought us to Africa and, in sense, had impelled me to follow through as an odd sort of pilgrim of history, in the wake of Vasco da Gama and his successors. I continued to watch the land until it vanished not below the horizon but burnt out like ashes into the fire of the day. I went to lunch marvelling at the indifference of everyone at table to the event, and resolved that no matter how far and long I travelled I would not allow so culpable an *ennui* to overcome my spirit at any first glimpse of land after days at sea.

At last we came to our first landfall after Port Natal. The drama for me was set the evening before when the *Canada Maru* felt her way delicately through the last coral water between Dar-es-Salaam (The Haven of Peace) and the Sultanate of Zanzibar. Almost on the equator now, the sky was heavy with thunder-clouds, black as Mohamet's coffin below but piled so high above it that they were pure and white in a sky of peacock blue. From somewhere behind them, the sun at the exit of the world enfolded all with wide arch-angel's wings of light, in an immense and final clasp of valediction. Just then we saw to the east on our starboard bow the sail of our first dhow, all gold in the last of the sun. It came to us like a bird out of *A Thousand and One Nights* pecking away at the dark blue fathoms between us, as it went up one roll of the satin swell and then over into the gleaming stretch of smoothed-out water before the next swell. Behind us, the wake of the *Canada Maru* changing course became more and more calligraphic. English helmsmen used to speak of their steering as writing their names on the sea. Our helmsman seemed to be doing so with the hand of a Zen master recording sudden revelation and illumination. My excitement was heightened when I saw that, small as the Arab ship was, it looked an exact reproduction of the craft that I knew from the earliest Portuguese and Dutch lithographs and paintings. Also, of course, it was still bound on the same sort of occasion then as its predecessors had monopolised for years between the Persian Gulf, the coast of Malabar and Africa. A light stirring of air out of the east brought a sharp, provocative scent to fall over our ship from the pagoda of its masts to the inner sanctuary of its captain behind the bridge. It was the smell of cloves which grew in islands of coral, emerald and palm just beyond the edge of the dark. These cloves, I was to find, spilt out from their bales for inspection on the floors of the market in Zanzibar so profusely that I had to wade up to my knees through them. Even in our ship the air was heavy as if filled with the scent of pinks and carnations. As a result we went ashore the next day, as if initiated in an order of centuries.

Yet there was little of history on show in the offshore island of Mombasa itself and particularly where the *Canada Maru* lay. She had been tied up at the new harbour of Kilindini to load a dull cargo of potash from the strange lake of Magadi in the south-east of Kenya. All around us were cranes, sheds, railroads and trucks of steel. Compounds of tin and metal, glittering piercingly in the sun, brought an ague of fever to the heat of the shimmering day. Beyond a scattered and sprawling township of galvanized iron roofs, brick walls and dark little verandahs behind mosquito-proof netting, any expectation that one might have had of a continuation of history in stone was killed. The influence of Arabia was visually absent except in the loose, white, ankle-length dress of tall men who looked more Bantu than Hamitic as they tacked like sails from one sandy street to another. The atmosphere of the East, however, was out in force along the streets lined with Indian stores and eating houses. The smell of hot curry seemed to be as much a part of the day as the sunlight. Every now and then a flash of colour from some Indian lady in a sari (one of the few forms of dress that defies change because of its perfection of line and colour) appeared briefly in a pool of silver light between one purple shadow and another, and was as fresh and immediate as an explosion of cannas in Cochin.

Yet what redeemed Mombasa was not town and modern endeavour, brave as it was, but the land and what issued so abundantly and passionately from it in vegetation. It was my first encounter with tropical Africa and the quick, electric-green of the thick, surging grass, brush, flamboyants, jacarandas, acacias already crowded with sun-beetles and crickets to salute the day with platinum voices. In a few hours Plomer and I had seen most of the town and hastened back to our cabin and its shade. There we were visited by the editor of the local paper. He was a courageous, independent Scot of forty who looked seventy, aged by the impact of the powerful, overbearing character of the land of Africa and its influence on his interests and expectations. It appeared that now he seemed to live only for the redemption of the violent exercise of its power and assault on body and senses during the day by the coming of night and alcohol. Resentment and disappointment were so deep in his defeated, romantic character that all his adjectives were pejoratives of the coarsest kind, uttered loudly and constantly in an abrasive voice. Plomer, who could never deal with violence of mind and deed in others, withdrew into himself and maintained a total and somewhat apprehensive silence which the basically insensitive Scot felt to be an immediate and undeserved cause of offence. I had to work hard to keep the peace, and succeeded in doing so only because even at my age I had known many people like him and understood how desperate human beings became when Africa did not allow their lives to fulfil the love and hope with which they had begun. Gradually, with the help of ice-cold bottles of Japanese lager, the Scot was reduced to relative calm. And when he heard that Plomer was a poet and novelist, he convinced himself that Plomer's withdrawal from the conversation was proof of great artistic sensibility. So he set out then to find common ground between them and talked about books.

There was a book, he said, which had just caused the greatest uproar in Kenya he had ever experienced. This was Norman Leys's *Kenya*. Plomer had reviewed it for *Voorslag* and welcomed it as an attack on colour and racial prejudice. I had not read it myself then and when I later did so, although I basically agreed with Plomer, I had reservations about sitting in judgement as Dr Leys was. I also felt as concerned for the accused and for the plaintiff. Undoubtedly, however, at that moment it was a devastating reappraisal of the colony's history and condemnation of the ethics of Empire and the men who served it. Yet, not interested in either illumination or understanding, it was particularly scathing about the settlers, their great buccaneer and titled pioneers as also the commercial community. So I asked the editor if I could buy the book in Mombasa. He assured me that no one in the town, or even in Kenya, would be so foolhardy as to put the book on sale, let alone be seen in possession of it. But he had a copy of it in his office and would gladly let me have it, indeed would welcome the chance to be relieved of the burden of a tiresome and provocative volume. In due course he sent the *Canada Maru*'s agent on board with the book wrapped in an African blanket.

I mention this apparently trivial episode at some length because it was one of the first straws in what Harold Macmillan was to call many years later, 'the wind of change', not only in Africa but in the European Empire everywhere in the world. Plomer's book, of course, was such another in a different and deeper level of the contemporary spirit. It was no 'wind'. In those windless days under the mediumistic blue of Africa in the year 1926, it was little more than a change of atmosphere which foretells the breaking of a drought before the thunderclouds mass and the rain falls. But to us it was like a postman's knock on the door of our time delivering an express summons of fate. Yet, I do not want to imply that these immense processes of change with which Leys's book was connected began precisely at that time. The notion of a precise beginning to living things and events is a highly abstract and arbitrary one. In a sense all, at any and every moment, is a beginning, and the most one can say, perhaps, is that when one refers to something as having begun at such and such a point, one is speaking only of one awesome minute (which Kipling once called 'unforgiving') and where a stream of endings are converted into beginnings as fast as they occur; then suddenly one can become conscious of a particular change. 'Alles ist übergang': 'All is transition', an inscription on a church high in a mountain valley in Switzerland, was to inform me. So in that sense the day in Mombasa was part both of a time of great transition and of personal beginning. I accept, therefore, that I was predisposed to project an overwhelmingly subjective sense of change into the world and the larger scene of life around me. Yet, this book was some objective evidence of how much more there was of change growing great in the human spirit of the day. But, like its mathematical opposite, it may have possessed no substance as yet but only position. For I still assert there was already something in the heart of the darkness of Africa and that age at large, which marked the end of the era in which Bartholomew Diaz and Vasco da Gama had been instruments of 'beginning'. And this 'ending' was being made more evident daily by increasing tangibles of the beginnings; of cataclysmic convulsions in the lives of men and their societies in my lifetime; the falls of darkness; the trembling of the earth; the tearing apart of the heavy curtains of the temples of the world that always precede the birth of something new and unforeseen . . . except by a few whose vision of the future is rejected and they themselves left to the cold comfort of their own tears. But what then was this objective evidence?

To give an example of something which deserves mention, there was for instance the presence there of the *Canada Maru*. It was, in its modest way, evidence of change brought about by the victory of Japan in the early years of the century in their war against Russia. The consequences of that victory for the European empires which were to continue to expand so confidently and brilliantly for another generation were, perhaps, too subtle to be understood by a singularly impervious age. But they were, nonetheless, lethal to Empire: as is said of the incurable illnesses of medicine, they were 'terminal'. The Japanese shattered forever the European hubris of a power and right over the lives of non-western peoples all over the world, which they held to be as absolute and lasting as it was self-evident. Since the peoples in the empires over which Europe did not make the inflexible distinction which it drew itself between matter and spirit, but held them to be one and indivisible manifestations, the power of the West over their lives was automatically regarded for centuries as proof of a spiritual authority greater than they themselves possessed. The worst consequence of European empire, therefore, was not the exploitation of their imperial subjects for greed and commercial gain, as is held against it by its critics today. Had that been all there was to Empire, it could not have lasted as long, and engaged so much of what was good and noble in the many who served it with devotion to the welfare of its subjects. That dedication was incomparable, and particularly impressive when measured, as it should be measured, in the balance of values of its own time. It gave, I believe, far more than it took, and hurt itself in the end far more than it hurt others, except in the one dimension of Empire. There, its worldly power, so great over their lives and so ambivalent in their minds, had a paralytic effect on the spirit of the governed. It was as if they were hypnotised out of being themselves. They lost the will and even desire to 'do' and 'be' in their own natural right. But the Japanese defeat of Russia broke that spell, overnight. The paralysis of natural spirit vanished, and from that moment on began with a swiftly accelerating surge to mobilize sullen and long-arrested forces against the authors of empire.

Already there was Gandhi, who had found his Damascus a quarter of a century before in the Port Natal from which we had just come. He had broken out of this paralysis of spirit and, with the diabolic cunning of which only the saintly are capable, mobilized the forces of shame in the British, and used their own innate decency as the ultimate weapon of their destruction. The time was coming fast when this kind of decency would be carried to indecent lengths, and when the Chinese exhortation to men to 'be modest also about modesty', should have been adapted in relation to the British to urge modesty in decency and self-criticism upon them. Norman Leys's book was startling evidence of how quick that poison of shame had been to act in the spirit of the British, and how far already this deadly, partial re-examination of their motives and manner of Empire had penetrated their judgement, even to the extent of identifying the revolt against Empire with what they sloganized as the 'class war within our own society'. It was easy enough to laugh at these simple settlers for their outrage over the book. As simple people do the world over, they recognised far more clearly the consequences and the deadly intent of a cold calculated disregard of their own humanity than did the intellectuals and leaders of the crusade against Empire in Britain and Europe.

More important even, this process of change was not confined to colonial Empire alone. At that very moment while we loaded potash at Mombasa, leaders of the Commonwealth were being summoned to London for the gathering that ended in the Statute of Westminster, which was as revolutionary an event in the history of Empire as the Great Durham Act of 1830. Nor was evidence of change to be found only in the British Empire. The year 1926 saw the significant and strangely overlooked revolt in Bantam in Java, which was part of what was held to be Holland's superior model of empire, despite a long, unfinished war in the Atjeh of Sumatra. The Governor-Generals of the Dutch East Indies were by the year making more and more use of their powers to exile people in 'the public interest' and making political prisons out of some of the smaller of those outer islands. Because they could do so without explanation, and without the bringing of formal charges and public trial, the world was not to know this until the end of the last War. In French Indo-China the incipient war of centuries between North and South which was to explode in Vietnam had not been resolved by a central metropolitan authority; and the whole of South-East Asia was uneasy, resentful and depressed as the air before a typhoon.

I could not pretend to have had a fore-knowledge then of the end to which these illustrations pointed. But I use them merely as evidence of a submission that my own awareness of change in myself was not unrelated to vast forces released in the world in 1926 to give impetus to the formidable energies accumulating on a wide front for the transformation of an entire world.

There is far more to the German concept of *Zeitgeist* than the antiseptic, highly sterilised rationalism of our day will admit. Nor is it German alone. The Greeks had it in their concept of the Platonic year; the Arabs and their alchemists in Astrology. The Ancient Chinese, too, held earlier than any others that time was seasonal and not merely a measure of distance travelled by the spirit of man. It was also an element with a character of its own which contributed significantly to the nature of all things born and enclosed in the same moment of itself. Not surprisingly, therefore, the German Spangler, however short in specifics but long in his intuitive vision of the *Zeitgeist* or character of the time, had recently delivered himself of his resounding warning in *The Decline of the West* – all the more awesome under a German title which reverts to the ancient Sanskrit name of Europe: 'The Place Where the Sun Goes Down' or 'The Land of Evening'. Even my own sensei had just planted the seed of greater awareness of these huge imponderables of life and time in me, through that simple Taoist proverb I have cited: 'At noon midnight is born'. Perhaps that is as good an epigram as can be coined for the essentials of all I felt in the atmosphere of time around me in 1926: it was a high-noon in which the midnight of today was being born.

Yet, I must add that even the instruments of the universal change implied were not conscious of what they were doing or had done. They confused the elemental urge by lusting after the very things they were destroying, just as the long-distance men who had encouraged the *Canada Maru*'s mission were already busily conceiving a dream of the greatest Empire ever. In Japan itself, the external catalyst for precipitation of fearsome change suspended in the national soul was about to be released. Two generations and some years of the great Meiji era were about to be relegated forever to history. The Emperor Meiji's successor was dying, and indeed was to die before I finally said farewell to the world of the *Canada Maru*. The cold of one of the coldest New Years was to begin under a new Emperor, and to bring into rapid being a new, dangerous, and, like those of Hitler and Mussolini, self-destructive society. But it was on the mainland of Asia, across the Sea of Japan in China, that the end of Western Empire was nearest of all. A young Chinese soldier, Chiang Kai-Shek, fresh from a Russian Military Academy and married to the formidable Mei-Lin, one of three gifted sisters of perhaps the most powerful Chinese family of the day, was collecting his forces at that very moment for his famous march on Nanking. In so doing he began the campaign against the great warlords for the reunification of a long fragmented civilization, as well as the abolition of the rich, self-regulating concessions the British, French,

Germans and Japanese had established on the most strategic areas of the coast, like Canton, Shanghai, Tsingtau and Tientsin.

In Europe, too, Hitler, the sleep-walking missionary and therefore the most dangerous of all, was starting his *Götterdämmerung* conversion of a whole people in the back streets of Bavaria, as if in total subservience to an ancient Teutonic mythological pattern which is the only one I know wherein the gods themselves have to be defeated by the forces of evil for the quenching of all but a very little light. Mussolini, already installed in Italy, had picked up this stirring of strange ambitions in Japan and spoken openly against the Japanese, referring to them not by name but as "a species of people who resembled nothing so much as monkeys who had newly come down from the trees". If they did not desist, they would find, I seem to remember him thundering at Milan, "thirty million Italian breasts and thirty million bayonets barring their way". The Japanese were duly insulted but the rest of the world did not, as I did, see it as the first open hint that he also had his sights on Abyssinia and a dream of Roman Empire resurrected by Italy as the pilot of his mind.

All these things came unbidden, as did the foam and spray of the Indian Ocean that we were sailing, to whet my imagination. And at times, most unbidden and prominent of these crowding premonitions of change, there rose the beautiful face and words of the prophet, newly arisen among the Zulu people whom I had sought out just before our departure, following my instinct despite the cynical disapproval bordering on an outright veto from my Cockney news editor. But I knew in my blood how profoundly Africa was an Old Testament country almost more in need of prophets than medical officers. Therefore I was aware how significant an event this could be. And how strange, therefore, that my last piece became one of the few things I ever wrote not to be published because it was, I quote from both News and Chief Sub-Editor: "All my eye, Betsy Martin and mumbo-jumbo." The face of the prophet I will always remember. It was transfigured and of compelling beauty. My final image of the prophet is of him telling how at midnight as a boy, a Zulu Samuel, he heard a voice that could not be denied, calling and telling him to go to the top of the Sacred Hill of Inanda and look up. He did so. He looked up, in another August night, to see with blinding lucidity five great stars fall out of procession and proceed, against the natural order, to move from west to east. He rushed back to his bee-hive hut full of fear at so unnatural a sight. Five days later he was told that on the morning after the midnight vision, five great nations, five great stars of humanity, had fallen out of the lawful progression of the universe and gone to war: Germany, Russia, Austria-Hungary, France and Great Britain. The First World War had begun. Many more stars were about to fall out of their courses, he warned me, and that, he stressed, was how and why he had been called on to prophesy and warn, for that was all a prophet could do. But warn to what effect? It was not for prophet or man to say in an age, he declared tragically, when no-one spoke any more of Umkulunkulu, the great first spirit. His praise-names were forgotten, and men now spoke only of things useful to them. How could I, a child of the same Africa, myself have failed then to conclude that no year for centuries had been of so meaningful a transition as this year of our absent Lord, 1926?

It would have been even more startling had I known what I was to discover only after the last World War, which was in itself such a gruesome confirmation of the objective aspect of my subjective feelings of change. But then, in the very Kenya which Mombasa served as port, Carl Gustav Jung had just terminated his sojourn among the Elgonyi of Mount Elgon and returned to Europe satisfied that he now possessed all the objective evidence that he needed to confirm his hypothesis of a 'collective unconscious' in man. This, in time, seemed to me at least as great a breakthrough for the human spirit as Einstein's 'Theory of Relativity' and those illustrious others who penetrated the secret of the atom and discovered a mysterious universe expanding in reverse. I mention these three re-directions of human seeking not because they were the only ones but because they are all, for me, part of one another. They are aspects of a potential of mind and spirit out of which will come, however much and tragically societies were about to crumble, the energies for a renewal of life and its movement, a recharging of the arrested post-renaissance spirit into greater and more significant forms of being.

Certainly all these shadows of coming events, which the frustrated Mombasa editor had cast over my mind by his talk of an 'infamous' book, were made specific by Mori, who suddenly appeared later on another afternoon when the three of us were together again, as if summoned to illustrate how, as far as he and his own people were concerned, no sense of paralysis of will to assert themselves was left. He was outraged and angry. His face was set firm and resolute. The bone and tense muscle almost visible through skin drawn tight as to be more translucent, sat like a 'Noh' mask on him. Even his beard trembled like an heraldic pennant in the tumult of swords, spears and arrows, clashing and clanging on ancient armour. The cause was proclaimed to us in simple, precise words struck like metal discs from the machine-tool process of transforming the hurt of centuries into a vehicle of a precise action.

Our ship's carpenter had died. Mori had not mentioned it to us because he did not want to spoil our day, and had gone ashore to see his agents to make arrangements for a funeral for the dead man. As he spoke, I assumed it had been to arrange for a cremation because I already knew how important it was for the Japanese to return their dead, one way or another, to the land of the rising sun. But I was wrong, and wrong in a way that showed how little I understood the underground symbolism at work in Mori and his people to elevate this voyage of a cargo ship far above a purely literal and lateral level. He wanted his carpenter to be buried decently in the Christian cemetery of Mombasa, and had been resolutely refused permission with the ostensible justification that, as it was consecrated earth, no pagan or non-Christian could be buried in it. To our Captain this was sheer evasion and camouflage for deliberate discrimination against the Japanese; discrimination all the more culpable because it was against the dead.

Plomer and I were appalled because to us such a ruling seemed profoundly unchristian. But for Mori it touched on a volcanic area of spirit of which I, through my daily instruction, had just begun to have an elusive inkling. It was simply that I was beginning to suspect that, in the Japanese spirit, death and not life was the ultimate romance: the way one died was more important than how one lived, and even the best of living was not valid until vouched for by the manner of dying. The dead, not the living, were the really happy spirits. This, in brief, must serve to point at great and complex fundamentals which, for Mori, made discrimination against the dead far more reprehensible than against the living.

It had apparently been no good explaining to him that Hindus, Muslims and Africans accepted this religious prescription without demur. The Japanese, he held, without yielding, were different. They had a right to be buried the way they thought conferred most honour on the dead. Yet why then deny the poor carpenter the right of all Japanese to be returned to the shrines of their ancestors? The answer was that the carpenter was to be buried that way not for himself, but his country. Almost unnoticed among the crew during his life, in death he had become a plenipotentiary of his people which, taking into account all these powerful associations and supreme status of death I have sketched so briefly, meant an ambassador at the highest level.

His account rendered out of a charged mind, Mori turned to our ravaged Editor. He had by this time the highest regard for the power of the press and appealed accordingly to him for help. The Editor, who combined a kind of Burnsian scorn of priests and their practices with an equal love of liquor, was immediately on his side. Within minutes we were back in the heat and dust of Mombasa. First we saw the Chief of Police who, allowing for vocational differences, was a replica as well as a pal of the Editor. He, too, fell instinctively in line, and, the Editor's language becoming coarser and more colourful as our numbers grew, conducted us with a fury of spirit on his own inner account, to storm the church itself. Within an hour, we reached an agreement which was a compromise in the eyes of the church, but a victory for our Captain beyond his expectations, and, a confirmation of the Editor's and Inspector's view of the hypocrisy of churchmen. The carpenter would be buried in the Christian cemetery provided a proper Christian burial service was read over him. For Mori, that a member of his crew in death would receive the highest religious honour available far outweighed the fact that he was not being given the burial demanded by his own religion.

I know that this may seem cynical and grossly expedient on Mori's part, but at heart it was nothing of the sort. It told me much about the Japanese attitude to religion and how naturally religious a people they are, but religious without any marked attachment to doctrine or dogma, and with an instinctive predisposition to find common ground in all sincerely held religious beliefs. Religion, in fact, was more a matter of feeling than of lofty thought or high technology. However much social and political considerations contributed to Mori's satisfaction, there was not, I believe, any fundamental contradiction in his seeking the best available religious decency in death for his carpenter. I was not surprised to hear from him that night how as an adolescent he had seriously thought of becoming a Christian himself.

The funeral service was duly held in the Christian cemetery the following afternoon. A cross-section of the crew, the ship's officers and Mori in their best uniforms, always fastidious in appearance and manner, were there, drawn up in a neat and compact pattern. In the presence of the dead, round a freshly dug grave raw as a wound in live flesh in the crimson earth of Africa, they were at their best. The sun, at three in the afternoon, was at its fiercest and its light, like a flicker of fire around them, raised from them a kind of radiance of the physical cleanliness which was indeed godliness to them and which influenced their states of mind accordingly. By comparison, the crumpled and decayed Editor, the Inspector of Police, Plomer and I, hot in collars and ties, uncomfortable and sweating, and the curate in an un-ironed surplice which was lifted, by a breath of air of the steep afternoon, as if for inspection and quickly dropped with distaste, were inclined to look uncared for and deprived of grace. Nor did the perfunctory, unconvincing reading from the prayer book by a self-conscious and embarrassed curate before this most attentive audience do anything to improve the European image. I felt this so much that I resolved to go over those great words from Ecclesiastes, St Paul, and St John of the Revelation with the purser and Mori in order that they should know what was also inspired in the Christian concept of 'decent burial'.

It all filled me with a sadness that was not just my own. When I looked beyond the sunlight breaking on a sea of green vegetation against the town-line, I saw the land of Africa itself swelling to the west like a tidal wave in the blue distance, drawing itself up to a great height where it joined and matched the blue of the horizon. There were no hills there to which we could lift our eyes; only a range of mountains of high thundercloud in which my native continent excels. The voice of the curate in such a setting was like a twitter of protest at the majesty of death. Our group, large as it was for a Mombasa funeral, suddenly became insignificant and vulnerable. Yet my sadness was not for us or the carpenter. He was, in a sense, lucky in that his going was marked and remembered and would be honoured on the lips of living men. No, it was a sadness as of a remembrance within my blood, of the long succession of anonymous dead for over four hundred and more years who had died unrecognized and unrecorded on just such a course as we were bound, with only clouds such as these to be raised like marble tombstones over them in the vast impersonal cemetery of Time.

I have not told all of this chronologically, but in the sequence of meaning as it possessed me. I must hasten, therefore, to record how in between funeral and departure Mori, Plomer and I had been up-country. We had gone 'out into the blue' appropriately enough in terms of that sweeping upward view of the main of land from the cemetery I have just described. It was the strangest of journeys which I have already referred to in *Venture to the Interior* and can only resurrect partially here in so far as it touches on our voyage and Mori.

From the beginning our Captain was determined to go to Nairobi in order to build on the relationship he imagined his service to the Governor-General had established. It was a legitimate expectation on the whole and, at the most, was designed not to press home an unfair advantage but to raise the prestige of our Captain and his country. If there was anything excessive in this design, Plomer and I felt that it could not only be easily contained by an experienced head of state but permanently defused by a formal reception and display of good manners. But the telegraphic exchanges between Mori and Government House made Plomer and me fear for Mori. The responses were, to say the least, evasive and discouraging and only a person of Mori's self-confidence and determination would have failed to detect signs that Nairobi, already suspect by the settlers, saw a danger in welcoming the representative of a nation under public scrutiny for designs on the fishing industry along the coast. And other unspecified ambitions were implied by the sudden letting loose of a team of film cameramen, journalists and travellers, in the Interior. Plomer and I, who suspected this, felt that if there were indeed such a danger, it would have been more honourable to have reckoned with it before and not to have made use of Mori and the *Canada Maru* in the first place. We did our best to persuade Mori not to go: but in the end he came too.

We went by a train that left Mombasa on a Sunday afternoon, and our departure was important to me because we arrived at the station just as a Nairobi rugby team and its followers invaded the station and swarmed all around us. These interested me enormously because they were not the Lord Delameres and other titled latter-day Elizabethan buccaneers, who were the villains of Norman Leys's piece, but were a more representative indication of what Empire in Kenya really meant. The players were the human element which the lovers of humanity usually leave out of their reckoning. Or, if they are included, are despised because of their human fallibility which cuts across plans for the desirable and ideal in life. There is no hatred as great as that which springs from the love of perfection. The cunning with which hatred of the muddled, incalcitrant human being is concealed in an abstract love of humanity, seemed to me, even then, so inventive that it defied general detection. Yet if I had had the new breed of sociological idealist with me there, they would, I believe, have had to admit that not one of these men looked as if he had done well out of Empire. On the contrary, they all had an air of having been unscrupulously exploited themselves, and would give anything to be back in the Britain on which they had turned their backs. It was obvious that the game of rugger they had just played against Mombasa could not have been for fun. Rugger was not intended for such a climate in the mid-day sun along the equator, where even keeping upright at a grave-side for half an hour was a trial, and had nearly caused Plomer to faint. It was not sport but extreme nostalgia that had brought it about, and to me there was something very brave and moving about it.

Most of them were ex-soldiers who had converted their pensions after the World War, and been induced to invest their lives and money in Kenya. They had promptly lost all they had of money and also the firm expectation of a future. Profoundly disillusioned in Europe by four years of war, they had come to Kenya to begin another and better life. But for the moment they were totally rejected by the nature of Africa, and so at the end of their material and spiritual resources. They had only courage and determination to hold on to, which the ritual of the game just played implied. But beyond that the laughter was too loud; the fellowship, too assertive; the general cheer, too good to be true. And the eyes of one and all belied them: they looked out of a spirit deeply, perhaps permanently, bruised.

In this company we travelled over-night to Nairobi. At a siding called Simba, the Swahili (the slave-traders' Esperanto) for lion, we all went into a small, overcrowded dining-room, where even the unstarched table-cloths were limp and sweating, to be given a meal. We had a hot soup that was more cornflour and browning than meat and vegetable stock: tough beef, Yorkshire pudding and roast potatoes; prunes and custard made from powder, all of which the rugger players swallowed down with whisky. Plomer, Mori and I had tea, over-sweet and red as the fine dust of Africa which the wheels of the train churned up from the track and sent swirling like a fog into every compartment – a dust which, at that very moment, spread a film all over the floor of the eating-room. Indeed, it was even in our food and tea. So I remarked to Plomer to the effect that, no matter how or what one tried, one could not keep the earth of Africa out of either one's food or philosophies. And yet the occasion for me was not sordid, for the same reasons that redeemed rugger in the midday sun on the equator from lunacy. That I was not alone in finding it so was demonstrated for me just then by the roar of a lion.

At once all conversation ceased and everyone listened with instinctive reverence as if to the voice of a god. The lion was close, and the immediacy of the sound came like lightning from its throat; the authority of the voice proclaimed as if on behalf of life itself, through the absence of fear and doubt in its utterance. Even when the lion's announcement ended, we remained silent long enough to hear another lion answer at length from far away. Only when that answer ended the primordial dialogue did the men gasp, as if coming up for air out of an unfathomed deep themselves, and start to talk again. The rugger players behaved as if there had been nothing unusual about it. But Mori was excited, and Plomer, whose first roar of a lion in the open it had been, was moved and uplifted until Mori asked him: "Plomer-san, I would like to buy some lion's whiskers to take back to Japan. They tell me they sell them in Nairobi. Will you help me?"

The wish, delivered with the innocence of a child asking for an ice-cream, roused Plomer's acute sense of the absurd, and I saw him struggling not to laugh aloud. I could almost see the relief from pressures that he found in the vision of Mori on the prowl in Nairobi after lion's whiskers! Certainly, compared with the grandeur of the sound, Mori's proposition was full of bathos. And yet to take a hair of the lion that roared at you back to Japan, in the sunken levels that nourish the Japanese imagination, was a way of honouring majesty. And as it happened, Mori's longing for lion's whiskers was about to serve us well.

We had hardly arrived at the New Stanley Hotel in Nairobi when Mori, as optimistic as ever, began a long series of calls to Government House without success . . . only a request to telephone again in the morning. That left him puzzled but not dismayed. Plomer and I were fearful and angry. We had, meanwhile, seen what there was to be seen of Nairobi and talked to numbers of an English community of barely two thousand. My own impressions were on the whole an intensification of those initiated by our encounter on the railway station at Mombasa. The pity of it all hit me far harder than the black evil portrayed in Norman Leys's Olympian sentence on the colony. The need for compassion and understanding was far greater than any impulse to judgement and condemnation. And with it all, a feeling of foreboding insinuated itself, and was made worse by an almost intolerable and unusual aching head and smarting eyes. The large, blonde Scandinavian housekeeper of the hotel, whom I remember with gratitude and affection, cured this part of my affliction with an aspirin: my first ever. The world of drugs and medicine has achieved such undreamed of successes and complexities that no-one remembers how eventful was the coming of the humble aspirin: for us in Africa it was almost as great a breakthrough as M & B 693. The tyranny of the headache in the Africa of my youth was to the human body what Chaka was to the Zulus; its abolition was a miracle. The relief I found that night at the ministrations of the house-keeper leaves her angelic in my memory, and is comparable only to the instant abolition of an almost unbearable pain in the war by an injection of morphia. Even the food in the hot dining-room of the New Stanley Hotel was transformed by it and left me relaxed and open to appreciate what was to follow.

Our sweet, a suet-pudding, was interrupted by a woman riding a spirited horse into the dining-room. She dismounted smartly, threw the reins at an unperturbed Kikuyu waiter, a double-terai hat at another, and then snatched the wine list from a table to select champagne for all around. The aspirins had made me more than ready to join in the toast to her in the tepid but sparkling liquid. That done I studied her more closely. She had two pistols, one on each hip, a divided, khaki riding-skirt, leather boots, and an open-necked, silk shirt with a scarlet kerchief, as if she were a cowgirl straight out of the circus of the Ringling Brothers. But her face was unmistakeably of England, and of the England of the post-war gentry who had suffered and proportionately lost more in four years of killing than any other section of the country. It was fine-boned and well-shaped, but the skin was beginning to wrinkle through time and dehydration under the sun of Kenya. For all her bravado and demonstration that life in Kenya was all 'high-school' fun, the excess of the gesture was a measure of stress and disillusionment within, as well as being the sort of defiance that is also a cry for help. Her looks were as faded and dimming as her years and hope. And how little impact she and her kind had as yet made on the inexhaustible and unrepentant Africa, biding its own dark time in the great night outside, pressing so heavily against the feeble flickering light of the room that one almost feared that the glass of the window where we sat would give way, was made plain in the early hours of the morning. Some of the colony's finest race-horses were killed in their stables in a back-street, not far from our hotel, by a pride of lion – presumably unaware that their whiskers were under threat. We had hardly heard the news of the killing when a message from Government House struck down Mori's own racing expectations. The Governor could not or would not see him, and no diplomatic euphemism could disguise the brutal nature of the announcement, made worse by an offer to pay Mori's travel expenses.

Mori took the news well and, indeed, fell as he had exhorted me to fall at judo in the well-deck of the *Canada Maru*. Had Plomer and I not known him well enough by then, we would not have known how hurt he was. He gave no external signs of injury, but we both knew that one of those apparent trivialities of life that can affect a whole trend of fate for millions had just been tossed at our feet by the impervious and bland bureaucracy of Government House. It was my introduction to a long association with Government Houses all over the world that made me feel much about them as I did about the General Staff in the War. The men in the field in war and peace seldom failed one's highest expectations. But at headquarters and capitals, both rather far from, and behind, battlefronts, the war that men were directing, or the country that they governed, was often not recognizable in the terms of the actual fighting, or of the provincial crisis that I had just left and on which I was now reporting (often in vain) to my superiors. I must hasten to add, though, that there were many exceptions to prove the rule because the exceptions then were beyond reproach, and became more and more frequent as both War and Empire neared their end.

From the moment the news broke, we kept close to Mori and took him with us everywhere we could think of going. We never mentioned the Governor-General or the incident again. It was as if he used his whole spirit to make a philosopher's stone out of this cruel little episode, and succeeded. He has not mentioned it yet after fifty-five years, and it is as if it had never happened. One wishes that more of the negations of history could be dealt with likewise. No wonder, therefore, that on the day itself, Mori was unusually silent and only spoke to abjure us not to forget the lion's whiskers, as if that were all that was left for him to seek in Africa.

The Indian tourist shops round about the New Stanley knew of no whiskers and smiled politely at the very idea. But at the offices of the R.A.C. a remarkable man called Galton-Fenzi, who might have been a P. G. Wodehouse character at his best, with an Edwardian moustache, big lively eyes, tall, thin and full of a certain romping Englishness that was to see the Kenya experiment in colonization through to Uhuru and beyond, gave us not only a clue but our first positive inducement to laughter. All was adventure and fun to him. Far from being depressed in spirit by the intractable nature of equatorial Africa, he regarded it as one of the most formidable of wealthy and disapproving aunts who had, at all costs, to be outwitted. He had just pioneered a road from Nairobi to Mombasa by car and, of course, it had all been the greatest of fun. He told us we should have seen the giraffe racing by his car blowing out steam at the radiator like a railway engine! "It reminded one," he declared with eyes large and innocent as a child, "so vividly of pre-historic times. By Jove! It was top-hole."

Plomer and I could hardly prevent the smiles of appreciation that politeness demanded from becoming outbreaks of uncontrolled laughter. The language even then was beginning to date but was not pre-history yet, and that one could still call on Jove for emphasis tells one as much as any social history how different our Then was from my Now. For it only made sense because European man's thread of continuity with the vanished world of Greece and Rome was still intact. Now, for many, history begins with the tiresome manifesto, the revolutionary hiccup of 1848, that testified to the onset of ideological indigestion from which the world is still suffering.

Galton-Fenzi took us to the Club which outwardly was a model of the English prototype. It gave visitors, as it was said of its London model, a feeling that a dead duke lay upstairs, which was belied by the presence of the eager and efficient Kikuyu waiters. Also there we met a clutch of men that left one certain that something unique and, perhaps, decently outrageous would be made of Kenya, for they were determined not to allow Africa to quarrel with them. There was, to use one of several examples, Grogan Caney, who had walked from the Cape to Nairobi unarmed at the beginning of the century, just to prove to the father of the girl he wanted to marry that, though penniless, he had other resources than money to offer. Nothing, I knew, could make such people go away. They proved that there was far more to the matter than the Norman Leyses of the world dreamt of in their high philosophies. I was to know him well because he stayed and lived on into Uhuru and his nineties. And I am glad it was he who put us on the spoor of the 'lion's whiskers', Mori's 'much-to-be-sought-after-hair', as he had it in Japanese. To this day I am not certain that what he bought in the end really were the lion's royal whiskers, and not, perhaps, just the hair of the substantial *haute bourgeoisie* of the elephant in the bush of Africa.

Whatever they were, to Mori they acted as balm on the wounds of his spirit and enabled him to go with some joy to a place where, since the day was clear, we could see the summits of both Kilimanjaro and Kenya. The view, for me, is part of one of the great dreams that Africa unfailingly evokes in human imagination. We stood there deep in long, lean, green-gold grass, on the upland plain, stretching to a remote horizon where the distance first made it blue and then finished it off with superbly spaced hills of purple, so that the quicksilver tides of air could ebb and flow swiftly in between and ruffle the tanning yellow tassles of the tall setara grass with dark catspaws of wind. The sky itself was blue-black, but north and south of us there were startling flashes of feathers of snow from the two sleeping volcanoes. They were as pure as the wing of the albatross, which is the greatest white hunter and slayer of vast distances in this world. Elsewhere, even on the Mountains of the Night of my native South Africa, snow in winter is a platitude. But here on the equator it was a miracle: as if archangels were coming and going in that dazzle of Antarctic white so high in that remote, arched, tranced and timeless blue of the interior. I fell silent, as I am certain the lone Greek wanderer stood silent when he saw the same snow in the year before Christ, and took news of it back to Tyre, from where it had travelled to Ptolemy and guided his cartographer's hand to include them as 'Mountains of the Moon' in his map of Africa. For me it was like a form of religious confirmation without need of a catechism. But before the dark fell on us and that little capital behind us, looking as if it had lost its way and was about to go round in circles to search for the place where it had entered that vast unrepentant scene, I remember that, in the long level light of evening, the Ngong Hills, well outside the town, wrapped a mantle of heavy purple around their shoulders, and drew a hood of gold over their heads, against the cool of the fast oncoming night. Not only did I have the feeling that 'I-had-been-here-before' about the hills, I had also a feeling of 'I-shall-be-back-again'. It was almost as if Africa at that moment made manifest Shakespeare's prophetic soul of the whole wide world. It was a moment, therefore, of profound hail and farewell. But because of Mori's hurt in spirit, we felt the sooner we left the better.

That night we hastened back to the sea by train, silent where before we had been cheerful. We sat all night upright and uncomfortable, each of us wrapped in his own cloud of red-dust, like men of the hill-people of Basutoland with their crimson tribal wool blanket over apprehensive shoulders, on their way to mine in precarious tunnels, miles below the surface of foreign earth. But at Tsavo I went out to stretch my legs and body, and also in the hope of hearing lion, since Tsavo was the place where their storm-troops, the great man-eaters, had once in my lifetime massed to hold up for months the building of the railway line, until subdued by the enigmatic Colonel Patterson. But none called for us that night. I saw only the Dog-Star in the trembling silence, now drawn white above the dust: two cupped hands full of silver-green light; and on our right the Southern Cross, dipping alarmingly low since I had last seen it at Port Natal.

Some days later, in the middle of the day, we put to sea again. Plomer and I, side by side and silent, watched the land foundering behind us. He watched it as if for the last time, and aware that he would never see it again, except for one perfunctory glimpse nearly fifty years later. But that could not count, since by then he would have lost his unique inner eye which had seen with such startling clarity and awesome comprehension. He watched it, moreover, with an absence of regret that must have hurt more than outright pain would have done, because in that uncompromising heat and light the view was an epigrammatic statement of the Africa that he had once known and which he remembers from afar in one of a number of poems dedicated to me. It concludes with the lines:

> *That was the Africa we knew,*
> *Where, wandering alone,*
> *We saw, heraldic in the heat,*
> *A scorpion on a stone.*

And so began the longest period we had yet spent on the open sea. Our course was due east and followed almost exactly the line of the Equator. As a result we moved through a sea without wind, without seasons and without time. So still, glittering and immense was the ocean that we thought of Coleridge's: "So wide and lonely that God himself scarce seeméd there to be". Were it not for the world we carried in the *Canada Maru* with us, which provided us with some wind out of its own speed, setting its own time and observing its own seasons, the experience could have been as fearful as it had been for the Ancient Mariner. But for me it was full of interest and delight. Even the simple fact that the sun daily came up fast on the *Canada Maru*'s bow and set dead astern, raised the journey into a dimension above a mere physical displacement from West to East. Since the sun rose with such speed I would be more punctilious than ever in performing my dawn ritual and would be out of our humid cabin in good time to see the light explode and shatter the darkness ahead. I was usually also in time to observe the deckhands come out of their quarters in the fo'c's'les to submit their ship to as scrupulous an ablution as the one they had just observed for themselves. Each one of them, as out of his own inner accord, never failed to turn about on deck, come to attention with arms fully stretched downwards, hands extended and pressed to their sides, and bow to the rising sun, before they became servants of cleanliness and order. That, of course, seemed part of the extra Eastward dimension to which I subscribed with such imponderable and inexpressible feelings. It was also yet another instance of a marked feature of the world of the Japan which our ship contained in miniature. Once the Japanese knew their 'position' they seemed in no need of orders. It was extraordinary how few commands were given in the ship even by our Captain; and when necessary, they were only uttered in low, clear-cut and almost confidential tones. An implicit love of form and order seemed to be in everyone, so that they ordered and observed the most complex of forms as of themselves. As a result there were no feelings of external compulsion, and one could enjoy and marvel at the atmosphere of liberation and freedom derived from it. But of how all this looked, seen from within by the Japanese themselves, I had only the most elementary indications.

These came from the little that my teacher told me about one of the most profound and pivotal concepts produced by civilization: the concept of

'Li'. Originally (and 'originally' in terms of the Chinese who first evolved it meant going almost as far back as the conscious spirit can go without losing altogether its consciousness), Li meant sacrifice and the ritual and ceremonial that proceeded from the exercise of sacrifice. But for my 'sensei' and those humble deckhands who began their day bowing to the rising sun, it meant outwardly the observation of courtesy and good manners in human relationships; and also ritual and grace for acknowledging the presence of the Gods which, as the relevant ideogram graphically implied, represented that which was *above*. Courtesy to man, and the ritual that was courtesy to the Gods, were one at source, so that not surprisingly within, it was a transcendental ethos which inspired the morality that alone could impose order on the competing elements of the human spirit. The importance of all that flowed from this in the course of centuries until it became a reflex of the Japanese spirit, cannot be over-estimated. To explain its width, depth and ramifications would take all the libraries of heavy volumes already devoted to it and, even then, a lifetime of living it, in order to be understood in the way that the most humble of men who are born to it understand. But at that time it emerged as a moving and, for me, sympathetic attempt at bringing human relationships continuously into harmony. Then the resolution achieved could be joined in the overall harmony of the universe.

When my 'sensei' warned me that not to know Li meant not being able to take one's position in life, I understood at once what he was saying. It made strange sense of my own early morning compulsion to contemplate the first light at sea, as if it were my instinctive way of centring myself for the day. Consequently I came to watch the deckhands at their sunrise ritual with feelings of kinship. When later I observed a clutch of cinema men and their photographers who had boarded the *Canada Maru* at Mombasa after a public relations reconnaissance of East Africa, I felt something else. As I watched them bowing to each of Mori's officers in the order of introduction at a reception for them in the saloon, as the deckhands bowed to the sun, they were, I realised, one and all taking position. They were doing this not just in the social order of the *Canada Maru* but also in that of the universe and all its stars. It was as if they were acting out Origen's great injunction: "You yourself are even another little world and have within you the sun and the moon and all the stars."

My moments of early morning contemplation as a result tended to start earlier; they became longer and ended only with the arrival of our tray of early morning tea which I could take from our steward and serve in the cabin for Plomer and myself. Calm as the sea was, the swell remained long and heavy, and its regularity and rhythm made the little *Canada Maru* roll, dip, rise and glide forward all in one accorded movement, as if part of scales of music made for the observance of Li itself, so that our ship would not lose its position in the convoy of the wheeling universe. Although the seas on our departure had been far heavier, they had upset Plomer less than this consistent unrelenting movement of a ship with the smell of home in its nose. He was now as little on his feet as possible, went to bed earlier and slept later.

I hoped in vain that habit would come to the rescue that his fortitude deserved. But far from improving, his immunities to the movement if anything got less except that, provided he kept seated or lounging in a deck-chair, he maintained his ease and high spirits. As a result I was saddened that the voyage could never mean for him what it meant for me, and that he missed most of the manifestations of a beauty that his love and gift of the sensuous would have made him appreciate perhaps even more than I did. The sea of wild silk and the colours of dawn; the magic lantern sky at sunset like a reflection in a mirror where the invisible were made visible, the unknown known, and even Li at times acquired tangible form; the dark, pirate sail of a dorsal fin of a shark keeping station on our starboard bow and, without the hint of a crease, cutting through the silk of the fine-spun sea. Or perhaps just the coming of the first lone sea-bird from some atoll of coral and palm far below the horizon taking on all that weight of stainless blue of sky and great emptinesses of space and sea, as if it were courage itself on the wing. I could rarely get him to the rails in time to see porpoises and dolphins playing games with the *Canada Maru*. But once, when a whole school of them came to us out of the East and passed close by before vanishing in the West, the arch of their backs like stitches of silver twist in the Indian silk of the water, he was able to watch it all with an intensity that I could not match, before exclaiming after they had left that immense scene: "What an astonishing creature, Shakespeare! There was nothing of importance he missed, even about dolphins. I have just been thinking in a way I could not before, of his Mark Anthony: ' . . . His delights were dolphin-like, they showed his back above the elements they lived in'."

Because of the absence of wind, so finely balanced was the *Canada Maru* on the razor edge of the line dividing the world, time and space and all their seasons into two, that we rarely saw flying fish. Their efforts to rise were usually aborted and just left a scribble of their effort on the water. But when they did get airborne, and the achievement after so much effort exalted me, I would shout the news to Plomer. When they had vanished in a sparkle of diamond water that hurt his eyes, he asked one day in a voice that I thought to be oddly forlorn for so confident and assertive a character: "Were they like Roy's flying fish? Do you remember?

> *And the flying fish*
> *In their silver mail*
> *Rose up like stars*
> *And spattered down like hail.*"

He was quoting from Campbell's 'The Flaming Terrapin'. And for all that it may have of exaggeration and excess, it is still for me one of the most exciting epics ever written about the sea and the wild wilderness of the dark main of Africa that we had known as boys. So in agreement with him I nodded emphatically.

Fortunately, after many days there came an evening that Plomer was to enjoy as totally and as actively as I did. It came on the night of our first full moon at sea. Although the goddess who ruled the Japanese from the beginning, and is also the source of authority for them, is the sun, yet their love is for the moon which makes light in the darkness that haunts their spirit, day and night, below the horizon of their doing and being. Even as they bowed to the sun that morning, they did so with a slight abruptness in order quickly to be able to turn their minds to the imminence of a full moon. Part of this concept of Li is that all should be received with courtesy and ceremonial in order to create a state of grace in the presence of the reality. Accordingly, there is no other culture which has such an abundance of colourful, imaginative and heartfelt ceremonies. These arise out of a devout land to mark the passing of time, just as do the milestones, shrines, wayside chapels and temples that stand on guard of highroads and by-ways of the Japanese spirit. Hardly a day is without some special time-meaning of its own, which has to be honoured by some act of recognition from a hypersensitive people. I thought at moments, from my studies and observations of our life at sea, that it was almost as if time were an important master to which a whole culture was subject. Thus it was compelled to be greeted with a ritual.

Again this was an approach to natural events which I could understand without reservation. In my own home in the interior of Africa, when it was still an unthreatened keep of natural life, we took these things seriously, as indeed did the animals around us. Not only would our dogs bay at a full moon but even one lion that I knew was so indescribably affected by it that he would go on roaring at it with a new voice until dawn. I knew fragments of the shattered, copper-coloured, hieroglyphic Hottentots and the apricot Bushmen, the first people of Africa, who in their last sanctuary of the vast desert places of our land, went walking in procession, clapping hands, singing and dancing, just for the love of it. Sometimes just dancing in one magic circle in the scarlet sand, round and round, as in a carousel of a passion of thankfulness because they feared that unless they paid homage to this glory of light in the dark, it might be offended, fade, and go away.

That was our nearest indigenous equivalent to the instinctive Japanese compulsions that inspire, in honour of the moon, the host of colourful, complex and profound ceremonies which are beyond our comprehension, as I have hinted. The European way was to celebrate the going down of the sun, and perhaps there is in that an indication of the divide between them and us: we were born in love with the sunset, they with the moonrise. I was to remain astonished throughout my life by the role of the moon in their lives and temperament. Perhaps it is all best left to the symbols that inform us of meaning which take over on the frontier where articulation fails. And there was one such symbolic statement of which my sensei told me that was to stand me in good stead. Appropriately it is contained in a piece of that noble and ancient order of the theatre of Japan which is called 'Noh'.

This particular play is as spare, simple and yet full, as is demanded of all that is best in this spartan discipline of theatre. It was about an anonymous woman, in the grip of tragedy too great to be named, standing at the rim of a deep, dark well. The moon rises behind her as she stares into this black pit until it is high enough for her to see its reflection at the bottom of the pit. . . . That is all, not because there is no more that can be brought to it but because it is enough; and enough for a humble spirit, like that of this woman in her anonymous lot, is almost too much. My sensei told me this after a day of trial for him and a particularly sustained assault on science by the Chief Engineer at dinner; an attack not unconnected, I felt, with the approach to fulness of the moon. And, my teacher added, on other similar occasions this 'Noh' play would come to his mind in the way that he had first seen it in Kyoto when the Momiji, the maple leaf, was turning red and the first blue of winter was beginning to creep like smoke through the evening. He had only to relive the moment when the moon found the water in the back of the pit, and all in his spirit would be well.

I have enlarged on this theme which I could easily have left at its literal level. But to me it was as important to pursue my instinct that, in this new world which I was entering, 'meaning' was rarely purely literal. It seemed to me that unfailingly it touched on a natural symbolism, so that everything made manifest served also to draw attention to a latent area of unrecognised significance. Without some such elaboration, what followed at nightfall after one of our hottest and totally windless days would have been inadequately understood; as would the air of anticipation which came to the ship like a catspaw of wind on a becalmed ocean, indicating that a trade wind was hurrying near. It had such an effect on conversation and behaviour around us towards sundown that it was transformed into the first breeze of an excitement which filled the sails and sent the world of the *Canada Maru* moving out of the calm in which it had lain since its last port of call in Africa.

I would have liked then to be in two places simultaneously: with our Captain, but also with his crew. But there was no chance of even a dash and a quick glimpse of the crew's quarters because, well before sunset, our steward appeared. His greeting and bow were both subtly exaggerated, almost as if deliberately forced upon himself to produce a balance against a certain intoxication induced by the occasion. With pride and an emphasis of grace he presented Plomer and myself with two of the largest 'Yukata', the Japanese under-kimono of light cotton, to be found in the ship. They were crisply laundered and ironed out into the prescribed traditional pattern. Dressed in these, with bare feet in the half-slippers which the Japanese use indoors, we joined Mori and all his officers massed on the starboard deck. They were dressed likewise, and words of lively conversation bustled between them like bees about to swarm around their queen. Behind them the evening was dissolving into the light of copper which preceded the precipitation of the inky darkness of night in that world without latitude. Somewhat startled, I noticed that where the light of copper was brightest our ship had altered course some points to the north-east. I was puzzled by it until Mori told me he had ordered the change in deference to the moon, for otherwise we would not be served notice of its coming in the east: nor would we be in a position to see it rise with light unimpeded by any rude interference from the foremast and bows of the *Canada Maru*.

Our slippers were then at once discarded and we were invited to sit on silk cushions. There was again a large ceremonial mat of pure rice-straw; yellow, soft and still with a subtle scent of the earth that had produced it clinging to it. In the centre stood several charcoal braziers, each with a little mist of steam carrying the incense of different food. The largest bowl of all, a superb black lacquer vessel, was filled high with whole unpolished grains of rice, flushed and red as with the repressed passion of increase, but of the right ceremonial quality. Before us, laid out with the art and craft of the past, was all that was appropriate for the Japanese table. We each had a beautiful little lacquer tray by our crossed legs, with ivory chopsticks at its side. Then there was an infinity of lacquer bowls, black and red with a single leaf or flower motif as if to invite nature to be also present at the celebration. Little porcelain dishes and minute cups of sea-blue and driftwood white were filled with substances where the colours had to match the diversity of tastes, and all combine to make a meal not only harmonious on the palate but also to the eye.

And soon our steward, in the dress and kimono of a feudal attendant, seemed to be transformed into a figure of elegance and grace disguising his homely face and awkward shape as he came, solicitous and unobtrusive, to fill our cups with warm saké. He had hardly done so when the September moon served notice in the East that it was rising.

An extraordinary stillness and atmosphere of expectation enfolded the ship. Behind us in the crews quarters in the stern, a shakŭhachi began to play with clarity and purity as if the notes themselves were made of moonlight. The purser, who was on my left, whispered that it was addressed to the souls of the newly dead who would be resting there on the rising moon on their way beyond. As he spoke, the moon lifted itself over the rim. Its light began to show fast on the ring of the sea. A second before the sea had been black. Now it glowed and was pierced together with a flash of light that ran all round the flawless horizon, until we were in position at the centre of an unblemished equatorial night. The light, indeed the entire moon, was of a rich red-gold, and it rose with the unhurried and even movement that is its measure in the world of the spirit. It began to push before it the heavy water of the sky pressing down on the *Canada Maru*, and then roll back the darkness from east to west. I had been told by Mori and my sensei that no two moons are alike, and each exacted according to its individual character its own measure of tribute from the heart. But that September moon ruled over them all, and received, whether one knew it or not, the largest tribute and responses of feeling. I could not, therefore, have been more prepared intellectually for the occasion. Even the 'Yukata' that I was wearing conformed: it had become a state of mind. Yet I had still been reckoning without the direct impact of this most august of imperial moons in the Japanese year. For it rose as if to a trumpet sound of victory. Like birds who sit on some tranquil water for hours without a sound being uttered or a single feather stirred, and will yet suddenly rise and take wing altogether, so we all, including Plomer and myself, rose as one, and made a profound obeisance to the moon's presence. On my right, Mori sighed, and a certain deep satisfaction rumbled in his throat before he exhorted us with reverence: "Look! It is Momiji, maple-red."

"Like the Momiji at Arashima," someone hastened to add.

"No. Like those by the Imperial Villa of the Ascetic Doctrine, Shigaku-In, at Kyoto," claimed another.

For a moment there was an excited clamour of counter-claims, until my teacher asked delicately, as a great connoisseur might have done, a question whose answer he wished to share with others: "But the colour of what kind of Momiji, would you say?" There were apparently many distinct kinds of maple which coloured the fall each in their own way.

The Chief Engineer, opposite, made the Japanese equivalent of a snort, shrugged his shoulders and remarked dismissively: "As if any of all this matters: the moon is just red and that is all there is to it. Why bring leaves into it?"

I would have thought his interjection deserved only to be ignored. But it shook that close and closed little world on deck, like an earth tremor. Everyone thought the matter so grave that it should be left to the Captain. Wisely, Mori let the silence speak first, in which the throb of the *Canada Maru*'s engines was like heart-beat. I could almost hear the rustle of moonlight join the brush of the sea along the bows of the ship. Then Mori remarked with total courtesy: "You know more about engines and what makes things like ships move than any of us, Chief Engineer-San. But do you know what makes the things of the universe move?" And before the pugnacious officer could spoil the evening by saying more, Mori added: "You know the ideal of the Samurai demanded that he should be able to compose as appropriate a poem to the moon as to the battle before him, or the seppuku after defeat in battle! Now, who will compose a poem for all of us?"

Mori looked imperiously around and then most suggestively, I thought, first at the purser and then, not inappropriately, at Plomer whom, I recollected in a relief without envy or malice, had elected in advance to be minister of art and literature. Wisely he excused himself as too inexpert for so essentially a Japanese occasion, and said he would participate with all his heart and observe the persons who knew best how to order these things. The way he did it made Mori and his officers accept it as an act of modesty that they prized so much, particularly in a poet whose reputation already proved he was a master and who might have 'felt it was his due to be first in this moon-field.

And so the stage was gracefully cleared for all those who had it in them to produce appropriate salutes in verse. They were, I could see, with the exception of the Chief Engineer, all involved and helped on with more and more saké, and more ritual dishes of good, whose symbolic foundations were provided by a raw egg broken into each of our cups and the addition of a sauce of a bean that turns aside evil. By this time the yoke of the egg in my imagination, warmed on rice-wine, looked like a full moon reflected in my bowl of resplendent lacquer. I thought of a Zulu saying: "Patience is an egg that hatches great birds: even the sun is such an egg." Little did the Zulus know, I told myself, that the moon too is another such egg.

So while we ate and turned over poems in our mind, the moon rose through the level of its over-flowing Momiji-red self, into a more precise yellow self, followed by the lucid silver manifestation which was once enshrined almost to perfection in the quietude and seclusion of the Gepparō, 'The-Waves-by-Moonlight' pavilion in Kyoto. Finally it became a calm unwavering illumination of what was left of the night, until it enfolded our ship with a tender feminine authority in a soft shawl of light. It left the sea with swift impressionistic transcriptions of its unhurried climb to the summit of our glowing world, attended only by a single star whose companions had all been lost on the way in moonlight, as other things are lost in darkness. No sooner had it moved into this final phase of its ascent than the poem of the evening appeared. It came from my teacher and I remember it, perhaps not with dead accuracy, but live as it lives with me still for the quick of gratitude to the man. I translate it roughly as follows:

On this ocean
without season

the moon writes
in So-sho
the symbol
for Autumn
and declares
it is full
because it is
about to die.

The reference to So-sho (which was, for my teacher, the swiftest and most immediate of all forms of calligraphy) was inspired, because the sea, at that moment, was spread out like yellow rice-paper for the brush of a Zen Master in the sanctuary of his temple to catch what he could of moon-meaning.

Mori was the first to express appreciation, with a characteristic grumble of sound from far down in his throat. He was followed by exclamations of delight and the clapping of hands, in an oddly schoolgirl fashion, from the rest. Yet I thought I detected the reservation that, as far as his appreciation went, it did not go all the way, because the poem did not express altogther what the moon did to him. Mori, after all, came from Kyushu, Japan's southern island of volcanic, earthquake-ruled, impulsive, rich and thrustful earth, that gave a passionate and somewhat inflammable character to all it nourished. I had been aware of something of this on our first meeting. And daily I had come to realise the strength of character which contained all this inborn and paradoxical matter with apparent calm and dignity of mind and spirit. I felt it all the more when I learned how much of the Samurai example had been embedded in his family tradition. He came from near the capital of his island, Kumma-noto, where his family had been considerable land-owners, and high-caste farmers for generations. And he had an account, like a map reference in his mind, provided by his grandfather of the great battle near the capital in the sixties of the last century. There in the civil war which followed on the restoration and the Emperor Meiyi's decision to open Japan to the West after an isolation of centuries, the men of Kyushu fought one another for sixteen days. All this, combined with the congenital passion of Kyushu spirit (which the Japanese called 'Jo-netsu-teki'), made Mori decide that the sword should join the maple-leaf to do full honours to the most important moon of the year.

I do not know where the sword came from. All I know is that at one minute all had been trying to compose Hokku, the homespun version of the Haiku that the Zen Masters of this most exacting seventeen syllabic form of poetry have made great and famous. The next minute, Mori was there on his bare feet, naked sword in hand and trying it for balance with eager slashes that seemed to cut through the moonlight as if through a cobweb sparkling with dew. He followed this with an imposing and powerful call of a warrior either preparing for war, or in battle, and who accompanied his movements with sounds that left me in no doubt that Mori was already at war to the death within himself. Though impressed by the precision, the power, and even a certain atavistic glory of this performance under the observant moon, I was, as I had been at Kendo, uneasy. This uneasiness was not dispelled by my own rendering of a Zulu war-dance that drew blood from my right foot. Nor did the folk-songs which Plomer and I were called on to render as a reply to their own chanting to the moon for help. Neither of us could sing in tune, and the effort in my own saké-heightened spirit sounded like an insult to the moon which by now had me totally in its power. That power, and the fact that all in the *Canada Maru* had found it necessary to honour it, ultimately consoled me, and eventually sent me happily to sleep. But it left a memory of uneclipsed good.

The days that followed brought increasing signs that the end of a phase hinted at in my teacher's moon-poem had begun. We found no wind as yet, but there was a dramatic change of sky. Towards evening the pile-up of clouds, as if in a fatal collision of elements in the atmosphere, became greater and more dramatic. They testified to an inspiration and moulding power of great upward drafts of over-heated air that could come only from a land tormented in conflagrations of the sun, which was committed more to arson than growth wherever it went in these extremes of latitude. The heights of which these clouds were hurled would dominate the blue so that, no matter how beautiful their constantly unfolding form and changes of colour, they began to induce a certain fear, as if their wilful begetter had made them also more in love with power than lawful creation. Fortunately they gave us time to become used to their presence on the evening horizon, standing like the marble-walls and battlements of a great forbidden city at war with the night. For some days they kept a certain majesty of distance, as did the horizon, and left a lagoon of the deepest blue intact above us. But then they began to outflank and encircle us, so that the awe they inspired became almost as substantial as their foundations appeared to be. There was a gravity that pressed down on the mood of all, so that at nightfall we became unusually silent, almost as if afraid to speak: or, if forced to do so, we spoke in respectful whispers. The opal sea itself was also subdued, and so depressed under the weight of what it had to reflect that it hardly stirred.

It would be so still that the pulse of the ship's heart, and wash of her bows slicing through the rainbow water of a twilight doomed to die almost as fast as it was born, became too loud for comfort. There were moments, indeed, when it felt as if we were about to enter a vast valley of Himalayan giants, great and fat with snow so loose that any sound loud enough to reach these remote heads would release an avalanche to crush the little *Canada Maru* and make a witches' cauldron out of the prostrate, worshipful and moon-obedient water. Their power was all the more awesome because they seemed to possess an immunity against the dark which neither the polished sea nor our ship possessed. For long after the *Canada Maru* was dressed over all in full-length purple for the evening, the sun from below the horizon would light up their heads and arm them with long shafts, arrows of fire and throwers of flame against the oncoming night with such abundance and force that, in the religiously observed silence of cloud of snow, it was as if the air were hissing with the speed of their flight over the last sea of battle for light. Even when the sunlight had gone, they illuminated themselves with the swiftest, widest and most incisive flashes of lightning that I have ever seen. Only when they themselves decided that they had completed their business with the world would they mysteriously dissolve of their own accord, some time before morning, and vanish into the east which was their home. By that time my own imagination had become so involved with their subtle substantiation that it transformed them into some court of time, assembled by the wind which forever accompanies the earth as it spins, for an urgent late-night session to dispense justice and deliver judgement on the day for breaches of contract with life. And when the dark had finally locked one and all below in their appropriate cells of night, we would be near enough to hear the fundamental note of the voice of the thunder that followed to adjourn a court from which there was no appeal.

The reverberation of that magisterial sound sent the sea trembling and breaking into a goose-flesh of silver from cold fear. And above the black tip of the *Canada Maru*'s long fountain-pen like mast, still reaching to write poetry to the stars on the one blank sheet of sky, it looked for a moment as if the great Dog Star itself had stopped pouring down its generous light on us because of a passing darkness of heart. But this too changed after a lotus afternoon when all fear passed for good with the last flake of light in the *Canada Maru*'s wake, and the thunder spoke again as the cloud was at its summit. Though one had no rational means of decoding the precise meaning of the awesome voice, there was no mistaking the feeling of relief that it brought as evidence of a power greater than provisional man.

And as for myself, I felt rebuked for not realising at once, when I had first seen these mountains in pageant around the parapet of our world, how familiar they were. Even though I had known them by proxy through my reading of Conrad, it should have been enough because it was the proxy of an artist who, through gift, experience, and loyal service to imagination, was an honorary native and freeman of those jewelled shores, flashing streams and hidden boroughs of emerald jungle beyond the cloud. Even the sea I loved so greatly seemed to sink into a new intensity of silence, as if it were dumb with reproach. Indeed, it became so still and clear, open and shining with light, that it was a flawless image of the mirror that Conrad had made of his heart and mind to reflect, as though it were the face of a stranger staring over his shoulder from behind him, the ultimate countenance of truth.

Conrad possessed for my own immature self then, more than any writer of a time dangerously deprived of instinct and intuition, what an inspired French observer of primitive man called a capacity for *participation mystique* in the world around him. I had a hunch that, whether artists knew it or not, and however civilised and far removed from the primitive they thought themselves to be, they had deep down just such a first being within themselves; held in trust, impartially, without censor of judgement and prejudice, as a source of aboriginal wonder which enabled them to share in the secret of being in its most unfamiliar, improbable and even abhorrent forms. This dark gift of participation in all things and manifestations of life around them produced the act of transformative wonder for the light of an understanding which would find nothing on this rounded and turning earth ordinary or mean, and so made of their craft the instrument of the increase of awareness which I suspected was the abiding function of all art. It was something of this kind that Shakespeare was always unerringly after, as, for instance, in a passage over which William and I, because of the fundamental difference in our natures, had been at odds that afternoon.

It referred to that sombre moment in *Lear* when the doomed King at last finds rough comfort like a rock in the sea of deception and the unreality of a world of worldly and self-seeking men, with the conclusion addressed to his soul, his daughter: "We shall take upon ourselves the mystery of things and be God's spies". Conrad, for me, had been such a spy in many a world beyond the established range of the arrogant and narrowly focused European awareness of his own day. He had been such a one even in the heart of darkness of my own native Africa, and forced a whole new world of unknown earth, being and human considerations upon our slanted and inadequate reckoning. But nowhere had this sense of participation in the strange, antagonistic and totally forbidden been more marked for me than in his discovery of this world the *Canada Maru* was now approaching.

My own pilot into these seas and those gleaming latitudes of well-nigh countless islands which a great Dutch writer – disguised under a Latin pen-name, signifying that he had suffered much – described as 'slung along the equator like a girdle of emeralds', had obviously to be "Lord Jim". It was significant how at the moment of recognition "Tuan Jim", to use his Malaysian honorific, was instantly at my side on the boat-deck like the visitation he had been for Conrad. I remember how even Conrad, who accepted his own charge of the mystery of things with utter obedience, was so haunted by the subtle way Lord Jim and this new world into which his betrayal of a greater self had led him, felt the need to explain how he had become one of what the writer called his 'family'. The conscious explanation was inadequate and in a sense superfluous, because it was all in the story to be known only as through the profound sense of participation which forced Conrad to report on him. He could, in the end, say little more to put his readers on a course which passed the understanding of his day, than that Tuan Jim had come to him 'like a cloud' and, when his own truth finally came to his side in the tangled jungle that was its temple, 'veiled like an oriental bride', went on to vanish 'inscrutable as a cloud'.

And there it was, for Conrad as myself, cloud could not be kept out of it. It was as if they were the supreme image of the unfathomable interfolding of the spirit of man and the hidden forces of the earth, sea and sky and all the other untouchables of the nature which produced and nourished man and which Conrad described as no one before him had done for me. It was so not only in *Lord Jim* but in many other tales of fever and unrest, recording the shattering impact on upstart European intruders among seemingly barbaric people, like that of volcano and earthquake on the earth beyond the cloud going scarlet ahead of us – the sort of Barbarians we might have need of because they possessed a noble and vivid nature of their own which they followed without question as they went like torches in carnival procession through dark and fearsome jungles, up to their knees in the moss and the twisted footpaths which conducted their lives. Indeed the effect of the scarlet cloud ahead just then was almost a repeat of Conrad's image of the crimson crack as of doom itself in the final cumulus formation which crowd around the end of his *Victory* leading with classic inevitability, as in *Lear*, to a death that is nonetheless a triumph and vindication of life. Accordingly, this moment in the *Canada Maru* was to live with me through the long, strange, dangerous and random years which were to follow, as one of personal revelation and intimation, no less intense for me than for Tuan Jim. It began with the lesson of learning how pursuit of my own craft and this experience of the sea, with a thrust of my own to the East, was also a search for my own truth. That alone is why I have had to record it at such length. Even so, it is very much an abbreviation, which needs perhaps the authentication of the fact that Conrad had prepared me so thoroughly for the experience that I did not need Mori's charts for verification of my conclusion.

All the same, I was compelled, for my own satisfaction, to go on the bridge, enter the little chart-room where, despite the spectacular rainbow crepuscule aglow without, some electric light was already focused like a dentist's torch on the map on which the officer on duty had marked our course. In answer to my request, he immediately made a point with the pencil of his callipers on the line drawn there when the sun was shot at noon. We were, I saw, as we whispered like conspirators against the night over the yellow chart, off the bunkering port of the island of Sabang, directly south and at a right angle to our course, and thus closing in on the western-most point of the island continent of Sumatra, where Lord Jim had found himself at last and was free to close his account on earth.

Though composed entirely of great imponderables such as these, my mood as I went towards my bath, which daily acquired more of its Japanese meaning to me, and dinner in our tiny saloon, was as if bowed down with great substantial weight. I was unusually silent. Happily no one but my teacher seemed to notice it. As I watched William and Mori, the sardonic and needle-sharp chief engineer and the talkative radio officer, whose hobby was Kant and Hegel, I was amazed how these intangibles that meant so much to me passed by them unnoticed. It was as if, through their functions, their imaginations were craft already fully loaded without room for more, like the *Canada Maru*. Even William seemed to have taken on cargo enough not to want more, while I in comparison was a kind of tramp travelling light, looking for trade of spirit and daily finding something worthwhile to stow in the hold of an imagination which had set out only with some ballast of history. Much as I had already found, little was comparable to what I took on board in that sunset hour. Hence there was inevitably much more of the same to follow in a progression of an almost logical series of growing intensity, for the following afternoon found us well inside the entrance of the strait to Malacca and receiving the first glimpse of the forbidden earth itself.

It was little more than a broad band of ink on the starboard horizon, with the highlands of the interior of Sumatra and the violent land of the Atjeh people of fire, hidden in cloud. As we were now inside that world ourselves, it rained heavily upon us, with drops like glass marbles bouncing upon the deck and darkening the early afternoon. There was a gloom deep enough to transform into a superb purple the lightning that stabbed fiercely and incessantly at everything around us, like the dagger the Malays call a Kriss, an onomatopoeic name for it, like a flame of the fire used to forge it. This purple, I was to discover and rediscover in peace and war, was the speciality of the equinoxial hours of all those days of the thunder and lightning of that world. But as the rain lifted and withdrew to the mountains, I had my first vision, with the help of Mori's glasses, of the immense flood of a jungle of high entangled trees and pagan palm rolling down to the edge of a violet sea. Plomer, who also looked at it all briefly through the same glasses, was obviously made uneasy by it. As was often his way to the end of his days, when caught out by emotions not welcome in his reckoning, he deflated it with a joke which made us both laugh.

"It is the first time, Lorenzo," he exclaimed, "that I have seen earth with pubic wood."

But on me it made an impression that caused me to remember it in a poem a generation later, which began:

> But sometimes still a dream like flame
> Burns through my country calm,
> And desires without shape or name
> Haunt me with vision of jungle and palm.

Finally, as the scorched earth of Sumatra was relieved of the pain of so much sun and the air cooled down by rain, a light breeze as of a natural thanksgiving came to us from that densely fringed shore. It filled the atmosphere on the deck by our cabin, already translucent against the fall of dark, with a complex of smells of immense swamps of mangrove and rotting vegetation, as well as the refreshed roots and leaves of profound trees and giant fern. But over all those there was a subtle presidence of bark spiced, as it were, with nutmeg, cloves and cinnamon, which reduced my anticipation of the future to proportion and restored the measure of the special incentives of history that had placed my people in a position to enable me to continue their pioneering into the final dimension presupposed in the original endeavour of a great European reawakening.

The next morning provided us with a more disturbing glimpse of what lay beyond the portals of Conrad's world of which William, travelling in a capsule of his own craft and working with Mori as usual until noon, was unaware. We were sailing now with not only Sumatra in view but the first of many small and jewelled islands. A morning still without cloud gave them a sparkling clarity in a diamond sea. I had read that, in terms of geological time, they, Sumatra, Malaysia, Java, the Celebes, Ceram, Timor and the thousand and one islands providing the long hidden universe of the peoples of Indonesia and South-East Asia with a causeway of previous stepping-stones towards Ambon and the Sprattley Islands on the rim of the Pacific, were the newly born of the earth. But even so, I was not prepared for them to look so young to an eye like mine, schooled in that oldest of antique faces of an Africa dark with the density of its remote origin. They had a gloss as of a new-born calf upon them. Their outlines were precise and clear, unworn and uncreased by time. Their vegetation was full of the thrust and eagerness and trust of a youth in love with growth, and still out of range of the deception and disillusion of time.

Yet long before noon this appearance of innocence was in trouble with the conflagration of day which by the hour made their thin-skinned earth look more and more vulnerable under the infinitely experienced and martinet sun. They, and the blinding sheets of air wrapped around them, began to tremble violently, so that they and their reflections and even the sea were shaking from one distorted outline into another, as in a universal hall of mirrors of deliberate hallucination. More disturbing, when I left my teacher to go on deck well before noon, was the distinct impression that this shivering and trembling of the view was set to the rhythm that came not just from the sun but also from earth so fresh from its smithy that its fires had not yet been drawn. It was as if those forces of heat were so actively at work on unfinished business underneath delicate clay that they could still shake all above it either with earthquake or eruption of unused energies and liquid rock through volcanoes of an incurable insomnia and numbers not equalled in the world.

In Java alone, just behind that melting glass of our window on the day, I knew from my reading, there were lined up for unpredictable action against the established shape and order of the earth, great un-sleeping volcanoes on the scale of one almost every thirty miles from west to east of a narrow axis some seven hundred miles long and barely eighty miles wide. Their presence and their impatient energies, like some colossal internal combusion generator idling below so as to be always prepared for another foray into the upperworld, were there by implication in the feverish beat and urgent throbbing of that overheated earth and sky. It struck me as the strangest and most significant of paradoxes that such beauty of scene, abundance of creation in plants and trees, rivers and mountains, and those formidable materialisations of cloud designed for their relief and growth when sun and flame became unbearable, should have concealed in the dark deeps underneath so thin a skin, a heart of such violence. I wondered how much of this accounted for the passion, the fever and unrest encountered in Conrad's reading of that vibrating world. I wondered even more because of a remnant of memory from my childhood.

We had, as one of the oldest racial ingredients in the rich mixture of peoples that composed my native community, Malays whose ancestors came from the centre of the world we had entered. They had told me that their name 'Malay' was derived from a word 'Malu' meaning 'gentle', implying that gentleness was their highest social value. The further we went and the more the light and view vibrated, the more this concept troubled me. We were by then in the narrows of the strait of Malacca, and the proliferation of ships making for that bottle-neck of burning glass of the tight pass into the Gulf of Siam, the Yellow Sea, the waters of coral and atolls of palm of the Philippines and Indonesia and so on to the Pacific and the roaring forties of the South Atlantic, was so marked and demanded so much attention from Mori and his officers on the bridge that William labelled it 'Piccadilly Circus'. It was profoundly impressive because it was the time when ocean-going traffic was about to achieve its climax before a fatal post-war decline, and I found high drama in the way Mori and indeed all those ships of many colours and flags that made a patchwork quilt of the day, avoided collision and argument and went in orderly procession to their berths in the harbour of Singapore.

It was even more impressive when one recalled that it was little more than four generations since the young Stamford Raffles, when barely as old as William, had laid the certain foundations in a great swamp of that busy harbour city and gone on from there during the Napoleonic War to become the most imaginative ruler, not excluding comparison with Dutch, Japanese, Hindu and Indonesian predecessors, that Sumatra and Java had known. Recalling all this from the bridge of the *Canada Maru*, as we wove and interwove our delicate thread of the only safe course through the complex skein of the massing ships coming and going, and watching the land shaking off its distortion in the mirror of the day, all these thoughts, feelings and considerations of history provoked by the thought of Raffles became rather disturbing. I asked myself repeatedly whether I was being subjected to a genuine premonition of the future or just a keener awareness than usual of this induction of history which insisted at all sorts of moments on inflicting a heightened assumption of individual responsibility towards it. Or was it just a hangover of the intoxication of spirit produced in my imagination by reading and re-reading Conrad at my most impressionable age?

I did not then and I do not yet know the true answer. Out of an instinctive respect that I have always felt obliged to observe for the natural frontiers of human awareness, I have never tried to push speculation beyond. But this is not to say I was not acutely conscious of how the natural walls of awareness for me had never been opaque but strangely transparent, as if bombarded by some X-ray light or cathode lamp of a mystery of flame from beyond. Nor could I ever ignore those strange events, like the working of chance and affects which the long arm of coincidence often lobbed over those oddly luminous walls, and which could not be left out of any honest evaluation of reality. All – I can say out of respect for the lucidity that we all strive after, and for the clarity of ignored feelings and other intangibles as much as of intellect, and those inexpressible emotions which the voyage and daily witness of the sea produced in me with increasing intensity from there on – all were proved fifteen years later to have been, among other things, a decisive overture to some of the most fateful years of my own time and life. Intuition is, I believe, a natural capacity in all living things to see around the corners of the future. Could I then have had some dim foreknowledge that I myself was to play a principal role in what was to be a repeat performance in contemporary dress of the piece of high history enacted at the beginning of the nineteenth century round Raffles in that theatre of the world into which we were sailing? I can only record the facts of that strange feeling of unease, and let the story speak for them.

Fortunately, I was brought out of this cumulus of personal preoccupations by the brief encounter that William and I had with the people of Malaya when we landed. Mori had been summoned to urgent business with his agents, and we had the rare experience of encountering alone a world foreign to us both. We were excited and uplifted by the rich mixture of races that mingled in abundance as great as that of the jungle which surrounded them and whose presence was always there by subtle implication as a kind of *éminence verte* rather than *grise*. We noticed even then how the Chinese tended to predominate and indeed dominate the scene with their inspired pragmatism, genius for improvisation and instinctive doctrine of work; how the gentle and native Malays moved with a dignified, decorative and submissive ineffectuality among them. We paid a ritual call, almost as if out of a certain *politesse d'histoire*, on Raffles Hotel and were dismayed and fearful for the future from what we saw there of self-indulgence, excess of prosperity as well as fat on the Europeans of all races who congregated there. Appalled, we watched them, laughing and drinking there at high noon, in a dangerous measure, on cool verandahs and in darkened halls, humming with blades of whirling fans that were part of their highly organised systems of excluding the steaming world without seasons and the all-powerful sun which surrounded them. The contrast with what we saw of the spare Chinese, slight Malays and ascetic Tamils, competing for survival in those narrow and overcrowded streets, was too poignant and stark to be ignored. But it was not until we drove across a causeway destined to be a tragic setting in Singapore's and my own future story, and penetrated for some twenty miles into the jungles of Johore, still an authentic part of the essential Malaya, that all those associations with 'Malu' and 'Malay' from childhood came disturbingly into focus.

It happened just where we ended our forward journey by a roadside clearing in the jungle of trees thrusting upward with such a passion that it was almost as if one could hear the sap from roots which mined the earth deep down for resin, drumming behind aromatic bark on their way up to feed those lofty pagodas of leaves gleaming in the sun above us. There, under an open stall thatched lightly with branches of palm, we bought soesaties from a slight Malay vendor and an even slighter and more delicate wife. The soesaties – spiced, sliced and grilled meat on a thin bamboo skewer – were still like those that their countrymen introduced into the national cooking of the Cape in the seventeenth century. While we ate them, dressed with a sauce that the Malays of my youth and their kinsmen there called a sambal (made of ground nuts and the most fiery of little red peppers known as Chubby Rawat), we drank cups of delicate Chinese tea, uncontaminated with milk and sugar, in a green shade that emphasized the ferocity of light and the vibration of the suppressed violence of earth under a day breaking in waves of flames on the Johore highway. Suddenly this fragment of memory I mentioned assailed me. A great paradox, which had been stirring at a submarine level of my mind when I was last standing on the ship's bridge, surfaced and over-rode all other impressions in a rush of overwhelming questions.

How did such greenness and fire, peace of scene, freedom of growth and tyranny of day, such gentleness of the Malay and eruptions of violence in their national character which were well-known and amply recorded, produce such a bright and vibrant appearance of harmony? The role of the physical scene in the sense of paradox was not beyond my understanding, in so far as paradox can be held capable of rational apprehension. After all, everything, one had been taught through science at school, had its opposite. The most obvious example, which intruded as of its own accord just then, came from a rumble of thunder out of a temple of cloud just visible above the jungle, where the blazing road led up into the sky and reminded me that the lightning which had produced it was composed of negative and positive charges of electricity. The principle of a positive and negative, of opposite poles in everything, was not in question. What was alarming were those implications of a disproportion and imbalance of natural energies to an extent of suggesting a fatal hubris in this immense portion of the physical world itself. Was this imbalance not inevitably bound to have tragic consequences again and again for earth, plant, man and sea, as it had done, for instance, when one of those volcanoes, the infamous Krakatoa, just below the horizon behind us, had within memory of Malays still alive, demonstrated catastrophically for all the world to witness? But even this alarm was capable of dissolution in an acceptance of the paradoxical heart of truth and a capacity in man to transcend opposites, as the lightning that made the voice of thunder and shook rain from these monumental clouds, for whom even the vast blue expanse of the immense skies of those latitudes hardly had room enough, transcended the negative and positive in electricity. No! The question that truly concerned me then, as it does today, was: did the clash of opposites so violently built into the earth around us determine accordingly the nature of the people native to it? Was some subtle and profound interdependence of the character of the earth and spirit of man such that he too unavoidably had to have earthquakes of soul and volcanic eruptions of vision? If not, where, for instance, arose the phenomenon of Amok, common among so gentle a people? What produced the darkening of the eye, the "mata-kelap", as the same phenomenon is called among their kinsmen in Java and Sumatra? Was it because of a disproportion of gentleness, of excesses of a graceful conformity and acceptance of the social conventions and decorous traditions of Malaysia, all constituted a hubris so great that it denied the individual the right to be his own awkward self? If that were so, might the result then not be that some long-rejected part uniquely of the individual would grow great with anger against the community which imposed so rigorous a self-denial, until it erupted in his spirit and compelled him without warning to himself and others to draw his kriss and kill every person within reach, even those most loved because they upheld so tyrannical a social system? Was this the cause of that shattering cry of 'Amok' which still went up on the calmest of evenings freed of fear of the sun? Was this a phenomenon due to the violences of the earth that nourished them, or to a trangression of natural law which decrees that all virtue and indeed even the greatest good had proportions and limits that had to be observed? Or was it, as seemed perhaps slightly more probable, a compound of both, an imbalance of human spirit encouraged by nourishment of the disproportion built in the physical world? I put it like this because already the awareness of the importance of the earth of Africa to my own state of being predisposed me to a belief that, however indescribable, there was between man and his native earth an umbilical cord of life-giving imponderables that no circumstance could cut, not even long years of exile. Most disturbing of all finally was the fear derived from my own reading of history, that the phenomenon of Amok, the "mata-kelap" or darkening of the eye confined to individuals in that world around us then, if not understood, could affect whole nations and civilizations, as I suspected already had happened in the First World War, so recently behind us.

I dwell on this abbreviation of the intangibles that were coming like thunderclouds over the horizon of my mind, as I did in the resumé of the quickening awareness produced in me by Conrad, because this was the real voyage on which the *Canada Maru* was taking me. I realised this, that Johore noon-day, with a start that was a stab of an awakening heart and mind and that made travelling into a new external world in the *Canada Maru* mean so much to me because it was helping me to go thereby into a great undiscovered country of my own imagination, which I could not have entered any other way. For William, I knew the voyage was an interruption and important only as a means of getting from Africa to Japan. Japan would mend the lines of communication for him again, but the journey between it and the severed significance of Africa had no special meaning for him. For me, however, the journey in between was even more important than our point of departure and arrival. The eruption of an immense potential of new meaning in life caused thereby was so great in my mind that I remained silent on our way back to the ship. For once William's keenness of observation and great gift of wit did not really reach me. I hardly heard him or looked further around me because the sound of music of gratitude to Mori who had made all this possible filled my senses.

So we came to sail on the last long segment of our voyage to Japan, out of the crowded harbour on the morning of a day that started brighter and clearer, if possible, than any before. It was as if the gates of the distance between us and Japan had been thrown open in the night and a broad highway of sparkling sea cleared of traffic to enable the *Canada Maru* to speed home. There was no doubt that the heart in her engine-room seemed to be beating faster, just as a distinct excitement quickened the pulses of everyone in the ship. Even the sun which had dominated our days ever since we left Mombasa was given its clearance. We were no longer sailing straight into a sunrise as we had done for weeks but heading almost north-north-east, leaving the sun to bow out of our play well to our starboard. The sky and the sea and a breeze fresh with a new idiom of temperance and moderation came to meet us, and yet we had not done with that world of Conrad for all our unknowing. Realization of how far the unpredictable rule of that world of his stretched came to us, though, in the most conclusive and typical Conrad fashion, as almost another particularly pointed working of the laws of chance and suspense which had so deep and disturbing a hold on him.

Many days after Singapore, Johore, Sumatra and Malacca were behind us, like things seen in a trance of fever from which we had been delivered, and we could just make out through Mori's glasses an outline of the nightmare cliffs that rise in places to thousands of feet sheer above the Pacific on the West Coast of Formosa, the unexpected happened as if to emphasize that a world of violence, even natural violence, has violent frontiers too, and that the great transitional demarcations between the inexorable seasons of life and time, and the static and kinetic in the dynamics of change also, cannot be accomplished without the help of great storms.

I had already thought the morning strange before then. There was something about it I felt I ought to recognize and that the something was somehow connected with Thor Kaspersen who insisted on continually coming to my mind after weeks of oblivion. Mori and William, in a spirit of increasing haste to finish the translation of *Turbott Wolfe*, had hardly got to work in the shade by our cabin when I felt a slight shudder in the *Canada Maru* underneath me and then a movement of a gentle rise and dip which steadily became more pronounced. Startled, I looked at the sea and the sky, saw that the water was losing its cobalt blue and that the air suddenly was sullen and veiled and beginning to weigh on me. Then I realized it was an exact reproduction of the spawning condition of atmosphere which drove Thor Kaspersen to aim *Larsen II* straight for the vortex of a cyclone spinning out of the Mozambique channel in the belief that it would provoke his heroic sperm whale to battle with his physical manifestation of evil in the shape of titanic squid, while the rest of the whaling fleet turned sharply about to make for safe harbour in Port Natal as fast as they could.

At the same moment a young apprentice came down the ladder from the bridge, a folded paper in his hand, and presented himself with great circumspection before Mori. I was struck, as so many times before, by how behaviour, in this floating world of Japan, was not really governed by concepts of quantity and preconceptions of smallness. There was an over-riding master value of quality straddling great and small alike, often raising the tiny and significant to a dignified height and lowering the seemingly large and imposing to a more modest level. The consequence, it seemed to me, was that all possessed an equality of dignity in the value of this world. As a result, though only the bearer of a message, the young apprentice, with a complexion as fresh as that of a girl from the mountains around Kyoto, approached Mori and presented his rice-paper missive as if it were an Imperial Summons. Mori unfolded it as if he had all the time in the world, read and re-read it. Had it not been for a certain tightening of the muscles of his determined and clear-cut jaw and a quiver of nerves at their apex near his high cheek-bones, which I had come to know, I would not have had an inkling that it was anything more than a routine message, particularly as he dismissed the apprentice not curtly so much as unambiguously without an answer to take back with him. He then resumed working with William as if nothing of importance had occurred. But to me it was as if I was at the beginning of a Noh play and was from then on to be drawn slowly deeper into some intensely dramatic symbolism of unknown meaning. Not believing that I had seen the end of the intrusion, I could not keep my eyes off Mori from where I stood at the boat rail some ten paces away. The air of hidden drama there was as thick as the yellowing veil being drawn over the sun. After some five minutes when, I guessed, they had reached a convenient break in translation, I heard Mori ask William to excuse him for a while. He did this in a way no more portentous than several times before, so that William went on working apparently unaware of the change in sea and sky. But to me, observing Mori's trim back disappearing not onto the bridge but to the Radio Officer's quarters, he appeared tense, moving like a person inwardly in great haste. He was not there long, reappeared and did not come below to join William but vanished through the entrance onto the bridge. My sense of impending drama was heightening because my apprehensions of the unusual appeared confirmed.

Mori was gone for some fifteen minutes before he rejoined William with an apology and, to my eye, more relaxed, as always when the relevant action a problem demanded had been taken. From there they went on to complete their daily routine, so that I might have revised my interpretation of the incident had I not soon afterwards noticed that the *Canada Maru* was coming slowly about to steady on a course almost due east again, bringing a sun now yellow and blurred almost to rest on the foremast's lookout ledge and well away from the parallel to Formosa that we had briefly followed that morning. I had no doubt Mori had taken evasive action, of which only my association with Thor Kaspersen prompted an explanation. Since my teacher was engaged in completing endless manifests and accounts for our arrival in Osaka, I was free to wander around and enlarge my watch of sea and sky.

I had hardly begun to do so when I met the Chief Engineer stepping abruptly out of his cabin. For the first time he was wearing engine-room overalls, as clean, white and starched as his naval uniform. He greeted me formally but without the sardonic facetiousness and insertions of the needle-sharp mind he used endlessly and rather diabolically against my vulnerable teacher. Yet he was all that my nickname for him, 'Science', implied. Had I heard the news, he asked me with a certain grave factuality? They had not long since received a radio-warning that a typhoon was heading perhaps in the *Canada Maru*'s direction. He said 'perhaps' because the typhoon was travelling with a 'little bend' and the trouble was that no one knew which way precisely the 'bend was bent'. And then his inner nature reasserted itself briefly and he joked: "Trouble also with typhoons: never know own minds, always changing bends, like women changing minds. Perhaps that is why Yankees always give them names of women?"

With an air of a joke successfully delivered, he went through the entrance to the ladder down into the engine-room obviously to make certain everyone was forewarned and prepared. On the well-deck between bridge and fo'c's'les, I went on to see that the first officer had made a most unusual appearance there with the bos'n and half a dozen deck hands. From then on until noon I watched them making certain that all the hatches were properly battoned down; extra wedges driven into all their clamping irons; everything loose stowed away or tightly secured with additional rope; forward anchors and chains strengthened and amply lashed with the best manilla cable; the ladders leading to the lookout post on the foremast tested, and indeed nothing however small left unattended but reappraised and if necessary reinforced, in terms of the onslaught that a typhoon at its most extreme could exploit to split apart the *Canada Maru*'s defences. Watching all this take place, fore and aft, in so orderly and largely unspoken a fashion, I was impressed by an implication of something more than the foresight and proper precaution of seamanship I would have summed it up as being in a European ship. I did not know at first how to fashion this feeling in words until suddenly it occurred to me with an unusual clarity. I was observing another manifestation of the subtle and comprehensive working of the complex concept of Li which my teacher had described to me and was constantly referring to because it was so much larger and deeper than rational and verbal definition. What the officers and crew were carrying out was not merely an appropriate exercise of seamanship but an observance also of Li at its most profound level: an observance of good manners in the sense that manners are good when appropriate and that man had to preserve his manners and be on his best behaviour especially against the anger of nature and its storm of wind and water, not just as a matter of survival but the more urgent one of bringing himself, his ship and storm into harmonious relationship again with the law of the universe. The dignity and the rhythm this realization induced for me in the behaviour of all from that moment on was almost like a prayer in action.

All this was confirmed by some brief words with my teacher just before we went down for our midday meal. He joined me briefly, his large brown eyes strained with overwork and yet inwardly illumined as with a refreshed emotion. Did I remember, he asked me, how at the beginning of our voyage he had been compelled to correct me when I took 'Maru' to be the Japanese for ship? He had tried to explain that it was a kind of suffix of the most ancient origin, and was intended to remind man that certain material things of his own fashioning were more than matter and carried a charge of spirit and symbolic meaning all on their own. In this way some of Japan's greatest knights and warriors had given names to their swords with this suffix of 'Maru' added, to remind them that their swords too were instruments of spirit which should be accordingly used. This practice was most profound and appropriate applied to ships because the ship was perhaps the most evocative symbol of spirit of all and deserving of the 'Maru' after its name. He hoped we would not be put through the ultimate test of having to go through the eye of the looming storm but, if we did, would I remember all the feelings and definitions associated with 'Maru' because he had a feeling I would have the best possible definition of the word since I would then see 'Maru' in action?

I was much moved because all this was so in keeping with the spirit which had made me designate him as the plenipotentiary of religion in the ship, and I could not help discussing with him the matter of the 'unpredictable bend' in the nature of typhoons which the Chief Engineer had mentioned to me. He suggested with unusual quietness that the voice of nature, as uttered through the spirit of ancestors so much closer and dependent on it in their day, should not be ignored in these matters. Their voice, however remote, never failed to help regulate one's behaviour even when asked to bear the unbearable. Then he added an afterthought, intent on removing any hint of rebuke to the Chief Engineer. It was not surprising, he said, that a new generation of Japanese raised in contemporary and especially Western scientific ways, should have forgotten the ancestors, their 'kami' and all that which was above them, their trees and mountains, and so their links with the gods. If they had not, they would have remembered that, in the ancient Yamato spirit and the first stirrings of the Japanese soul, one of the greatest god-figures was born of the breath of Izanagi – 'the inviting male God', and became a personification of storm, known as 'Susa-nowo', 'the impetuous male'. There was a vast area of life, he concluded, where the direction of such thinking was still a better compass than could be designed by the Chief Engineer-sans of the world. And there it was again: we were back at the heart of the Li which held that, in regulating one's behaviour, one helped to regulate great universals. I would have liked to ask him more; but we were joined just then by William whose presence gave our exchanges a different turn.

I asked after William's morning and whether he had been told about the probabilities of an encounter with a typhoon of unknown propensities and temperament.

"It is most extraordinary, Lorenzo," he replied, truly amazed. "Despite the interruption caused by your typhoon, it was the best morning's work we have yet done and sent us speeding to the end. Mori is really an incredible creature, doing all that with all this on his mind." He waved an expressive hand at the rising sea and yellowing sky.

"So you know it all then?" I interrupted.

"Yes I do, and I feel it too," he answered with a laugh I thought brave. "But I did not know until right at the end. That unbelievable creature did not let on until just now: then he excused himself for having been apparently so rude as to break off in the middle and told me laughingly, as if it were a small matter of no consequence, even absurd in comparison with what we were doing, that he had to attend to the 'little', yes he called it 'little' problem, of a typhoon."

With that we went down to lunch for the first time without Mori. The fiddles were already set, although they were not yet strictly necessary in terms of the movement of the *Canada Maru*. It was another illustration of how thorough had been this transition of behaviour everywhere in the ship, including the galley, into an attitude which could not fail to honour the challenge of the impending storm.

After a shorter and less conversational lunch than usual, the swell on the sea had risen so that William was forced to take to his bed. Despite the most courageous efforts not to miss meals in the saloon, he had to keep to our cabin for some two days. For him the experience was decisive in one way. During those days of suffering acutely in a cabin shut as a fortress against the storm, I believe the resolve was born not to journey from Japan ever again by sea and, when he ultimately left it, do so by Trans-Siberian railway over land, with travel by sea confined to a channel crossing to Vladivostock. As a result I went through prelude and storm very much on my own. To this day I have been grateful to the circumstances which arranged it so.

I had already experienced one of the greatest storms on record in the small naval sloop *Protea* off those fearful waters near Cape Agullus. But there I was never alone; moreover, I was made to feel part of the crew, sharing fully in their stresses and above all the one profound moment of dark and purest fear when that great black hole in the sea appeared to open underneath our slight bow and a range of water rose above us to erase the brown of the moaning and groaning sky with a midnight black. It was an unforgettable experience, not without its moments of revelation of incomparable beauty, and this typhoon, I felt, could be even more important since it would be my first lone and undimmed encounter with one of the greatest patterns of storm, and uniquely of Conrad's world.

At first I watched its portentous approach wandering freely on our little beat of the boat-deck backwards and forwards between port and starboard, but forbidden to descend to the well-decks and go on to the bow and stern, as I longed to do. I soon understood how right the Chief Engineer had been about the capacity of typhoons to bend unpredictably. Our change of course, timely as it might have been, was not putting us altogether out of reach of a storm which seemed to me to be still overtaking us with the stride of a giant. The swell was continuing to rise and already making mountains out of chocolate water, dressed more and more frequently with a gust-whipped mane of foam, but on the whole maintaining a steady, sombre and inexorable rise of rhythm which I knew from whaling along the fringes of cyclones with Thor Kaspersen. Soon the sky too went from yellow to sulphur and then to brown, deepening fast as the sun climbed down from its strut on the prancing foremast and seemed not to sink so much as be sucked down towards the heaving water. Then there came that strange, almost sub-sonic, finely sharpened pitch of moaning from the sky, as if in great pain. Hard on that the first real wind arrived, huffing and puffing, and then climbing steeply up the scale into the first dimensions of a gale.

The gloom at once increased, the moan rose to a shriek which became one long sustained scream mounting, falling and remounting to higher reaches of the tonal scale. The ship rose higher and plunged deeper to shudder with the water shipped. Although steering at first stern to the main storm, it was soon forced to change course north of east so as not to be pooped by the mountainous following water ascending sombrely behind us. As we changed course, I had time only to see the air producing dark flying objects; and to recognize that they were twigs and leaves torn off living trees and plants from some island earth at the centre of the storm, and watch them vanish, streaking fast to the south-west, when Gengo appeared at my side. He had to shout to make himself heard, and it was most marked how he managed to shout politely against the noise of the storm. The honourable Captain, he wanted me to know, apologized profoundly for disturbing me, but would be greatly honoured if I would be gracious enough to withdraw inside the ship. For the first time he addressed me not as Post-San, which everyone had come to call me, but with the higher degree of honorific, 'Post-Sama', a sign which, small as it was, I took as confirmation of the intensified realization of the importance of courtesy of spirit between men in the face of a storm great enough to provoke its opposite.

At first I tried refuge in our cabin. However, it was in darkness. The portholes facing the bow and sea had already been screwed down tightly before lunch and now had their metal covers clamped on as well, and curtains drawn. The door to the deck was locked and our cabin trunks were roped down. Everything loose had been stowed away; all by our infinitely thoughtful steward. I saw all this briefly as I switched on the electric light, but a sign of distress from William made me turn it off and leave. I could do so promptly through the doorway leading to the bathroom and short passage to Mori's cabin, as well as the inner companionway to our little saloon. I then tried the saloon. It was empty and had bolted, barred and shut out the external world even more firmly than our cabin. Yet I was not unobserved. Almost immediately our steward appeared and presented me with a large bowl of steaming beef-tea. As he did so he smiled in a way which illuminated his homely face, transforming it as well as exposing his teeth of gold. It has always amazed me how moments of crisis and extreme stress and distress can be almost mathematically measured by the importance attached to trifles so as to reinforce through attention to externals the conscious qualities of will and measure needed against the upheavals produced in submarine levels of a subjective spirit. The teeth of gold and the dentistry involved in their insertion became an image of the worth of the norms of the human round, and reassurance that the rule of experience, scrupulously honoured as I had seen it observed all day, would assert itself over the abnormal in the storm and re-emerge no matter what the odds against it for the business of life as usual. In addition, his smile was so full of natural and straightforward solicitude and contentment in his order of service that I smiled gratefully back at him. But it was significant how often I was to think, at all sorts of odd moments without predeliberation, of the gold of his smile as if it were the metal to be transfigured out of lead in the alchemy of the storm.

I sat there so absorbed in paying attention with all my senses exclusively to the effects of the storm that the seconds and minutes were forgiving and the hours sped. I listened to the way in which the noise of the storm was countered by the response of the *Canada Maru*. It began to creak all over and at times, although I felt the metaphor disloyal, to squeak like the mouse trapped between the sharp claws of the storm of which my whaler off Port Natal had so often reminded me. Yet the dismay that might have been aroused by the sound was redeemed, for me, by the great shudders that shook the ship and ran right through from bow to stern before driving home with the thuds of thunder of water breaking on our fore-deck. They struck me more and more as an unperturbed bracing of muscles of a confident athlete capable of soaring out of the deepening troughs and clearing even higher hurdles of waves.

All these affects were joined by the noise of the wind which made a harp of the ship as it had already made tribal drums out of the pounding waves, and drawn music of cosmic form from the taut sky. After a while I thought I was beginning to discern a pattern in all those wild, barbaric sounds and violence of movement that led to the certainty that even so great a storm was not beyond the law. There came moments of resolution when all those sounds and movements combined in one superlative achievement of singleness and harmony, and a vast calm was incorporated into the heart of the typhoon whose voice would then rise high into the turbulent universe outside like a song of truth and the freedom from falsehood that was fear; proclaiming an abiding theme in the opera of all living things on earth. I was so moved and uplifted by all this that I hardly slept that night so as not to miss a note, at times clinging to my bunk to prevent myself from being thrown from it because of the greater extremes of movement of the ship.

Most impressive too was the simplicity imposed on us all by the necessities of the storm. It was, for me, another profound aspect of the experience that the storm seemed to demonstrate how, despite the complications of circumstances and complexity of demands of life on earth, truth, rooted in sheer necessity, united and simplified all in the end. Even our food became of the simplest. All set rules for eating were abolished. I was alone most the the time, rarely saw any of the officers who remained near to their stations in touch with the bridge, and when I did meet anyone over a dish of rice and what became for me the nectar of this strange feast of the storm, bowls of boiling hot beef-tea, our conversation too was of the most economical. No one referred to the typhoon except indirectly by asking after William's and my own well-being, and expressions of delight that I was up and about, by a polite exclamation, hissed between their teeth and invariably followed with messages of solicitude for William. It was not until nearly noon of the day after the storm overtook us that Gengo apppeared in the saloon with an invitation from Mori, whom I had not seen yet, to join him on the bridge. I accepted at once, and with gratitude, for I was full of longing to have confirmation from my eyes of this invasion of elemental beauty which presided over the rest of my senses. Already it had seemed to me for an hour or so that the storm, although risen to yet a new height, had maintained itself there as if at its own summit. But when I came onto the bridge I was not nearly so certain.

I was met by Mori who had been on the bridge all night. A look as of inspired command on his clear-cut features erased any suggestion of strain and fatigue to which he must have been subjected. He welcomed me with a warmth which I had not experienced in similar measure except at our first meetings in the beginning. He moved an immaculate hand of a polite greeting at the wild scene almost as if he owned it, and asked rhetorically what I thought of it? Before I could answer, he explained that they had been in radio contact all the time with weather stations and all ships in the vicinity, some of them in deep trouble. He was, he said calmly, reasonably certain now that we had avoided the eye of the typhoon. It was, he believed, now travelling away in an exceptionally narrow and powerful vortex, south of our course. He proposed, therefore, riding out the rest of the storm, with only enough way on the *Canada Maru*'s engines to keep her head on to the seas. How wise he was became evident immediately, even to me. Wherever I looked a sea of a substantive colour compounded of royal purple and incorruptible green, was transformed into Himalayan ranges of water with manes of snow raised along their crests by the wind, and blown away in this spectral mist of finest lace which one normally only sees streaming along the necks of the high mountains of the earth. Over all, the high pagan music, of which I have spoken, was higher and wilder than I had yet heard it.

Mori and I had to shout to be heard, but all fortunately spoke for itself so eloquently that speech was unnecessary even if I had wanted to talk. In fact, I had never felt less like speech. The beauty of that sea and its accompaniment of wind was miraculous. I had never seen anything that moved me and produced so great a need and fall-out of stillness at heart. It was, in the recollection of the tranquillity to come, a moment of almost religious confrontation and experience of an assertion of ultimate truth that was an onslaught on all that was false in life and space and time. Nothing that was not true in conception and execution, one knew, could outlive the presence of the uncompromising power and glory, implicit in the scene. It was, to use the inadequate language of the rational word, as if what I looked on was a manifestation of grand design in the universe which seemed to ennoble to the same high degree that it induced humility. I went on to stand there silent on the bridge at the back behind Mori and a quartermaster, his feet placed well apart and his hands firmly on the wooden spokes before him, like an archetype of his vocation, incessantly winding, spinning and unwinding the wheel again as if it were an instrument of fate to keep the *Canada Maru* unflinchingly face on to the seas. I stood there for hours to witness at the end what proved to be the climax of the storm in the darkening minutes of the long afternoon. I had ignored all suggestions that I might be tired and cold and needed to go below. I could not have deprived myself of any part of such beauty. It was as if I was discovering in myself this capacity of mystical participation of Conrad which I have mentioned. Indeed his story *Typhoon* was present in my mind, and the resolution, when it came, was not unlike the great crisis for his travel-stained old master of the tramp in his tale. It was brought upon us by yet another steep ascent of the waves with the suddenness that is at the disposal only of potentials of cataclysm in rare circumstances. They rose and fell with an acceleration of speed as of an avalanche from Everest itself, and were hurled at us by the wind in a succession of sevens, each seven with at least one summit higher than any that had preceded it. The quartermaster replaced his feet even wider and more firmly apart, and Mori's back seemed to straighten more and go taut as if to say: "Here it comes."

The waves by then were so high that they had ceased to break over the ship. They seemed to move in a wall which was pierced at its base by the bow but not broken apart over the poop and well-deck, to reach still intact higher and higher and nearer and nearer to the bridge. As each wave did so, the water went like a dark green shadow over the day and ship. The *Canada Maru*, after each visitation of water, seemed to recover with less and less resilience than before, and to sink lower into the midnight blue trough in between. These, even I knew from what I had been told, were the moments of greatest peril, not so much because of the weight of the water which forced the ship down but the speed and power of the wind that drew such a taut, tight and tense cover of atmosphere between one summit and another, almost like a sheet of steel, which could, as typhoons often did, lock the vessel in the trough and prevent it from riding upwards again. A wall of water, the highest we had yet met and dominating a succession of giant others I could see wherever I looked just then, came sweeping towards us like a tidal wave, as if about to push us down and do just that to the *Canada Maru*. It moved over the bow without a break in the precipice of its front and a confident hiss like that of a mythological serpent about to strike. All went green, blue, purple and black, as it hit the bridge and everything around us, before it travelled all over the rest of the ship behind us. The *Canada Maru* sank down with immense gravity of movement, slowly but it seemed endlessly until we felt locked under that cover of air, dense and heavy as lead. We seemed to hang there long on the rim of doom and shadow of a valley of a fathomless sea, unable to move and apparently incapable of mounting to meet this monumental water's successor already looming high and wide and demonically handsome ahead. Yet somehow, overwhelmed as I was with awe, I was not afraid. Far down underneath me I felt the steady heartbeat in the engine-room and a tremble all over the hull and deck that made me certain, however plausible the appearances against it, that the ship was still alive and willing. And indeed the *Canada Maru* proved before too long that it was not only willing but undismayed. It began to rise slowly, the foremast vibrating as if it were about to break under the tensions of piercing the capillary air, stretched taut between the peaks of water but not enough to bow and snap. Somehow the ship managed to rise shuddering and high enough to divide the next wave as it went like the rush of a great wind over us. The *Canada Maru* once more began to slide down and through its successors rather than sinking as before. From there on, more and more, it began to ride the prancing seas.

It was clear that the ship, the observance of the rituals demanded of it, its Captain and crew, were vindicated. The realization brought tears to my eyes; not of relief, I am certain, but because it had all been like some sort of transcendent metaphor in action of a meaning to all, however enigmatic and obscure, even in the utmost of storm. I went below then like someone leaving a theatre in which an authentic piece of life pitted against anti-life had achieved its catharsis. I saw William and sipped some more beef-tea like wine at communion. By nightfall it was clear the worst was over, and in the morning the wind had receded enough into another quarter and the sea moderated so that the *Canada Maru* could swing back onto its proper course and follow resolutely on after its breakthrough of the kind of storm which for me patrolled so appropriately this other frontier of the world of Conrad. And in the sense that all frontiers in reality are barriers, it still seems to me, as it did then, that being compelled to break through as we did was an assertion of universal design. The storm was part of the great law-abiding necessities which demand, for instance, that even the practised round of seasons cannot serve the change of one into another without storms to aid them. Certainly what was beyond speculation on the second morning after the typhoon first came to examine our credentials for crossing the frontier, was the clarity with which a new ocean and world and time was now open to us.

For the first time William sat up, was hungry and his confident self again. He was ready and eager to contain the experience behind us with characteristic laughter, as laugh he did when I told him of the final wave of absolute water hurled against us; when I remarked that the rush with which it went over us was like the swish of the skirts of the typhoon itself and that although we may have failed to keep well to its outskirts, as we intended, we were able to stay within its 'inskirts', however perilously near the centre.

"But what a how-do-you-do, Lorenzo," he exclaimed. "What an incredible how-do-you-do. I do not think I would ever again want a repeat performance."

CHAPTER FIVE

Full House

AND so we moved on northward into cooler waters with the same long, smooth, graceful and swinging stride with which the *Canada Maru* had carried us before the storm. We moved from one hemisphere to another; from a western world and darkening continent of sunset to another of islands flung at a rising sun and the light of a strange new dawn. The air became clearer and more precise by the minute. Finally, on a morning of unparalleled lucidity, it brought us through a window of crystal distance and over water of the purest blue, to see land itself in the tips of two of Japan's main islands. This tangible witness of reaching home, though it was only a darker blue pencil line on blue water against a peacock sky, produced a relief from nostalgia in everyone around us. This we could observe only with pleasure but not measure. For while William and I were left on the perimeter of their considerations and content in the natural justice of being there, there was compensation too in the freedom this gave us to deal with the approach to land, after our own fashion. And this land soon established itself firmly enough for us to recognize something of its underlying character.

It is true, we had moved into another season and back to our proper place in the orderly progression of time. This is denied to that world of feverish brilliance held in the blinding equatorial latitudes divided between Capricorn and Cancer, which are without season or apparent shift of change. Yet it was startling to me how obvious it was that the character of the earth itself remained the same. It looked as young as ever with the same sheen of the new-born upon it. The outlines of the hills came out of the water as clear-cut, and not yet worn down by erosion of time and the friction of care. Though the quality of the atmosphere around us had lost all traces of the high fever which afflicted the kindred earth, there was a constant and unmistakeable tremor in it as of a tuning-fork. It was the first warning symptom telling us that we were to experience more volcanic earth with veins flowing freely with lava, circulated by a heart of fire, and full enough of power to shake the thin skin of soil which nourished so bright and plausible a green above it. For me it was as if the recognition passed like a shadow between me and the rising sun. It still posed the overwhelming unanswered question. As I looked at these people of the world of the *Canada Maru* that I had so taken to my heart without any reservations, I sought once again for a lucid answer. Were they too deep down like their earth? All their graces, this harmony of behaviour, this calm and dignity of welcome and evenness of disposition which had made one feel as if one had arrived at the centre of the world which Baudelaire had referred to as *Là, n'est qu'ordre et beauté?* Were all these subject also to unpredictable eruptions and darkening of the human eye as among the Javanese? Even more important, were they too charged with the dark potential for irresistible collective surges of millions towards annihilation which, in the individual, was the dreaded Amok of Malaysia? The answers I sought were not available. I was too immature and still too full of the hope and trust of youth. But in so far as portents serve as pointers to ultimate diagnosis and answer, I was happily not too young to observe, and even to dare to interpret, some of them. So I dressed and prepared myself accordingly in an uninhibited excitement for our landing at the port of Moji in the early afternoon.

There we did not go alongside but dropped anchor to the sound of finality of arrival which the process always excites in me. We came to swing around our captain's 'hook', as the sailors called it, in a deep blue water with a sheen of the levelling afternoon sun upon it. The air was cool rather than cold. But already it was far enough away from summer to show it was close to another long session with winter. The clarity was impressive, and, just across a twist of channel opposite Moji, the port of Shimonoseki which gives its name to the narrow but historically fateful strait into the inland sea of Japan, showed up as clearly as Moji itself. Apart from the quays, the cranes and mechanical aids to contemporary ports and railheads, there was obviously little of the modern world in their appearance. It was still barely two generations since the guns of Commander Perry's ships had compelled Japan to open itself to commerce, and to traffic with the outside world from which it had put itself in rigorous Purdah since the seventeenth century. The buildings were still of a traditional modesty, unexpectedly compact and close to the earth, almost as if they had grown out of it rather than been imposed upon it. This impression, through glasses, showed them to be built of wood and windowed and flanked with thick translucent paper. They introduced me to the clear and ultimately unchanging feeling that in Japan all builders were not architects so much as carpenters of genius.

Partnership with nature in all forms produced a feeling of fundamental harmony which was proclaimed to the sky by silhouettes of pines, green-black against the blue. Many of the greatest of these were isolated, twisted and aslant with torment, as if shaped by natural design for the lone, inner endurance through which they had to overcome storm and wind. In this way they became the instant image of love of life and courage, and so remain in Japanese imagination in the role of supporters of heaven that they represent in the symbols of Japanese writing. However, we had little time to do more than take in an over-simplified impression of the land because we were quickly reproved by finding that Mori, characteristically, had made the most meticulous arrangements for our reception.

We were summoned to the saloon for a formal introduction to a special representative sent from the owner's headquarters at Osaka to meet us. There was also the head of the Africa section of Japan's Foreign Office. I do not, alas, remember the young diplomat's name. I do remember his courtesy, fastidious concern to make us welcome and to look into our needs. As the day and evening progressed, his increasing warmth glowed through the conscientiously cultivated diplomatic front. Yet it was as absurd as it was disproportionate and a warning as to how conservative, even at an early age, the human senses are, and how easily deceived by appearances. William and I were amused by his dress. For a while we avoided each other's glances in case it released the laughter already rebelliously under restraint. He had bought plus-fours of an exaggerated cut which were the diplomatic *de rigueur* for the occasion. As a result, I was more quickly impressed by the Osaka Shosen's Mr Tajima. It was not only because he spoke such fluent and idiomatic English but also because of an unpretentious, curious, well-informed, lively and cultivated spirit. He was also obviously on the best of terms with Mori, teasing him with warnings of the most severe censure from on high for once again having breached the norms prescribed for Osaka-Shosen captains. This time, he warned with mock portentousness, his Captain-san had gone too far once again by so impetuously inviting 'illustrious' personages to Japan without adequate preparation. But, as far as diplomacy was concerned, Mr Tajima saw himself already as the natural partner in the role of guardian to Plomer and myself.

So Mori hastened to discharge the many duties imposed by arrival in his first port of call in Japan. And for once he relegated most of his normal functions to his first officer, his 'mate-san', so as to be free as soon as possible to join us again. Mr Tajima and our Foreign Office shadow (looking more like a golf-professional than a diplomat) took us ashore in the company's launch.

All this was accomplished with such dispatch that when we came to stand by a waiting car, having landed without customs or immigration formalities, Mr Tajima asked us if we would like to go to our hotel first for rest and refreshment or see something of Moji. William and I had in fact brought nothing with us for a night ashore, so we both simultaneously, and without question, asked to see the little town first. This answer obviously pleased our guardians. Mr Tajima dismissed the car imperiously and hastened to explain why he was so happy at our decision. The people of Moji were celebrating an important festival. I do not remember which, because I have enjoyed so many of them in Japan. I know in fact no other country to be so rich and blessed with festivals. My teacher, for instance, had a book which listed some four hundred and fifty festivals without counting ones of purely local or minor concern. I only know that this festival exhilarated and uplifted me to a degree which not even the greatest among them, like the smouldering Gion festival in Kyoto with perhaps more of the Roman quality of gravitas to it than any other and which is of a sombre beauty as vivid and evocative as a dream, ever achieved.

All that my teacher had prepared me for was suddenly alive and in action around me. Coming as I did from a Calvinist country where the austerity was maintained to fanatical degrees, perhaps because of an unacknowledged doubt and sense of insecurity through contact with a vast continent of natural forces and a predominantly primitive human context, I realised at once how 'festival hungry' I had been from birth. The festival and ritual experienced in the *Canada Maru*, such as the ship's own improvised salute to a harvest moon, had continued to haunt me; but that had been a mere hors d'oeuvres to awaken in me a longing for more. The longing had been fed since birth on the bleak idiom of a puritanical education which had, for centuries, declared a war on ritual, instinct, image and symbol. Happily I had been preserved from total starvation because of my love for the indigenous people with their stories, legends, myths, and dedication to natural meaning. As a result, I was caught up in the tumult of festival of a small provincial harbour town glowing with a special autumn light of its own. I was not just an observer but a participant, both in ritual and spirit, to the extent of becoming a subject both of it and also of startling self-revelation. It began with an indescribable sadness for my native country which I had left so abruptly with an eagerness that might seem reprehensible to its exacting Calvinistic prescriptions. I uncovered a natural envy in myself of people whose spirit could be so full and overflowing with a sense of occasion that they had to express it continually in festivals of the kind in which we were sharing. I thought them, at that moment, to be fortunate and rich, where we were poor. They seemed blessed in such abundant apprehensions of the need for continual balance and constant renewal of harmony between human beings and the natural and universal forces that dominate circumstances and direction on earth. I marvelled at such vivid manifestations of instinct which appeared to run through the unknown and darkness that beset us all, like a gleaming river in the national disposition and carried them on through life predisposed to know that, as they were within themselves and retained their intuitive reverence for nature and all the great imponderables of the universe, so would they attract to themselves what was appropriate for the quality of their being. A surge of feelings associated with what my teacher called the 'kami' (all that was above and between earth and universe) rose naturally like an intoxication within me and added illumination, as from another world, to the light of a lengthening afternoon which was already of singular beauty.

What helped, of course, to raise the experience to so great a height was the light and bubbling mood of the crowds in the narrow, winding streets whose conversation was like an uprush of fountains around us; and also the beauty with which they had clothed themselves. They were just wearing their best clothes but, in my ignorance, the entire town appeared to me as if turned out in costume of taste and colour of a special design of great and significant antiquity. As far as I could see, only our own foursome were dressed in European clothes, and never had what passed for good taste and fashion in Port Natal looked so inappropriate and vulgar. I felt embarrassed and finally deeply ashamed of our appearance. I feared we would offend the sensitive sensibilities of happy, colourful and devout throngs around us. I found myself turning for reassurance to the appearance in the crowd of some men who, without offence or self-consciousness, wore European hats above the kimonos that flowed over and around them. But the crowds themselves never showed any inclination towards objection or resentment. On the contrary, they made us feel welcome despite our appearance of having just stepped out of a display window of a colonial outfitters. The glances directed at us were just naturally curious, and unfailingly produced lively comment and speculation wherever we went. Instead of affront or hostility, there was finally only acknowledgement and acceptance of us as yet another source of excitement.

Soon at ease, we followed our guardians, wandering from festival to the fairground that Moji had become; its streets and alleys like brooks, rivulets and streams flowing full of women in their national dress. The oldest ladies tended to be in kimonos of silk, formal and dark as if woven from the shadows of age. Matrons were in deep imperial purple and splendid sashes which were embroidered with full flowers, and were their true glory. Young wives wore bright but not ornate colours of their own choosing. Young girls flowed by unashamedly in full flower of all kinds, and then unbelievable masses of children led solicitously by the hand of one that was older than themselves. The little girls were like bouquets of zinnias, anemones, cosmos on the deep-dyed daïs of daisies. The little boys, addressed all round me with a punctilious "Mr. Little Boy", were more soberly kimono-ed, but already had their hair close-cropped in the encroaching modern fashion. In fact the children were, for me, the greatest delight of all. As affection and tenderness fell like lamplight over them from the heights of feminine heads held on elegant necks under shining lacquered piles of true black hair, they all stared around them, with an intense and profoundly solemn regard. Their fresh features were impressively still, almost immobile if not tranced as their faces were entrancing. It was only by observing their huge eyes, drawn and slanted as if to the design of the inspired hand of some Uta moro, that one was aware of the keen activity of response, reaction, observation and evaluation that went on within, with the utmost seriousness. For me they made art of the popular afternoon, and music out of the clatter of their Geta, their raised wooden shoes, which resounded at times like a massed tap dance on the streets.

Added to all this was the fact that there was no house, shop, stall or street that was not in appropriate costume too. Hardly an apex of the roofs of wood, turned a satiny blue-grey by time, was without a banner, pennant or streamer of designs of heraldic colour printed on wild silk, brilliant cotton or just on glowing paper, which was all that the majority could afford. No-one, however rich or poor, I was convinced, had failed to dress their property appropriately according to the limit of their means, as if in obedience to a prompting that what they had built and fashioned had become more than matter and infused with some of the quality of spirit implied in the use of the antique suffix of 'Maru' that my teacher had been at such pains to explain to me. Although there was no breeze but only a light air that was no more than the gentle movement of the afternoon turning slowly into evening, banners and streamers swayed against the blue roofs. They seemed to impart a movement even to the sky and impelled it to join in the rhythm of celebration. Below, shops and stalls set out all they offered for sale with yet more colour in neat and orderly patterns. Almost always they seemed to form a symbolic design to maintain the flow of communication from what was below to above, and entice from above what was sought for so keenly on earth.

This impression was encouraged by the appearance every now and then of priests, invariably alone on an inward course of their own. Sometimes one would go by urgently in the white Shinto ceremonial dress with scarlet cape of the most positive and startling dye thrown over it as of the image of the blood of life and nature itself. The red was all the deeper by contrast with a black austere headpiece with its lacquered backing of a long, clean-cut and abstract shape, holding it in position. Others were in white, almost painful to our eyes under so clear a light, but again unusually precise in outline, starched and stiff on the shoulders, and also with a black cover over their heads, though less dramatic than those of Shinto priests. They were guardians of some Bhuddist temple nearby. Indeed at the entrance to a road branching off to the temple itself, higher up the hill of an Euclidian skyline against which Moji leans so confidently, we saw a priest with an ampler black headcover held in position under his chin by a single cord woven of purified fibre, and a cloak of the translucent yellow which represents the Transfiguration and Resolution sought in the Bhuddist way. The cloak was thrown over another vestment of that startling white which spoke so eloquently, in Bhuddist imagination, for the purity and singleness with which the seeking had to be done.

All this and much more made the occasion so comprehensive and united in mood that I was reminded how my Sensei, discussing religion in Japan with me, had observed how the voice of Bhudda had tolled in the spirit of the world like 'a bell hung in the sky'. This teeming segment of colourful and aroused humanity around me was experiencing something of the universal belonging which the Lord Bhudda himself had expressed in words constantly recalled in Japan, that he had 'met a thousand people on the road to Delhi and they were all his brothers'.

All these impressions gathered force and rhythm as we went up the hill and achieved its climax at the end of a narrow lane. It was long and twisting in the manner of all lanes that have taken over from footpaths. Soon our lane was filled almost to overflowing with people. Most of them were women, all of an unqualified and shining femininity. Many were leading children by the hand as if to a new school. The children were as large, lacquered and solemn-eyed as ever, and carried a great diversity of bags of sweets, dolls and toys. Suddenly lane and town ended. We faced a road which led to an opening in a copse of pine, already colourful from the sinking sun. The road wound unfailingly up to our first Tori: the sacred and triumphant gateway that later I was to see everywhere in Japan. They are of an unknown and much-disputed origin. But what is beyond special pleading is that there was no view of the land as I knew it then that was without its own Tori. Almost always they were rooted in a place which to my eye had some implication of quality with nature and so suggested constant communion, however intangible, with all that was above. There were, I knew already, at least one hundred thousand Tori recorded in Japan. A score or more were of monumental proportions, but the majority were modest structures and many were almost home-made affairs. I was to see them in the deep country rough-hewn by peasants and yeomen farmers out of trunks of trees, far from straight and upright as in the greatest of the national Tori. Some were of an ancient stone that was already so much a part of the earth around them as to be difficult to discern. They and their rustic wooden counterparts touched me deeply. Their testimony of the pervasive dedication to the rule of modesty was a lesson that was already fundamental to my experience in the *Canada Maru*.

Japan was the first civilized country that I was to explore in depth that presented me with landscape after landscape without the spire of a single church to draw the eye simply up to the sky. But the initial feeling of loss was soon abolished by the sight of these Tori. They added to the meaning of spires somehow, as the spires represented the essential striving of the religions in the west towards a mobilization of all the responses of the spirit to reach above and beyond its roots in the earth. The Tori was an image of religion as direction towards a gate through the dark walls of unknowing which still hems in the human spirit as it did in the beginning. The Tori was there 'so that passing through the god-way, darkness is taken from the heart as darkness of night is lifted at dawn'.

Solemn as all this must sound, it would be inaccurate to suggest that this aspect of the festival was grave and sombre. The mood was high and light; the gaiety, excitement and happiness of the crowd was infectious. It was the best possible introduction to the spirit in which Japanese, with very rare exceptions, celebrate all moments of natural eventfulness, even those connected with ancestors and gods. Even the great All Souls day of the land, the Festival of the Dead, was one of the most joyful, always flowing with happiness and the special excitement of ultimate anticipation. I had not imagined anything so happy as our progression to this modest Tori, passing through it and following the path among the pines to the unostentatious shrine. There the quality of the light was matched by the clear and all-pervasive scent of the pines. The partnership of vision and smell was highly intoxicating, all the more so because it was uncontaminated by the smell of humanity. No wonder I had been delicately warned in the ship that the smell of the foreigner was a problem for the Japanese. They did not find it pleasant, tended even to be offended by it, dismissing foreigners as 'people who smelt of butter'. Butter was a food which, in the early days after the country's release from isolation, caused such revulsion in the Japanese that in country districts I was to see people, for whom European eating habits were still unknown, finding it difficult not to be sickened by witnessing our modest consumption of butter and cheese. It immediately rebuked, as it restored to its place, the general European abhorrence and arrogance aroused by the smell of the indigenous peoples of Africa. I sincerely hoped that William and I, so heavily outnumbered, would exude nothing to blur or stain that clear, absolving scent of pine, ardent with vital resin and fresh with the smell of grass, leaves, moss and earth.

As a result, I myself came, somewhat chastened, to stand drably with my three companions in a line of many dream-like colours before the humble but immaculately carved pillars of the shrine which supported the roof. There we followed Mr Tajima's injunction, clapped our hands twice to draw the attention of the enshrined spirits to us, bowed profoundly to them and then, with our hands raised and together in front, we joined in the act of prayer without words in the line of people with us. I could do so only without preconception of any precise intent, in the way that my old Zulu prophet had described prayer to me, as a way of asking the first spirit not with the mind and words but with the heart. My asking, I remember still, was just an urgent longing to know at least what I was intended to pray for in life. . . .

It was most marked how, on the way back from the shrine, everyone seemed to have dropped a burden and be in even higher spirits. The same mood was intensified down in the town below us. Its centre was yet more crowded, the bubble of conversation, the rattle and chatter of Geta louder and more impetuous, and a sort of collective fever of celebration rose high in the cool evening atmosphere. Brightly coloured paper lanterns with symbols in sepia and midnight ink brushed upon them to announce what the shops, restaurants, tea-houses and stalls had to offer, now hung in long lines on either side of the streets, trimmed and ready for lighting. Somewhere near the centre in a sort of market-place, I noticed, among all the tumult of movement and babble and talk, what appeared to be a place of calm and orderly arrangement of men, women and children, squatting or sitting on their knees, apart from the rest and deeply absorbed in something seemingly invisible. I was strangely drawn towards them and compelled to ask Mr Tajima to take me there.

The not inconsiderable group of people, so absorbed that they did not notice our arrival, were there at the feet of a man sitting on a yellow mat talking in a low, clear voice. He was dressed in a golden kimono, held with a broad sash woven of green and red round the middle. It was a far more abundant garment than usual, and lay with ample folds around him that overruled any shape or movement of his body within, and fell wide to the ground to disguise even the way he sat. In this sense he was more like a monument of singular authority rather than the man himself. This authority was immeasurably increased by the head and face above the dress. It was the face of an old man with features of a cast so old that it seemed beyond measure of antiquity that I possessed regarding the history of Japan. He looked, in fact, like one of those philosophers, statesmen, poets or resolved servants of the earliest emperors of China, serving, in exile from the people they loved and all that they valued, on the frontier of some remote province among the barbaric subjects of their imperial masters. They did so with such absolute commitment that some of the most moving and healing poetry of classical China before its age of 'troubles' came from their philosophical brush to convey a quality which seemed personified in the man now talking with such hypnotic power to the little gathering. His skin was like an ancient parchment, covered with innumerable creases and lines as of sensitive writing describing a long record of complex experience of life, and so exacting metamorphosis of its hurt, injury, conflict and, perhaps even most demanding, the pull of its pleasures. It was the face, indeed, of someone who had made his final peace with chance and circumstance, and so could speak without impediment or interruption because the words that came to him were not so much his own as those of finalities and necessities of life speaking through him. And so, as if to complete the authenticity of the image that came to me, he had a long, thin, grey beard as in the earliest paintings of the pioneer sages of China, while the hands that emerged from the wide sleeves of his coat were elegant, the fingers long, palms broad and used to illustrate his meaning, eloquent not in terms of the words they accompanied so much as of the rhythm of a spirit conducting a sacred rite. Somehow I seemed to know him and his function before Mr Tajima whispered to me, "He is a travelling and professional story-teller."

I knew it because the look in his eye and the tone of his appearance, despite dress and dissimilarity of circumstances and place, were familiar and dear to me. I had met it on the faces of men charged to pass on the stories of Africa from one generation to another without help of the written word, in the belief that, if their story were ever to be forgotten, they and their peoples would lose soul. I asked Mr Tajima if I could be left there with the listening group. He was at first surprised, looked searchingly into my eyes, was reassured and then so pleased that he blushed with the incapacity of a generation of Japanese to control an endearing phenomenon which happily was as common to all as it is uncommon among us. I thought he turned happily away to join William and our spirit of diplomacy in plus fours, after saying he would come back for me soon.

I was instantly offered a place on a mat of rice-straw among the listening group, without a lessening of their concentration on this antique story-teller's tale. It was a slight but heartening indication to me of the belief I already had in the power of the story to conserve, increase and unite, which was natural in someone who longed to be a story-teller himself. Ever since I could remember, stories had a way of being more real to me than what passed for real life. Their eventfulness surpassed and transported the importance of the partial realism which people around me regarded as the one and only way of being 'practical'. At that time this belief was in a sense still an unproved and a far from properly exercised and tested emotion. It was sustained mainly by an instinct that my own life was beginning to make sense only in so far as it followed and evolved a 'story' of its own. Since I still hold more than ever to that concept and have been its apprentice for many years, I must at all costs avoid orchestrating that moment in Moji with hindsight. I must leave it to speak for itself through the ease my spirit found in being instantly at one with the group for whom the limpid story that they were hearing was, for the moment, far more real than the celebrations approaching a climax around them.

I could not follow the words of the story-teller exactly because he did not measure his delivery to my inadequate preparation: and the language too sounded like an ancient one, in keeping with the primordial nature of his function. I understood just enough to be held entirely in that 'Once upon a time' atmosphere which the story transforms into a timeless *now* wherein past and future are instantaneous, abolishing the sequence and hierarchies of ages and eras, and establishing such an acute propinquity in the dimension of art that Homer, the writers of Genesis and Exodus, Virgil, Dante, Mallory, Shakespeare, the Brothers Grimm, as well as the Hottentot, Bushmen and Zulu story-tellers of my youth, were in one moment present as close neighbours and as if looking over my shoulder at the scene with approval. Yet all I gathered was that the climax of tragedy, already casting its shadow over the tale, was coming fast.

It appeared that we were in a once-upon-a-time castle of a lord of lords of antiquity. This lord, sickened by the feuding and conflict that led to incessant fighting and killing among followers and subjects, had forbidden provocation of any kind which could lead to still more feuding and was, at the moment that I came to listen, just beginning to feel confident that he was succeeding, where so many others had failed, in delivering Japan from its blood-soaked past. But unknown to him, a potential of more feuding and bloodshed was alive and active in one of his most powerful followers. The listeners, of course, were already deep in the secret and increasingly fearful of its development. The fear was indeed so tangible and pervasive as to be like a premature darkening of evening around us. This subject could not overcome his jealousy and hatred of another as powerful as he, least of all because the illustrious and sensitive person was in agreement with his overlord's ruling and so resolved to observe it loyally. This Japanese Iago thereupon nourished his jealousy with a cunning as great as his zeal. Implicit in the tale already, I discerned, was the power of an abiding pattern of the negation in the human spirit, active as ever to this day, which makes men jealous and fanatical with hatred of a quality of spirit that they are incapable of matching. The realization of the horror and pity of all this came in a gasp of fear from us all.

We then heard how 'Iago' plotted to time an attendance at his lord's court on a day which his spies informed him this rival had been summoned to appear. He hid, unperceived, within the precincts of the castle. When his rival came by on his way to the most august presence of that ancient world, he stepped out and laughed at the surprised man. A cry of horror so intense that I was shocked came from the listeners, although I had already had an inkling of how ambivalent and dangerous a role laughter played in the mores of the land. Here, laughter was obviously intended to be something that could not be overlooked or forgiven.

So the innocent victim drew his sword and the inevitability of the action and the sheer tragedy implicit in it, although yet to come and still unspecified, held the crowd bonded to a sombre and irresistible foreboding. The villain also drew his sword and, when the guards rushed to intervene, maintained successfully that he had done so purely in self-defence against an action that he had done nothing to provoke. For if ever proof of defence were difficult, indeed impossible, the story-teller made plain, it could not be more so than in the interpretation of a laugh which the villain insisted had been one purely of welcome. The overlord had no option. The crowd knew it and suffered for him, as for the victim, to condemn to death a man he had valued above all others, with the one merciful provision that he could die honourably at his own hand and commit the Harakiri, or the Seppuku as it is called ritualistically. All this was accomplished by the story-teller in dignified fashion but in sparing detail. Many eyes of the listeners near me were bright with an unworldly light of approval of the tragic ritual prescribed for the ambivalent moment when life, through death, is preferable to death through life. For it was elementary that lords and samurai had to know 'when it was right to live and when it was right to die'.

A woman next to me now began to cry without sound; more out of a strange relief than sorrow at the description of the preparation for the suicide – the bath of purification, so that body and spirit, which deep down in the Japanese spirit are interdependent aspects of the same reality, would be free of dirt for the journey on into the beyond, and the despatch of the condemned lord's servitors to fetch the purest parchment of rice, ink and virgin brush to write a poem of farewell to the earth.

So we came to the final act, which I had to follow less through words and gestures and expressions on the faces for whom the world of festival and fair were utterly abolished. There were more tears and it was all that I could do, raised in a more indulgent discipline, not to cry without restraint. But I observed how vividly the stage was set for the appearance of the samurai followers of their dead lord. Convinced of his innocence and in great danger themselves from the increased power of the unscrupulous instigator of the tragedy, they decided to disperse immediately and pledged to meet again in disguise and secrecy, to plan an appropriate revenge.

At this point I became aware that my three companions were standing behind me, and must have been there for a while, showing signs of increasing embarrassment if not impatience. Mr Tajima whispered, I thought with great reluctance, that we had to leave if we were not to be late at our hotel. I had no option but to withdraw as quietly as I could because I felt departure just then was an offence to both storyteller and listeners. Deeply distressed, I turned away and faced the town where the light was beginning to shine through walls, windows and screens of paper, and from hundreds of many-coloured paper lanterns. The sun had set but over the sea spread out like a cloak of silk on which the tender feet of the gentlest of evenings could walk without danger of con-tamination; the sun had just set on its way towards the worlds we had left behind us. Yet the west was still bright and full of light. The pink deepening slowly, we climbed up a steep, quiet lane where small Japanese homes, their backs to the streets and world of men and their true faces turned to a garden, if only in miniature, stood shoulder to shoulder on guard in a common distrust of ostentation together with a love of privacy and communion with nature. We arrived at the entrance of our hotel as the epic pink turned red. For a swiftly censored second it intruded on my feelings with an almost imperceptible question; whether such a red were not also the colour of a personal forewarning?

To our delight, our hotel turned out to be an inn of long tradition and local distinction. It was called the Mountain Pine Inn, and was situated above the house line of Moji where the hill was covered with trees. Obviously any association with pines could be nothing but honourable and, with the fall of evening, their scent was emanating as an unadulterated element of purification all around and over us. The scent was almost tangible, as if the pines were spraying us with it in an atomized form of the finest of perfumes which we left behind us only in passing through the outer entrance to the inn whose sliding doors were drawn wide apart. We came thus to stand on a floor of flags of sandstone before some long and broad steps leading to the proper entrance into the hotel itself.

We were obviously expected because, the moment we entered, the lady in charge, in a dignified kimono, unbeflowered as her position of authority demanded, but with two attendants in brighter colours, appeared above on a smooth dark floor of wood. They sank on to their knees and bowed with their heads of lacquered and piled hair almost bent to their laps, and as they did so our lady of the inn uttered words of welcome of such a high degree of politeness that they were beyond my apprentice's grasp of the language.

Slight as it was, this essentially feminine nuance to our welcome was of great significance to me. Nothing, particularly at that time, was the subject of greater misunderstanding than the role of the Japanese woman. I had been told on the *Canada Maru*, with feelings which could not hide the injury caused by the words, how old Far-Eastern European hands reiterated constantly a favourite saying of their hard-bitten kind, that one must never trust a Chinese woman but could always trust a man. On the other hand, one should never trust a Japanese man but always a woman. In so far as trusting Japanese women were concerned, there could be no objection to the observation except to the extent that it was rooted in a mistaken assumption that the women of Japan had developed this moral quality, out of a total and dangerous submission to their menfolk.

It was not something demanded of them without qualification and consideration. If withheld, it did not imperil their chances of becoming wives, mothers, or occupying any accepted roles in society however insignificant, even those of the servants of the inn on their knees before us. The misunderstanding, the world of the *Canada Maru* maintained, was encouraged by exaggeration of how, in times of near famine, female children were heartlessly abandoned to death from starvation in order to increase the chances of survival of the male. Ignored was the powerful role Japanese women had played from the earliest times in Japanese history, in politics, war, society, religion and most strikingly in the arts. As examples, I had already heard some of the finest poetry composed by women like Higitabe No Akaido, Kasa No Iratsune and Ono No Komachi. I was looking forward to reading the most famous work in the literature of Japan, the six volumes of the Tale of Genjii by the Lady Murasaki Shikibuku, written towards the beginning of the troubled eleventh century, and also the witty, malicious and acutely slanted work of sharp observation of the life at court in the 'Pillow Book' of her jealous rival, the Lady Shei Shonugon. The truth was that women had always occupied an important place in the life and civilization of the land, and this feeling of the significance of the feminine in man was most convincingly demonstrated by its apotheosis and translation of the sun, which is so masculine and Apollonian an image to us, into the person of the goddess Amaterasu who occupies the highest place in Japanese mythology.

All this is modified, it is true, by the influence of Chinese civilization in its patriarchal historical essence, and to some extent driven into the background if not at times underground. But the roots of the significance of the feminine went so deep into the origins and evolution of national character that, even in their moments of greatest submission and apparent sacrifice of their self to men, women made out of the grace and totality of surrender a new significance and source of authority in life that the careerists of their sex beginning to emerge might have envied, had they experienced it.

Long as it is, this recapitulation of the meagre extent of my preparation for my first meeting with Japanese women is essential for understanding how we were, there within the translucent hallway of our inn, on the verge of crossing not merely the threshold of an hospice but a frontier of a world dominated by women. Wherever we stayed from there on, with rare exceptions, we stayed in inns, all ruled over by women with an efficiency, grace, elegance and a sensitive and happy consideration which, as far as I was concerned, raised my appreciation of their essential and positive feminine quality. Also its significance and power in Japanese society rose to heights I had not remotely foreseen, despite my preparation. This transition to yet another vital insight seemed appropriately symbolised just then by the fact that we were following the example of our guardians, undoing the laces of our shoes, to step out of them before going up the wooden steps to put on clean leather half-slippers and then entering the inn proper. Superficially this could have all been passed off literally, as our thickly-soled European shoes would undoubtedly have bruised the soft yellow mats of rice and straw inside. But almost as important as the practical considerations in my own mind was the metaphorical necessity of preventing any dirt of the physical world from contaminating the world within the hospice.

It was accepted everywhere that cleanliness in the Japanese spirit possessed high religious meaning. Dirt on the person was still regarded then as an insult to gods and Kami. That is why I had been told that no person, however mean or poor, would face the end of the day, unless prevented by force of circumstances, without a bath. And so we followed the two graceful young ladies, going like coloured paper lanterns before us, along a soundless, translucent corridor towards a pre-ordained bath. At the far end of the corridor they drew apart two sliding doors, so well fitted that they opened without a sound, and showed us into a room, unusually large by the standards of the Japanese who, by instinct, avoid excess of space and have an inspired gift of making much of little even in living-rooms.

A beauty which, without obvious substance to support it, was as impressive as it was pervasive. The room was without furniture and had no decorations, ornament or painting on the walls. In that sense we might just have been shown into the Japanese equivalent of a Carthusian or other puritanical religious order. Yet the walls of paper, two of them without intersection and wood just solid enough to support a slanted ceiling, to hold the thick paper smoothly within their frame; the others latticed with slighter wood as if they were windows of a sort and at one with the sliding doors inserted among them. Yet this paper was warm as any paper I had yet seen, and glowed with a steady, unwavering luminosity because even light would have been too bold a word. The floors were covered with deep mats of straw that glowed even more than the walls, as if still affected by the gold of the harvest which released them from earth and sickle for such privileged matting. They spun a light not at specific points as of an electric bulb, lamp or candle, but as of a full-moon glow all around us. This light, and the quality of craftsmanship that fashioned the materials, continued for me to form a single statement of simplicity and beauty that any addition, however tasteful, would have diminished. In a mere provincial inn it far surpassed anything I had imagined or foreseen, and was breathtaking by comparison with the best hotels, let alone the sleazy ones, on the waterfront of Port Natal, whose recollection intruded like an uninvited guest of crude and uncared for disposition. To convey what was profound and complex simply, instead of conveying the simple in a complicated way, which depressing thought had been the great problem in my youth and inexperience, seemed to me superbly achieved in that room. It was so still as we entered that even the slip-slap of the soles of our inadequate though purified slippers provided by the inn, sounded loud and profane.

The moment we were inside, the two girls bowed to us and excused themselves. As they vanished, Mr Tajima told us that they had been appointed specially to attend to us and provide all we needed for our stay. He added that, although mere country girls and probably illiterate, they were highly trained and would not fail us. All Japanese inns of standing, he emphasised, prided themselves in providing so slight a service for their guests. They only regretted that even so the service could not be high enough.

His suggestion that two such graceful and prepossessing young ladies could be illiterate did not shock me. I already knew from Africa that people could be profoundly cultured without the capacity for reading or writing. Already some of the most civilised persons, and indeed a few of the greatest gentlemen I knew, were coloured and black, men and women, who could neither read or write. By contrast, some of the greatest barbarians and envious spirits of my acquaintance were men of the utmost fluency of word and pen, adroitness of mind and insatiable devourers of books. However I was not allowed to carry the thought far because Mr Tajima went on to excuse himself, and foreign affairs, in plus-fours, saying that they would rejoin us later with Mori and the rest of the invited party.

As they left, our two young ladies returned carrying towels wrapped in sealed paper, miniature safety razors, tooth- and shaving-brushes, the appropriate soaps, toothpaste and hair combs, all sealed likewise, and, finally, fresh laundered Yukata, the most ample and stately kimonos, of black silk with fine gold stripes to illumine the material, and broad sashes of black knitted silk for each of us. They indicated that if we would be so kind as to condescend to divest ourselves of our honourable livery, they would conduct our noble persons to our august bath.

But we had hardly undressed and were standing there naked, undoing our towels, when they re-entered. I extracted my own towel in a fever of embarrassment which prevented me from looking at William for guidance. But I did catch a glimpse of the faces of the two young ladies. They were totally without embarrassment or marked emotion, except perhaps bewildered by the nature and speed of my reaction. They went on standing there looking at us steadily and providing me with a mute but indelible lesson on how innocent and frank was the Japanese attitude to the nude. I remembered how in the *Canada Maru*, during a discussion of the importance of the daily bath, we were told by the Radio Officer about the discipline of Hegel, of a custom in the villages of the deep country where he had grown up. There in summer, he said, it was common practice towards evening to put large wooden tubs full of water out in the streets and light charcoal fires underneath them to warm the water. When the tubs were warm, neighbours of all ages and sexes would emerge naked to bath in them. Friends would place their tubs side by side to gossip and soap and scrub each other's backs. And every village, he had added with a laugh, had at least one tub with a Foreign Secretary in it. The purser had added to this the observation that in Japanese homes the bathrooms were situated in places of honour with the best view available on to gardens and into nature, and never combined with lavatories as with us, in tucked-away corners in unconsidered and ill-ventilated spaces of our buildings. In this way, body and mind absolved from dirt, a person could commune with himself and nature without need of any other ritual. This recollection, so tranquil with a primaeval innocence as of the garden of Eden before the first disobedience, helped immeasurably to calm if it could not overcome incorrigible feelings in myself as to the privacy of my own body. However, once our towels were firmly wrapped around our middles we followed our attendants barefoot out of the peace and luminous beauty of our room.

More sliding doors at the end of another silent corridor were in due course drawn apart, and we entered another world, blurred and warm with steam. Our young ladies bowed and indicated with a graceful flurry of their sleeves of silk that the honourable bath was waiting for us beyond steam and mist. They held out their hands to take our towels and we had to strip, which I did sideways, and stepped smartly through the frontier of steam. The bath was more like a small, indoor swimming-pool, fed by water from a nearby hot spring conducted through bamboo pipes. The water was obviously very hot to produce such dense steam, and had I not been so embarrassed by what I saw in the pool, I would have taken a warning from the fact that the eight members of both sexes, whose presence disconcerted me so, were in varying degrees of carefully calculated submersion, so as not to scald themselves. But I was too put out for any considered action and promptly dived into the pool in order to, as Shakespeare had it, 'hide my native semblance from prevention'.

It was shameful behaviour, of course, and an exhibition of the worst possible manners in an honourable bath. It was not surprising, therefore, that some time elapsed before shame would let me smile over such uncouthness. According to William, who made a carefully controlled entry into the hot water which earned him praise, I raised a tidal wave in the bath which swept up many as yet unacclimatized torsos, breasts, shoulders and even unprepared Adam's apple, and scalded almost everyone in the bath. It says much for the deep reliance on manners and courtesy in the provinces in Japan that my entrance seemed to be understood or, even if not understood, instantly pardoned.

When certain of my concealment in water, my confidence somewhat restored and my eyes free to look around, I saw no reproach on the faces of the scalded participants but only looks of keen enquiry and welcome. We bowed to one another as low at the water permitted, exchanged formal greetings and paused for a while, as if spirits all round were pulling back for a new leap forward. Then the oldest and bravest approached us to ask what good manners demanded, where did we come from? How old were we? Were we married and if so to whom and did we have children? If not, were we thinking of getting married? And what might out professions be, and so on and on to more intimate matters like whether the hair on William's chest was real? I was too young to be similarly blessed but the matter was of great interest to them because the Japanese were not a hairy people. William's affirmative answer, which barely concealed a certain consternation of amusement in me, raised a hiss of astonished appreciation all round. As a result, when we at last extracted ourselves and were safely conducted to our room by our young ladies I could not help christening William the 'honourable sir of the much-to-be-envied hair on the chest'.

"Rather that, my dear Lorenzo", countered William quickly, with no need to search for words, "than the 'impetuous male Susanowo', which will surely be your Japanese label for evermore."

In this mood we dressed in our Yukato and Kimono, our young ladies showing us how to deal with our shining sashes. That done, they stood back and, in the most charming and feminine of ways, expressed approval, with a markedly more prolonged and elaborate appreciation of William's appearance. It was without doubt justified. Happy and at home as I felt in my own kimono, the result must have been respectable rather than inspired because it was so taken for granted. The effect on William, however, was much more impressive. Tall, slim and broad-shouldered as he was, the kimono sat on him with immense elegance and conferred a certain air of aristocracy upon him which my tailored clothes, so inadequate on his person, could never have done, even had their European idiom been the fashion in Moji, which it was far from being. Yet there was something else; a profound nuance to the overall effect I had never observed before and even then, when first felt, I could not name beyond an intangible suggestion that a flowering dress was more welcome to his being than a man's tailored suit. This and his extra inches added a dimension to his appearance, making it by far the most prepossessing of all on the occasion to follow.

This began soon after our two invigilators approved our appearance with the pleasure of young mothers taking their firstborn to a birthday party. Mori, Mr Tajima, the company's local representative and the rising young diplomat all appeared together. They were in high spirits and all dressed in kimonos like ours except for the diplomat who seemed cast in his plus-four mould as if in cement. But he quickly apologized for his appearance saying he had to board the late night express for Tokyo so that he could report on his mission in person to the Secretary of State first thing next morning. We had all barely bowed and shaken hands when our young ladies reappeared with towers of cushions leaning perilously on their arms, despite a firm embrace of their foundations. These they duly arranged on the mats for us to sit, legs crossed in front, an exacting position which we fortunately had practised adequately on our voyage. Cushions were followed by low and small black-lacquered tables and trays of the same material of a fine archaic lustre placed in front of each of us. The trays were of the utmost simplicity, beauty, precision and a taste that seemed incapable of error. As beautiful for the same reasons were the lacquered bowls, platters, porcelain receptacles, cups and even the wooden chopsticks laid on them with scrupulous care which demanded just the appropriate spacing between each object as well as distance from the edges of the tray necessary to complete harmony of object, volume, eye, mind and palate. Even the colours of the receptacles appeared chosen as carefully as a painter his subtler tones, necessary for relating the world without to the imagery of his own within. My chopsticks, for instance, were of smooth golden wood, ends placed on a porcelain rest of snow white. But the platter itself, shaped like a maple leaf, was an epigram of all the colour of the fall which was gathering pace outside towards the fullness of its final flame. My bowl was a deep, restrained yet positive green. My porcelain cup had an impressionistic abstract of earth and vegetation of aboriginal Chinese blue upon it.

All in all, as far as I was concerned, they did everything that was intended to create a spirit of harmonious anticipation, until it was shattered briefly by the appearance of a man, as homely in feature as our steward in the *Canada Maru*. He was dressed in black evening trousers, white mess jacket and bow-tie, like a bar attendant in a liner. He carried in his hands an outsize cocktail shaker which reflected the mellow gleam of our room with hard and alien platinum flashes. Yet despite his incongruous look, he seemed to raise a shiver of excitement in our hosts, perhaps because it was still an era wherein the cocktail was a kind of badge of all that was new, trendy, fast, daring and definitely post-World War One. Even where William and I came from, to label someone as a member of the 'Cocktail Set' was almost a social form of death-sentence, and the drink itself was permissible at the most in ones or twos at home. But at large, on the unbridled social scene of the Charlestoun twenties, it was the nearest equivalent to Pot.

Mori, who, in the best sense, found his own personal meaning in a vision of himself as born to be a bridge between ancient and modern for his country, told William with an unusual note of satisfaction, if not awe, that the newcomer had just arrived from San Francisco where he had been studying the art of mixing cocktails for two years. He was reputed to have become expert, so expert that he could mix more than a hundred different kinds without necessity of referring to recipe books. He was, as a result, discussed and for the rest of the evening accepted by our hosts as one who would govern without question the orchestration of intoxication of our senses like a conductor who knew by heart the score of an opera whose prelude was about to resound around us. It was for me, slight as it was, another example, to add to the many others, of the natural thoroughness of the Japanese in their approach to the matter, no matter how small, which they had decided was needed in their national system and esteem.

Nonetheless, though impressed, I was alarmed by Mori's proud pronouncement that he would demonstrate his art that evening by mixing a different cocktail for each new course. My recollection of our Full Moon Festival in the *Canada Maru*, and the number of courses its inevitably epigrammatic galley produced, was still as lively as it was deep. The certain fore-knowledge of the many more courses in so well-found an inn, and on so special an occasion, that would be set before us, made me see William and myself under sentence of doom to an extreme of intoxication, unknown in our experience and undreamt of in our imagination. The result was that I sipped my first cocktail as if it might strike at what was sober in my make-up, like the deadly black mamba of my native country. It was, however, disarmingly familiar and deceptively reassuring: a Manhattan, mixed, I thought, to perfection. As we sipped it, the cocktail artist left the room and reappeared shaking his mixer even as he bowed to us on entering. I saw then the reason for his disappearance and became even more certain and afraid of the doom awaiting us. He had brought with him so much material for innovation, and even improvisation, that it had to be kept in a large room exclusively for his own use. There he made up the magic potions of his black art and refilled his mixer with new witches' brews in sufficient quantities to fill cups already over-flowing with doom, to speed the race of East and West in the room to a Bacchanalian oblivion. I could go on enlarging on this gentleman from San Francisco and his subversive contribution to the evening but it should be enough to say that he did the duty Mori and his country expected of him, far beyond and above the normal call of duty and exigencies of his service.

I came to measure his subtle success as this part of the evening pursued its irrevocable course without scruple or remorse, by the behaviour of the diplomat's plus-fours. There came a moment when it struck me that plus-fours and stockings on the legs of the slender man were beginning to part company and became incapable, as it were, of recognizing each other any more. Intoxication, I thought to myself, as a warning and in the manner of a medical diagnostician, was setting in, and from that moment the growing gap between stockings and plus-four straps became a kind of thermometer to register the rising fever of alcoholic affliction. The severity of the affliction can be judged by his final appearance, or disappearance, in whichever way a sobriety that I did not possess totally, yet miraculously had not lost either, suggests. An hour or so after the inevitable photograph of us all had been repeatedly taken, he stood there, between sliding doors on uncertain legs, and balancing like someone on a tight-rope, trying to bow low to his audience with the courtesy diplomatic good manners demanded. His stockings were by then round his ankles, and one plus-four leg, apparently more sensitive and caring than the other, had followed its own stocking down full length in a vain effort at reconciliation; the other, affronted, had gone aloof and somehow got itself haughty with affront in position above the knee. Yet to his credit, he completed his bow like the forked twig of a diviner pointing at a new-found point of water. All around him must have appeared crumbling and illusory, for his sartorial unity was shattered like an empire in dislocation, and its constituent elements in a state of unilaterally declared acts of independence had severed all diplomatic relationships with one another. Yet the spirit so sorely strained was still intact for, although we lost him then forever, he went into the night with a dignity of absurdity and impressive disregard of appearances which remain endeared and endearing in memory.

Meanwhile the processes of innovation of courses and innovation of cocktails were locked, as it were, like drowning persons in each other's arms and pulling all under with them in a manner I need not describe because the essence of it is plain enough and does not warrant more detail, clear as it all remains to me, despite the sustained onslaught of mixed spirits. I had to endure to the end. Something of much greater importance to me, meanwhile, had happened and continued without interruption from the moment a Manhattan was followed by an impeccable Dry Martini. Our hostess had announced herself at that precise moment with three Geisha and their three assistants in her Mediaeval train. On the faces of our own attendants, who had not left our side except in the bath I had so dishonoured and to whom this was apparently a sign to take up a less prominent position by the wall behind us, there appeared, like a shadow of a butterfly wing, a strange and complex expression I could not decipher with much certainty except that I was sure it was a sign that they did not relish their displacement. But if I were inclined to think the causes of their dislike egotistical and composed of elements of envy or rivalry, I was quickly corrected. The simple country girl beside me gave me a tentative, almost frightened glance, took courage with an effort that made her blush, and whispered to me as she bowed beside me before withdrawing. It was, apart from all other considerations, exceedingly brave of her. The Lady of the Inn, for all her grace of dress and manner, I felt certain, was of a formidable character and would have regarded such an approach presumptuous enough to warrant the girl's dismissal.

I could not be certain of the words because of the exalted degree of politeness in which her spirit conjugated its meaning as well as the fact that Japanese women still spoke in matters peculiarly their own, an ancient vocabulary not used by men. But it was a warning that was a plea utterly objective in its concern for my welfare, to 'be careful'. This concern from someone I had known for a few hours only, and in the most formal of contexts, was a relevation of the speed of the Japanese reflexes of feeling which they seem to me still to possess with unequalled reserves, but also of the importance they attached to following their feelings on and through and to themselves with truth and precision, no matter what the consequences. I was so touched that I wanted to show it with a gesture but, alas, I did not know how to do this appropriately and in a way that would not violate the need for discretion implicit in her whisper, nor the subtle nuances of degree and station involved on these occasions. All the same, I still regret that I did not try, as I would have done if I had attached as much importance to truth of feeling as this humble girl did and so not allowed the demands of my European education for justifying action always with reasons in preference to feeling or emotion in deciding the issue for me.

But the rising tide of celebration almost at once swept me on and away from any self-examination. The formal entry of Geisha and attendants demanded our immediate attention. It was, for me, an enchanted moment. All European preconceptions about the role of the Geisha in the social life of Japan, and the popular tendency to look on them as courtesans and prostitutes of a most select kind, had been erased by what Mori and my teacher had already told me. I cannot pretend that even then I knew and understood their profound and complex role. I do not understand it even today, in the way it should be understood, because understanding was not a matter for knowledge so much as experience. But knowledge such as that imparted in the ship, rooted as it was also in experience, helped. I knew that they were professional hostesses and entertainers of men on special occasions such as this. I knew that they were chosen carefully not merely for their appearance but their intelligence, quality of spirit, wit, culture, and gift for dance and music and much else beside. I knew that they had to go through a training and discipline derived from traditions grown over many centuries and of an exactitude not unworthy of our own more austere monastic orders. I knew that they did form liaisons with men but by no means automatically and, in general, for a balance of many considerations over which their sponsors, to whom they were bound for years by exceedingly exacting contracts, had little if any influence. I knew that those who became truly distinguished in their profession exercised great influence over remarkable men and in most important areas of society. I knew that the most illustrious of all had been sought out by politicians and powerful men because association with renowned Geisha would promote their careers and increase their power. I also knew that even in those early years their presence at these essentially male occasions was becoming too expensive for ordinary well-to-do people, and the process had already begun whereby today only those who can do so on expense accounts, and for reasons of state, can afford reputable Geisha parties. I knew that their arduous training and their magnificent clothes and upkeep were so costly that they had to mortgage years of their future to repay their patrons. But all this was elementary stuff, insufficient for an understanding of all, and inadequate even for my own after one glimpse I was to have before I left Japan, of a certain freedom not attainable any other way, which they found in bondage.

Our introductions, bows, responses, almost religiously ordered with an atmosphere of remote and enchanted symbolism, prompting all from the wings of a stage set for traditional theatre, were obedient to another sort of protocol in which I thought I discerned Mori's fine hand of command. He and William were allocated the senior Geisha and her attendants; Mr Tajima and myself the second in rank, and diplomat and agent the third. Number One was called Teruha, 'the Shining One'; number two Chiyono, 'Eternity'; and number three Tamako, 'Jewel'. I had been told there tended to be three types of faces in Japan: one of a more Mongolian strain, another tending towards a somewhat Polynesian cast, and the third and most prized, as the oldest and most authentic, the type of the original Japanese who conquered the Ainu aboriginals of the island, which was called Yamato with all the mystique implicit in the word. Tamako's face conformed to the Mongolian, almost Chinese type; Mr Tajima's and my Chiyono to the Polynesian. Polynesian-wise she was rather plump and, although of pleasant expression and features, podgy enough to deny her the look of eternity which her name was designed to uphold. William's and Mori's Teruha possessed the classical Yamato features and all their subliminal capacity of legendary evocation: fine-boned, eyes wide and their slant a slow but precise curve which allowed them to show a light as from afar and enveloped in her name, while her nose, delicately drawn, had a subdued yet clear suggestion of a Roman arch. All in all her face and bearing were of a natural aristocracy, making her one of the most attractive and interesting women I had yet seen, so that I was resentful of the middle-class estate that was my lot for the evening and imposed on me a position of compromise between outer-Mongolia on my right and the aboriginal order of the feminine in Japan on my left. My reaction was all the keener because it was unexpected and found me with no immunities against it.

It was the first time I had been powerfully attracted by a woman of totally different race and alien culture. The attraction was all the more disconcerting because it seemed to be at work on two levels, one without and one within, to such an extent that I could not be certain whether it was due to an inner image or just an outer one or two separate faces totally coinciding in the objective and subjective levels of reality but in every detail forming an irresistible one. Today I would try to explain it perhaps on the lines that, as a total stranger of an undiscovered femininity, she was an externalization of the unrecognized caring and feeling potential in even the most masuline of men which the Greeks always personified as a 'woman' and called the Psyche in man. She was, of course, to me, a person of great beauty in her own right. But the beauty was enhanced by the power to reflect, for me, an aspect of myself of which I was totally unaware and which I began to experience for the first time that evening as the most powerful of emotions. Though it is probable that an element of exaggeration in my reaction may well have been due to those demonic cocktails appearing for consumption now like mass-products on a conveyor belt of spirit compounded in a cocktail factory designed according to the most advanced technology of the day, to make too much of it would be an injustice to the innate qualities of Teruha and my own perception. Indeed my attendant's warning had already helped me to decide to beware of my intake of alcohol.

I am still amazed, as I can be distressed, by the animal cunning which rushes to one's side when the prevailing expectations of reason fail one. I heard myself, as some stranger, explaining to Eternity, with the occasional help of Mr Tajima, that it would be a grave breach of the best European manners not to share my glass with them all and that if I did not I would offend the spirit of my ancestors. As a result I managed to make Eternity, her attendant, my own young lady demure in the background, and even the Lady of the Inn – seated like a Manchurian Empress on the side – drink most of my cocktails, while I confined myself to the less invidious saké, the rice wine which Chiyono poured into my cup with strict regard to a commandment of her evocation that the cup of a guest was kept neither empty nor full. It was a process I could subject largely to my own measure, and although I drank more than ever before in my life, I managed to slow consumption to such a pace that I acquired the kind of uninhibited vision and magnification of perception that had the intensity of those transfigurative experiences which I felt certain were the cause of the breakdown of law and order and finally the fall of Empire in Thebes, produced when Bacchus first appeared in the classical world. It was, perhaps, reprehensible of me to achieve this at the expense of the bright-painted acolytes of Chiyono, and even Chiyono herself, who soon lost the look of eternity in her eyes and became increasingly obsessed with the here and now. But I felt that they must be better equipped than I to deal with so professional a hazard. Chiyono rapidly became less formal, her eyes more brilliant and her sensibilities more aroused. In consequence she became more single-minded and, to my embarrassment, she became especially so about me. For ironically, ultimately I was only interested, as far as the ladies went, in Teruha. As for Teruha, she refined this irony further as if on behalf of an inflexible fate.

She and Mori clearly knew each other well and had a great regard and respect for each other. Mori, from the moment of their introduction, had made himself an accomplished go-between. He had put himself totally at the service of herself and William. She was, I could see, instantly taken with William and I had to witness for the rest of the evening increasing evidence of how this attraction grew. It was, in essence, a repetition of the same pattern that had raised William, as it had demoted me, in the order of Mori's interest and affection. At this point I thought irony had achieved its ultimateness. But there was a final, almost lethal refinement to come. William made a convincing show of being interested and was, as always, scrupulously courteous, but without warmth or enthusiasm of heart. This a sensitive woman recognized at once; and yet could not accept. The clearer Teruha's knowledge grew the more she applied her great gifts, art and beauty, to awaken deeper responses from William, but without success. Neither of us knew then but some five years later I was to walk the streets of London all night long with a near suicidal William; I learned only then that he was incapable of being physically attracted by any woman.

The same sort of single-mindedness meanwhile increasingly afflicted me. I never failed, whenever Mori allowed me an opening, to try and get Teruha to talk to me and, at the same time, doing all I could to avoid any emotion to an 'Eternity' whose mortal objective was now clearly established beyond any even unreasonable doubt. Teruha, in turn, was courteous and not unfriendly but always disposed of me as quickly as good manners permitted. Nothing could show the final outcome more clearly than the official photograph of the evening which I have mentioned. In it I am hardly visible and almost forced flat on my back by the weight of an ardent embrace in the arms of 'Eternity' and her attempt at a kiss, whose mention, let alone public demonstration, in the Japan of the day was regarded as a social outrage. William, on the contrary, is seated, upright and dignified, looking straight before him with uninvolved eyes, his right arm, on Mori's command, dutifully rather than affectionately round Teruha's shoulder while she, her eyes bright with illumination of character that justified her name, looks out at the camera, resigned almost to the point of dismay. It was as if already her spirit knew the outcome and was resorting to the acceptance framed in one of the saddest though most frequent expressions I was to hear in Japan, especially from women, "It could not be helped".

I may seem to have made a great deal of so brief an encounter, but so much are the subtle fabrics of circumstance which set the tone and rule the life of the imagination woven of intangibles such as these that the emotions of the occasion are still fresh enough to have the power to hurt and will, I believe, always do so. I know too that they wove a shroud of melancholy through the evening that would have affected and perhaps spoilt all had it not been for the remorseless flow of cocktails, feeding the enjoyment of all the others. And, of course, there was as well the quality and novelty of the entertainment provided by the Geisha and pupils.

First of all Teruha, dressed less ostentatiously than the others but with the fastidious taste that was expected of the senior Geisha, left the place between Mori and William and seated herself on a cushion in front of us. The wide sleeves of her kimono, as she flicked them to compose herself, rustled like an air of darkness in the silence caused by anticipation of her performance. She sat there for a moment still, eyes looking inward, until her attendant brought her a samisen, whose strings she stroked tenderly as if to wake a delicate child from sleep. Then she played to us. The first sound raised a kind of gooseflesh all over my skin. It was so like those stringed instruments overheard at night in the Indian quarter of Port Natal on my lonely walks down from the hills to the sea, as they rendered the god Krishna's lament for his lost Sita. This new music, simpler and much purer, sounded even more as if plucked alive and bleeding from nerves of the theme itself. Moreover the recollected theme was not irrelevant to my feelings just then because they too smarted under a sense of loss, perhaps keener than that experienced on the southern tip of India when the gods still walked its earth. For the cause was the denial of any chance there, as I saw it, of a human relationship worth the price of losing it. No life, I thought rather bitterly, was so poor as one incapable of acquiring things of mind and heart which were worth the grief of losing. In that state I envied the stricken Krishna; however, not for long, because Teruha, her samisen and singing, put an end to any indulgence of personal feelings and made me feel petty and mean. Even the power of alcohol was no proof against the mood induced by her performance.

She sang in a voice which I, with my long Indian induction behind me, found as moving as it was beautiful. It must have been assiduously cultivated, of course, yet it sounded utterly natural, with no marked indication of having any musicological preconceptions imposed on it but issuing directly out of the feelings represented by her name together with a spontaneous expression of their own innate design. With a significance that I was not to appreciate for many years although the weight of it was already full upon me, she sang first of a great Buddhist saint, not in the light-hearted manner that was normal on these occasions, but almost as a tender thanksgiving to his example, which I noticed was too solemn a trend for the liking of the Lady of the Inn whose ideas about the evening, I am certain, were closer to those of the gentleman of San Francisco. I am certain that Teruha was aware of this, but she was indifferent. As the most distinguished Geisha in the region, she was honouring the inn and us, and not the other way round. Without arrogance, she went on to use the self-evident fact for her own, not ignoble, purpose.

Much of the art, philosophy and imagination of the people of Japan is profoundly preoccupied with mood and states of mind at the expense, it would seem, of the external realities that we put first. Truth to mood and fidelity to state of mind, if there are priorities in the human regard for truth, often come first, and some subtle new mood was in command of Teruha. For she followed this Buddhist theme by more music in the same direction, with a delicate intimation of denial of the popular version of her kind, namely the life that the Japanese call the 'world of flowers and willows' and 'transient things', terms inspired by the Buddhist influence on the national character. Her music by evocation and suggestion, more than direct statement, indicated that she too had immortal longings. For she sang of her own world of Geisha as impossible of definition and capable of being understood only through experience. She sang of states of heart that had their seasons, and bloomed unrecognized by man to fall like petals of a flower of a beauty that made grace out of its own decay. She sang of the moment when memory of the beauty itself was enough and the pain of renunciation awoke a realization of things that had always been and would be, even though the 'floating world of appearances' vanished like mist before the sun. Her voice at times trembled as if in pain, and the samisen matched so truly that the decline of subtle flower and fall of petal into grace was so moving that the impact was as much a visual as a musical one.

I longed to know it all word by word. I saw Mori absorb all with genuine approval and clear indications of how this justified his devotion to Teruha, his vision of her right of place at William's side and her claim on his own mind. But I could read them only as signposts to meaning. What stirred me in the main was the deep impact that music, words, gesture and quality made on me. I was sad when she ended with an air of authority that seemed totally unexpected in a person outwardly of so unassertive, delicate, frail and poignant a beauty. But there was no doubt; she had done her part of the show for the evening, and was happy to take up her place again with William and Mori. She did so obviously not a moment too soon for the Lady of the Inn, for Eternity, Jewel and their pupils were immediately ordered to dance and sing for us.

They were dressed in kimonos far less restrained than Teruha's. They looked like flowers and did their dances like flowers, swaying in the air of some summer evening. In fact it was more a stylized movement of their feet and bodies and especially their hands, with fans in their palms and their lacquered heads. There was no fat even on Eternity's performance, and I forgot that she was in fact rather plump. The kimonos seemed to arise spontaneously out of the heart of the feminine, and unfailingly to serve it with an enhancement of beauty of rhythm of movement of their feet and bodies. I saw before me with fresh insight the ideogram for Geisha: gei – culture, sha – person. As a result, although more lighthearted, at times in a deliberately comic way, they made their own unique contribution to the spirit of the evening. Indeed, I felt disappointed when suddenly it was all over. Possibly it was the result of the saké I had drunk, but I had no idea it was so late.

But suddenly there we all were saying goodbye to one another, bowing and re-bowing, exchanging thanks, promises of meeting again and proclamations of our determination to defeat the destructive alliance of time and distance in their perpetual war on human relationships. Then all at once our room was strangely empty and large with unused space. Only William, Mori, Tamako, Teruha and Chiyono remained. I thought that they too would go, and held back only for a more elaborate parting of such principals which politeness might demand. But they showed no sign of going and still went on performing, though in a minor key, their role of hostesses and defenders of men against the ennui of spirit which was held to be their weakness.

They practised all sorts of endearing little diversions and attentions on us and teased us by proxy through Mori in the manner which would appear to make it a form of affection, as if for little boys almost who had to be cured of their reluctance to go to bed. Their sallies and counter-sallies with Mori raised some more laughter, until suddenly a disconsolate Tamako announced her goodbye. I did not and do not yet know why. But that traditional goodbye of Japan, the sayonara, the 'if it must be', particularly from its women, the deep bows, whispers of the irrevocable word, a rustle as the last air of summer of the long wide sleeves of their kimonos, always from the beginning has had the power of upsetting me as no other form of 'adieu' that I know. It had, for me, a finality about it that made it not so much a human gesture demanded by custom as an act of human beings about to set out on a journey from which they would not return. A silence fell over us all. Outside in the street I heard clearly the urgent 'Kara-Koro, Kara-Koro', as the Japanese describe the sound of their wooden Geta on streets, of some inordinately retarded person hurrying home. Far away, beyond and yet above the 'Kara-Koro' there rose the sound of incomparable purity of a limpid shakuhachi. But I seemed to be the only one who did not want to accept that the party was over. Mori was ready and impatient for bed: so too was William. But why then were Eternity and, most of all, Teruha, of

such immaculate manners and circumspect a nature, not releasing
them for sleep and saying that 'goodbye' of theirs, so full of fate? Yet
Teruha for the moment was still absorbed in a profound reappraisal of
her reaction to William, her eyes wide and great upon him. Eternity,
undefeated, was making me uncomfortable with her warm glances. So
what did this delay mean?

The answer came as soon almost as the question. Our two attendants
reappeared followed by the dauntless Lady of the Inn. They were
carrying leaning towers of quilts, silk coverlets and cushions. Super-
vised by their implacable employer they simply laid out quilts and all,
in five parallel rows against each other. But with two of the raised
wooden head-pieces that Japanese women used for sleeping so as not to
undo their elaborate and intricate hairstyles, at the top of two rows.
Three piles of soft cushions completed the remaining rows. The only
explanation, it seemed to me, was the late hour. Eternity and Teruha
were too far from home to be sent out into the night on their own. The
only hospitable and considerate thing possible was to bed them down
with us for the night. That I had no other thoughts might be an
indication of how young and innocent I was. One glance at William
suggested that I could even be exceedingly ignorant and naive. He was
uneasy almost to distraction, and I do not know what he might have
done had not Mori taken over command and dismissed the Lady of the
Inn and our own attendants.

They went graciously with an air of a delicate mission accomplished,
except that my own attendant, in coming out of her bow to me, raised
her eyes more plainly than before, looked me straight in the eye, as if
full of fear. Their expression was another form of pleading direct from
the heart, since even a whisper now would have undone her. Should I
be more careful than ever? She clearly had an experience of this world
of willows and flowers that did not match that of Mori who, as if once
more in his position of full command of a ship, alotted us our sleeping
places. I was to sleep on the extreme right, Eternity next to me. He
would sleep in the middle with Teruha next to him and William on the
far left. This disposition of sleepers accomplished with naval precision
and authority, seemed to resign William to the inevitable and console
him with the thought that apparently there was safety in numbers.

That done, Mori lost no time in divesting himself of his kimono. He
did so as if cocktails and saké in mixtures undreamt of in any alcoholic
philosophy, had not impeded his movements. But he was flushed with
spirits as if his face would burst into flame and lost no time in getting
down on his bed dressed only in his Yukata. The two Geisha did
likewise, helping each other, I thought, to undo the long, wide and
glorious sashes wound round their middles. But I did not look to make
sure, and so, with a haste that was perhaps an unseemly conclusion to so

finely conceived a celebration, I took up my own sleeping station in Mori's convoy for the night.

I do not know about the others. I had the impression they all, including Eternity who was disturbingly restless by my side to begin with, were asleep before long. Exertion, a long day, over-eating and too much drink, would have seen to that and should have also laid me low with sleep at once. Yet I could not. I was obsessed with a picture, despite the dark, like a magic lantern slide projected on my mind. I saw the five of us lying there in parallel lines, and the vision haunted me of all things with a mathematical association. It could have been an indication of how only mathematics of the spirit can rescue one from those extremes of turmoil and chaos of new impressions imposed on one by such a day. But I thought it came more out of a love I always had of the principles of mathematics and their universal validities. These principles had never struck me as of purely scientific application. They held for me also a great philosophical and religious significance. All the axioms of mathematics I knew, for instance, were utterances of devout spirit to me. And the axiom that haunted me at that moment was the one which proclaimed that parallel lines meet only at infinity. The definition implied seemed to me to gather together all the profound little ironies of the evening: we were laid out thus because, with the exception of Mori, our lives were conceived on parallel lines that could never meet, not even perhaps in infinity.

I found it almost unbearably tragic particularly at an hour to which T. S. Eliot referred as 'when midnight shakes memory, like a madman a dead geranium'. And the sadness kept me awake until a vision of the professional story-teller's face came to deliver me. It came with such a startling clarity and over-riding authority that I realized that encountering him and his story was, despite all, by far the most important thing which had happened to me that day. The resolved look on his face resolved my own tumult and tensions. A 'once upon a time' feeling took possession of me. 'Once upon a time' was 'now'; 'now' the beginning of another story, and the story my own. That sent me into a sleep without need of dreams.

I woke early, content at heart but in severe physical discomfort, if not pain. I had an acute headache and was unbearably thirsty. I was about to creep out of bed quietly and go in search of water when I heard a whisper beside me. I looked sideways. Eternity was awake. She was sitting up between her quilts, her face sullen with sleep. So was Teruha, to whom she was whispering over our sleeping Captain. It was something to the effect that she feared they had not pleased us and that we did not like them. Teruha shook her finely shaped head like a black tulip in a wind and told her firmly that she thought her wrong. She herself was convinced that we had appreciated them so much that they had been shown respect in a form seldom if ever encountered in the exercise of their profession. I do not know if Eternity was convinced and still had more primitive ideas of the respect due from men to women. All that mattered was Teruha's vindication of my measure of her quality.

Almost immediately afterwards, Mr Tajima arrived to rouse us and we had to bath, shave and dress and hurry back to the *Canada Maru* at such speed that I was still thinking of Teruha and her reply to Eternity, when we lifted anchor, and the *Canada Maru* braced herself with an over-all shudder for the effort necessary to push aside the waters that race with such power through the strait of Shimono-Seki, as though to prevent those who had not earned the right by failing to endure from entering, as it appeared that still, clear morning: the Temple of the innermost sea of Japan. There was indeed for me a stillness and light and beauty of natural sanctity over the land beyond the strait. William and I stayed together on deck for the first time in many weeks so as not to miss a moment of this delicate unfolding as of a great illuminated scroll. Again the youth of earth, the freshness and clarity of line, the houses of wood and thatch holding on so naturally and intimately to the slopes and folds of the unblurred hills, as children to the kimono of a great mother, struck me first but with greater force than on our approach to Moji. It was one of the most beautiful and original landscapes I had ever seen, and even the earth within itself looked profoundly cultured rather than cultivated. This impression was powerful enough to create a vivid fantasy of an earth that was artist enough to paint itself. It did so, however, not in one vast and transcendent sweep of a landscape of itself but in countless paintings in oil, water-colours, black and white on surfaces of silk and most of all a seemingly endless series of woodcuts so that one saw at once why this art was an inevitability of the relationship of land, sea and man, and such an appropriate expression of its visual character.

It was, therefore, to me not a single painting by an artist of genius but an all-comprehensive collection of inspired renderings of significant fragments of itself. The trouble was that every fragment revealed to us was eminently 'paintable'. All the themes great and small of Japanese visual art throughout the long troubled centuries behind it were there: the hills with the clarity and authority of line; depth of sky, blue with distance; segments of tormented heads of rock in the placid water, holding aloft pines distorted by struggle and suffering for survival and afflicted, knotted and doubled with the arthritis of time itself; or an island, no more than a triangle of sand-stone, still flying a green of pine like a flag nailed to the mast of a sinking ship of war in battle; a sacred gateway of a double Torii dark as ink on a parchment of a glowing sky pointing the way to some hidden peninsular shrine; tiny bays and coves, raised to visual significance because they lapped the foundations of cliffs and crags, sculptured by winds and storms into enduring symbols of life's contempt of defeat; and always at the heart of all, a reference to that unruffled vision of water full of light, and at peace with its role of impartial reflection of truth of land and sky like the mirror secure in its simple frame which is worshipped at the centre of the greatest of Shinto shrines. Yet it would be false to neglect the intuitive question which had accompanied me all along like my own shadow and was brought forth, however tentatively, to temper these impressions: could all that beauty not be too good to be permanently true and leave out of its reckoning some warning implicit in the refined trembling of light on a day without wind? Knowing what one did of such volcanic and violent earth, could it be only skin-deep and this fairest of scenes cosmetic rather than aesthetic in the most profound sense of the term? The question, however unfair and much as it puzzled me, happily did nothing to spoil the morning which had other diversions than watching to it.

Of those the happiest for me was a sudden diversion of the *Canada Maru* from its true course. It swung round sharply to aim directly for the coast long before our arrival at Kobe, which was our final port of call. It went so close to the shore that I feared for its safety. But when so close that we could see fishing boats on the beach and some figures of people all as if in a classical woodcut, the ship turned parallel to the shore. As it did so, from one of those small houses, a mirror began flashing at the ship. Excitement was implicit in the speed and reiteration of flashes, and confirmed by a blast on the ship's siren that sounded sacriligiously loud in such a dedicated setting. It was an incident against all conventional precepts of the vocation of a ship's Captain; but another heartening illustration of Mori's stand always to be his own man. It was his way of telling his wife, wielding her largest mirror so passionately there, that he was on his way home. It was also proof, if any were needed, of how essentially romantic his disposition was, not only in affairs of the heart.

And so, free to resume our objective course, we hastened on into the highest of noons and were able to berth at Kobe some hours later. Kobe was our first modern city and showed uneasy signs of the uniformity of unprepossessing design of modern buildings and the proliferation as of single cells of concrete that passes more and more for 'architecture' in the world. Beyond that I do not propose to enlarge on a description of that aspect either of itself or of other cities to follow since it is so self-evidently a platitude of contemporary life. What conveyed so early in a more startling manner the depth of penetration of the contemporary world into the awareness of Japan was our reception by the press.

My friends, Mr Shirakawa and Hisatomi, as well as the ship's owners, had organized the press on a royal scale for the occasion. The moment the gangway was down, journalists and cameramen rushed up it in numbers that I had not previously seen, even on the occasion of the Prince of Wales' visit to Africa which I described earlier. William and I were interviewed and questioned at length and in depth, photographed and re-photographed all over the ship, so that Mori had an album which was thick with news cuttings and photographs of the event. The subsequent reporting, as summarized to us, also suggested that the reporting had been as intelligent and relevant as it was informed and well written. The headlines too were large, black and imposing.

One headline stands out in my memory. It disconcerted William as much as it embarrassed me for him; indeed so much that I have never mentioned it to anyone and so refer to it for the first time publicly here since, alas, it cannot now embarrass him. He must have given Mori an account of his legendary Arden connections with Shakespeare of greater substance than he meant Mori to derive from it. And Mori, part of whose mission and indeed battle for vocational survival it was to invest our visit with as much prestige as possible, could not have resisted making the most of it with his own highly developed sense of drama. As a result, the headline proclaimed in boldest type across an entire page: 'Descendant of Shakespeare arrives in Japan'. But what justifies recording it is that I still believe, as I did then, how much the elevation of this piece of information to pre-eminence in the newspapers of the day indicated priorities in national values, absent from the world of newspapers that I knew.

Soon after, we had to say goodbye to the *Canada Maru* and Mori for a while, and get busy at once about the main object of our journey. We had to hasten to see and learn as much as possible of Japan, the new Japan, Mori tended to stress; in the little time allowed us. I call it little because, much as I hated the necessity, I had no option but to return to Port Natal in the *Canada Maru* if I were to keep faith with Mori and my newspaper. William, though not yet decided but under pressure from Mori to do so, could if he wanted stay on indefinitely in Japan. The problem for him was that such a decision would commit him to an adventure of great risk and uncertainty. He had no promise or prospect as yet of work to keep him, which was essential because we neither of us had any money to spare. But Mori, with his imperturbable faith that Providence was always benevolently involved with his plans, thought that between arrival and departure a way would be found for William to stay and write the book of books needed to interpret the true Japan to the world.

Our respective goodbyes, therefore, were uttered with a difference which was increased by William's relief at having done with a heaving ship for some three weeks at least. As a result he followed Mr Tajima, in whose keeping we now were, in a more positive mood than my own, despite my eagerness to set foot on the main island of Japan. Yet my inner reservations did not last long. For the first thing Mr Tajima made us do after booking us in at, to my regret, a modern hotel was to climb the hill behind Kobe.

He did so, I believe, out of an instinctive conviction that the proper introduction to Japan should be through not man but the Kami, otherwise what followed could be more vulnerable than ever to normal error and misunderstanding. Also, no hill at that moment could have been more suitable for such a prelude than the one chosen. It was in itself a natural temple of earth and fragrant pine. Its summit supported one of the most famous shrines in a land of shrines. Pilgrims came there in hundreds of thousands, to commune with the spirit of nature and of ancestors who mediate between earth and heaven and made them and all that was below at one with the Gods and their universe. Moreover, the hill was steep and constituted a test which was in itself a symbolism of behaviour as to how the spirit had to endure to the end if it were to achieve the harmony that gave the human round on earth the direction which is the 'meaning', held to be that which has always existed through itself. 'Meaning', so defined, seemed to me that afternoon significantly at one with St John's declaration that in the beginning was the word and the word was with God.

We could not be climbing so steeply, I seemed to feel, if this meaning sought by centuries of pilgrims were not also with the Kami and, through the Kami, with their gods in the beginning. Moreover it was as if centuries of pilgrimage and searching had aroused in all something above a compassionate obligation to help humanity in its obedience to such an intangible prompting of spirit. It had inspired it to hack and build some thousand and one steps in the side of the hill to ease the ascent! But even eased that way, it was an extreme physical test for both of us. We arrived almost at the end of our powers at the summit, as the light of the sun began to level out.

Once there, however, all fatigue of mind and body fell from us as easily as the first leaf of maple that I saw falling in Japan. It flickered as it fell with the colour which makes the Japanese speak of maple-flower, obviously pardoned and blessed, for being there so small and growing where no maples as far as I could see were supposed to grow. Below us we saw what looked like the whole of the inland sea, more beautiful than ever under the lengthening light of evening, and of a stillness as that of its Gods in a moment of deepest meditation. It made us hasten to present ourselves in reverence to all that the shrine and moment evoked, and to report our arrival to its spirits. The difference it made to my own state of mind was as profound as it was unforeseen. All the sense of rebellion stirring in me against my small ration of time for discovery of this new world, stretched out as in an act of worship below me, left me there and then for good. I gave the matter no further thought, and was as prepared as an alien could be for what had to be. But as if designed to prevent any disproportion and inflation of unworldly presumptions by a reminder of human mortality, I set out in the morning somewhat stiff in muscle in spite of the fact that I had behind me the benefit of constant exercise imposed by my Ministry of Sport. William, who had had none, could hardly walk and was in great pain, which he bore without complaint and so well that we two alone knew of it.

After such an initiation there was a satisfying logic of spirit as well as history in Mr Tajima's decision to take us on immediately to Nara. Moreover, we travelled by road and could share as if with the emotion of the earth itself, as it broke out of steep mountains and deep clefts to present itself open to the physical world as it had been in the mood of man who first created it. It is true that the land about it was still neither flat nor plain and preserved the nature of a valley; but a valley with the most gentle of slopes designed not to divide and imprison the earth, but to protect and contain it. Accordingly it was wide enough to keep some low hills at their proper blue of distance and spread out towards one solitary mountain of unassuming proportions. Moreover, the mountain at that time was covered with a dark, brooding forest held to be sacred; according to Mr Tajima, it had not been touched by the hand of man since the beginning of time – not a single tree cut nor branch wrenched from a trunk to be brought back by some Japanese Aeneas as proof of divine guidance to wavering followers. Mr Tajima, of course, was not, he declared, a superstitious person. But most people still believed the forest was a home of Kami whose voices could be heard on the wind moving over it as they consulted one another about everlasting things. At night the foxes barked uninterruptedly with an incurable melancholy in voices reserved for the Kami. Only the sacred deer who shared the forest with foxes, birds and gods moved freely between it and the world of man as emissaries of concern of the world beyond for living things on earth. I was aware, as Mr Tajima spoke, of more emotion to his rendering of the popular regard for the forest than his worldly-wise attitude would have admitted.

But already we had seen enough to marvel at in the wealth of forest and trees preserved in this instant cross-cut of one of the areas most densely populated with man. I marvelled at the fact that the Japanese had not destroyed their heritage of trees as had other men who were continuing to do so everywhere in the world. I marvelled because no other civilization seemed to me to have been throughout its long history so exclusively dependent on forests and trees as this one, because it was a civilization largely founded on wood and paper. Yet the phrase 'a land of paper houses', applied to it by Europeans, did not seem adequate. It had far too many implications of the picturesque and quaint in it to do justice to a most intrusive so-called reality. Worst of all, it seemed based on comparisons as odious because they were inspired by unconscious assumptions of superiority. If to accept what life presents, however imperfectly, at once and without question, as being the only material out of which man could transform both matter and himself, was a cardinal virtue, as so many a Zen Buddhist that I was to meet did then, this blend of loving presentation and creative use of a forest basic to civilization in Japan appeared to be an achievement of the highest and most original order.

So far wherever I had looked from ship, hill and road, all houses, bridges, to things of everyday use – windows, parasols, soup bowls, platters, chopsticks and trays – all were made of wood and paper. Even water clocks and handkerchiefs, let alone books and newspapers and, most movingly of all, shrines and temples; all were similarly constructed. I remember how struck I was at our inn in Moji to find that such plumbing as there was, had not been installed with lead but conducted by pipes of bamboo. Whether Mr Tajima knew it or not, I was convinced that the sight of that ancient forest meant far more to him than even I, with my love of the bush of Africa, could fathom. Even so, there was more to come. . . . This tentative inkling was raised to absolute conviction when we saw what Nara had made of wood and paper. The achievement was all the more impressive because what we saw, Mr Tajima stressed, was a mere fragment of the ancient capital of Japan. Although the year was still only 1926, Nara had long since declined into a small country town, famous only for such shrines, pagodas and temples as were left over from an abundant past. And it was sparsely spread around an immense park where we were compelled to go as part of a religious ritual, to feed the ambassadorial deer that were its citizens.

The decline, of course, had started centuries before when Nara ceased to be the capital. But it must have been hastened with the fervour with which the Japanese turned their backs on their own past when the irrevocable decision was taken some two generations earlier to emerge from its isolation. At first it constituted a grave threat to traditional values and particularly to things of religious value and veneration. In the Kyoto which succeeded Nara and became an even greater and more lasting achievement of Japanese civilization, for instance, a famous pagoda, neglected and in dangerous disrepair, was for some years put up for sale. Happily no-one took up the offer before the moment of madness passed and the rediscovery of a forgotten national self arose again so ardently that it was already in danger of swinging to the other extreme. So precious, however, was the fragment of Nara which remained that Mr Tajima described it at a length which I cannot repeat. Nor shall I attempt to dwell on all that we saw because it is done so already in libraries of books dedicated to the subject. I am compelled not to step outside my own living experience, small and brief as it was. The impressions, feelings and facts that were attacking my powers of expression all the time were formidable enough without complicating them with the statistics of history. I had to keep all that happened as simple as I could in order not to be overwhelmed, and try to follow the example of one of the most impressive lessons of art and religion in Japan and wherever possible use the small and insignificant to evoke the great and meaningful complexity of the unifying whole.

I tended therefore to make a kind of Haiku orchestrated in prose of what this approach to Nara did to me, evoking, I hope, an experience rather than deepening a gathering of wayside knowledge. It was enough, therefore, to hear Mr Tajima tell us how farmers everywhere in this valley were still ploughing up stones that were part of the foundations of many vanished palaces, castles and temples of Nara. He recalled, apologizing for not knowing the precise words, how already in the eighth century, within a decade of the founding of the capital, a classical anthology, 'The Manyōshū' or 'Collection of Ten Thousand Leaves', included a poem of the great Ono-No-Oyu, which salutes Nara as 'The Capital which blooms in fullness of glory like an unfolding knot that binds the petals in the heart of a beautiful flower'.

That was, in a way, enough for me. I do not think I would have liked to be presented just then with a Nara complete and active, as the capital of a country coming to desperate terms with its tumultuous and destructive past. All that administrative social and political activity, the buildings and institutions that went with it, could so easily have obscured if not obliterated the significance of Nara. It seemed to me that one of the great services of time to man is that it strives to separate the real from the unreal, false from true, and the imperishable from the perishable. When Shelley speaks of Athens as 'a crest of columns gleaming on the mind of man', he is in a sense using instinctively the Haiku way that I had in mind, namely evoking more of the still living essence of what Greece did for the spirit of man than cartloads of books of facts about the complete Parthenon and city state of Athens could ever have done. I thought of this just at the moment because, as we came nearer in a light of autumn that could not have been more Euclidian in its attention to precision of line and form, the silhouette of Nara was like a frieze of pagodas, not glowing so much as in full flower on the tallest of stalks, at a new dawn on a far frontier of the spirit of Japanese man and his devout and beautiful earth. I have to stress the 'devout' because the nearer we came the greater grew the multitudes of sacred stone lanterns of all kinds flanking steps and lanes to show they were signposts on a way of light to complete illumination and understanding in the keeping of the shrines and all hidden places of natural sanctity.

Wherever I looked there was evidence that hardly a part of this earth and what grew in it was not cared for in spirit and revered in heart. Most remarkable of all was the consummation of this abundance of diverse and minute attempts by man to be obedient to an essential part of his contract with life. The spirit which moves him and gives him meaning must not only be lived but made to be seen and graced with a beauty of tangible and appropriate form. The gathering together and summing up of all in one single final movement, as all the notes drawn together from the instruments of an orchestra in the last utterance of a great symphony, was accomplished for me by what was left of temple and pagodas in the scene before us. No wonder the Japanese were still speaking of one great Nara Pagoda, the Yakushi-ji or 'Temple of the Master of Healing', as frozen music! But since we were coming from the direction of Kobe, the process of consummation of all in one transcendent whole was begun for me, not by it but by the Horyu-ji, the Temple of the Flourishing Law, the oldest temple in Japan. It presented itself, not like so many of the great religious buildings of the land concealed in secluded valleys and mountains and clasped in a Shinto embrace of nature, but open and exposed in a countryside without secrets under the autumn sun. The avenues and trees that led to it, impressive as they appeared, were unusually ordered and not only naturally but consciously in harmony with temple and pagoda.

This harmony was all the more pronounced because the curves of the roofs of the temple buildings like the Middle Gateway, Golden Hall and Room of Sermons, as well as those of the tall pagoda of five-stories, looked at that moment to be not the solid work of man thrust upwards from the earth, but like something alighted from the sky like the wings of great birds folding as they found their nests before the fall of night. It was my first pagoda, and the shock of its beauty and grace as it stood high in the blue of sky and sun, illuminated all the talk I had heard of its symbolic, studied representation of Buddhist thought and its role as a visual diagram of its metaphysics of the universe. Elsewhere in the world of Buddha, I knew from many an illustration, the Pagoda and the Stupas that were its prototypes were bound to the earth in mortar and stone. This pagoda of the oldest surviving temple in Japan, and perhaps the oldest wooden building in the world, was impelled upwards with all the naturalness and simplicity organic in the stalk of a flower to enable it to explode in blossom like a firework in the sky. It had, as it were, special dispensation from the laws of gravity to soar as no other form of building I had ever seen; and the harmony and release it found in the process seemed all the more natural in that it was a composition entirely of wood. I thought it one of the most beautiful forms ever to issue from human imagination, and full of additional meaning because all special pleading and interpretation of mind were swept aside. Like the great spires of the West, it was an immediate assertion of the urge in man also to reach up to the sky, not just by climbing mountains and building towers of Babel, but through movement as in a singleness of voice in the searching music of his mind and heart. It made all I had seen more complete because where the Torii had accompanied Japanese man from so far back and provided him with a gateway to seeking on earth, the Pagoda had joined it as if in obedience to a new commandment. But a commandment to what end? I had no immediate idea and knew only, from the emotion stirred up within me, how important both vision and question were to me, as we approached the building.

This Temple of the Flourishing Law was to be partially destroyed by fire after the War. But we were lucky to see in it that day scores of original works of art and frescoes of Buddha and Bodhisvattas. Early as their hour of creation had been, the love of simplicity, lack of ostentation and rhetoric and devotion to purity of line so dear to the Japanese were already at work, raising them all to the highest level of art. I could hardly have had a better preparation for the answer and climax of revelation. These came to me not in the Temple itself but in a small convent close by it. The Temple, though modest by our own Greek, Roman and Gothic standards, was big for Japan. It was an essentially masculine and majestic creation, accomplished without diversion and afterthought according to a preconceived plan. The convent, however, was small, somewhat overshadowed and out to please the temple. It was, nonetheless, full of a feeling that seemed to have made it essentially feminine and, without hesitation, prepared to sacrifice a regular pattern dictated by reason of mind to changes of course and form determined by reasons of the heart. With a courage of feeling that seems to me uniquely of the feminine, it seemed to know that, in its own irregular way, it did something for the great temple which it had not been able to do for itself.

Once allowed to enter the convent (only by grace of the skill and sincerity of Mr Tajima's advocacy), in a silence that was total, and the cleanliness, the fastidious circumspection of rooms without stain, speckless corridors and floors without blemish, where even our movements on tip-toes sounded loud and lacking in the respect that was essential, we were compelled to an obedience of the atmosphere of the convent more exacting than any masculine commandment could have procured. And that too, for me, was an essentially feminine achievement because at the heart of the feminine there seems an understanding of the necessity of atmosphere for increase of the human spirit; and this is just as great as the fall of rain for growth in the world without which symbolizes it, and which the masculine tends not to possess. Nothing could have been more appropriate, I felt, that this intangible element to envelop us as finally we came to the small temple which was at the heart of the convent. In this heart of a heart, we found what we had been seeking, indeed what all sorts and conditions of men and their civilizations had been seeking since their beginning and were still seeking: a clear and full statement of the meaning of meaning, and stated with a beauty and authority I had never seen in any sculpted form. Its impact was immense and so unexpected that I believe I would have lost all inner self-possession and burst into tears had I been by myself and not supported by powers of restraint that come from being a member of a group. Behind the altar, flanked by all sorts of little feminine things and objects of reverence, was a figure of Buddha in an attitude I had never seen.

It had over it a glow of soft lamplight as if still not yet deprived of the diffusion of light with which it had dawned in the imagination of the artist; at the same time it was without loss of inflection or blurring of line because of its clear, immediate and conclusive presentation. Priests, artists, scholars, were arguing passionately then, as they still do, which Buddha the artist intended the figure to represent. For me the argument was support of my own feeling that I was looking at a summing-up in the most sublime form of all that Buddha in his various incarnations meant for man: a state of being no longer blinded by desire and ambition of worldly power but which had seen through illusion, 'the floating world of appearances'; a state that had renounced the provisional; and discovered the meaning even of suffering to attain enlightenment not only for itself but all 'sentient things'.

Complex as all this may sound, the Buddha before us was achieved with the greatest simplicity and economy of line, so that all this diversity and complexity were gathered together as in a single chord of music. But more than all this, the artist has succeeded where so many others had failed. In preventing this figure from abstraction in some remote dimension of infinity beyond 'the vortex of becoming' wherein we were caught, it made divine love human in an unbelievably tender and essentially feminine realization. That it should therefore have found its way into the keeping also of a convent and not a monastery, had a fitting quality about it. For if, among the aspects of the many Buddhas summed up in the figure before me, I had to pick on one as in command of all, it was that of Kwannon, the feminine apotheosis of the merciful, the benevolent and all-compassionate. Serene and resolved as the face was, it was lit with love and human solicitude. A smile formed lightly on the face could have been that of a mother watching a child falling in its first effort to walk; seeing its defeat already as the beginning of a final victory over the impossible. Old and wise, yet it was also young, and child-like, as if in the midst of its own translation and transfiguration it knew that on that far frontier of the spirit it was still a child of eternity. Most tender and human of all for me, perhaps, was the movement of the slender arm and long hand so that the tips of two fingers could feel the face to reassure even this metamorphosis of flesh and blood that a smile born of inner certainty was truly forming there. It still had to endure some element of loving doubt if only so as not to lose contact with fallible humanity in search of soul.

I knew then, I believed, why Nara was so significant a moment in time for the Japanese. It is because this representation of Buddha could have come out of a vision of truth which is imposed on men only when they have stood fast against the onslaught of death on meaning in their spirit. I thought of Michelangelo who carved in words, as though cut out of his most sombre stone, a poem in which he said something to the effect that 'whereas death killed all men, the thought of death also made them'. Here I was facing an equal of his in vision, if indeed not someone beyond even him in execution. The meaning they both served was the same. It was only when man looked death full in the face that the mortality which is imminent in the final regard releases him from all excess in his proportions, and in the surrender of egotistical presumption which follows as night the day, unlocks him for the experience of compassion for all living things, 'from ant to Emperor, whale to cat', as the Buddhists of Tibet put it, which is the sign of his conscious return from exile to the all-belonging which had been his point of departure and is then his Home.

I enlarge on this because it was here at Nara that Japanese civilization made its stand at last against death. Up to then whenever an Emperor or Empress died, the Capital packed up and moved away from such infected earth. For centuries fear of death, not so much in a physical sense, because even then no people had been braver in battle, but as death in the spirit, had made it nomad. But here at Nara in the eighth century it had decided to stand fast, look death in the eyes, and begin a dignified dialogue with it. The deliverance from fear and flight from death which followed, produced an incalculable emancipation of the Japanese spirit, already so confident as to hover on the face of this Buddha in that smile, subtle and enigmatic as in da Vinci's evocation of the eternally feminine. The dialogue that followed this stand elevated it to heights unforeseen and undreamt of in its past, and raised the sights of its imagination to the range of infinity.

So far-reaching have been the consequences that the dialogue continues even to this pragmatic and expedient day; and for centuries it put a whole culture in danger of having the kind of romantic 'affair' with death that most others prefer to reserve for life. It was as if Torii, the sacred gateway which had kept a whole people company so far back and for centuries, facing only one way, now faced two ways. Seen only from in front, it had been like birth: an entrance to a way of life on earth, ended abruptly by death. Now, seen also from behind, it was another entrance to a way through death so that in a birth of death it served a life greater than either. Though it continued to be pledged in the world without and to tangible witness of the Shinto promptings of the national spirit, it was now joined by the image of the pagoda on this new frontier where man needs more than the knowledge that he has brought it there in order to carry on beyond. A signpost of the sky rather than the earth, it commanded that it was no longer enough for men to live horizontally: they had also to live vertically. The conclusion made music in the imagination of an era in which pagodas dominated that skyline of Nara.

Appropriately I saw them last at dawn. Unlike the evening before, the outlines of their five stories now were like wings of great water-birds lifting as if to take flight upward with the rising light of day. Around them the trees, acolytes of the sacred forest, crowded in close support. Though the dawn was bright, the day was stormy, and a wind, already high, was rising. To my eyes it looked as if trees and pagodas were beginning to sway alarmingly. Mr Tajima reassured me. Trees may have been uprooted again and again by earthquake and flood, but no pagoda had tumbled yet. That too seemed a sign of the things Nara brought into focus for me, as the Japanese would say, for good.

And so not unprepared, but neither fully equipped, we went out to meet the fulfilment and increase of all Nara promised that was Kyoto. I have been back to Kyoto several times since and have not yet come to terms with all it gave and still gives. But on that brief encounter with a unique city that was to be the capital of Japan for more than a thousand years, I was totally overcome and incapable of forming any concept of the reality. I was compelled just to let it all happen to me, holding my breath like a swimmer tumbled by a wave of the sea more urgent than its predecessor . . . and yet managing to hold his breath long enough to be swept in a boiling surf back up on to the shore. The natural setting also in itself was awesome because we were back there in a world of passionate and dramatic nature, which refused to be denied, with woods sweeping from their base in a deep valley to peak over one hieroglyphic mountain after the other. Again the view kept on presenting itself as a series of landscapes. But by this time the great Sesshu had penetrated to the heart of natural matter and the potential of storm within it, as no painter of the external scene has ever done for me. Happily Kyoto's involvement with the mind of man had been so long, personal and intimate that all the stories and legends attached to it helped to keep the impact in an order that stood between me and chaos. What indeed could have indicated the position of Kyoto, the 'City of Purple Hills' and 'Streams of Crystal', more appropriately, for instance, than to call the two highest mountains around it: 'The Mountain of the Cave of Love' and 'The Mount of Wisdom'. . . . The first honoured its Shinto and aboriginal debt to nature; the second its Buddhist progression.

This last, I was told, predominated and was represented by numbers of temples and monasteries hidden in forests and valleys of the western approaches. But closer in and all round this city dedicated at its founding to 'Peace and Tranquillity', these two themes were continued more openly and enlarged and enriched not only in temples, monasteries and hermitages but gardens, villas, palaces and pavilions in numbers that were hardly credible and which made me aware of how bleak and impoverished in this dimension of life was the world of my own origin. I felt I was in Japan now almost as an orphan in a storm of riches. I have only to have a roll call of the names still waiting like actors in the wings to be called on to the full stage; as for instance those of the Temples of 'Enlightenment', 'The Blue Lotus', 'The Essence of Unlimited Light', 'The Three Treasures', 'The Great Science', 'Benevolent Harmony', 'The Dragon's Repose', 'The Calm Light', 'Serene Quietude', 'The Celestial Dragon', 'The Absolute', 'The Pure Fountains', 'Western Fragrance', 'Gratitude', 'The Miraculous Law' and the imperial villa of 'The Ascetic Doctrine', 'The Poets', and finally the 'Palace of Noble Fragrance'. Any one of these could have held one's attention for years.

The temples, above all, were huge complexes, not just of places of worship. They were used also for instruction, colleges and universities, living museums, and homes of vast and as yet uncatalogued collections of treasure of art, books and manuscripts as well as guardians of a heritage of gardens and parks. All these used the vocabulary of Nature as poets and seers use words to prevent men from becoming of the world while remaining in it. All these things, and more, were there to pass before my amazed and bewildered eyes; and also to move me to become a nucleus of future understanding significantly through my African, rather than my European, self.

It did so because all this imposing achievement spread out so abundantly in and around Kyoto seemed to me a result of the deep, ineradicable reverence for the small. The sense of infinity even in a grain of sand, blade of grass or fallen petal of flower, was the pulse of Japanese civilization. Here I was intimately at one with them because as a child my imagination had been permanently aroused by the reverence of the first people of Africa for the small and apparently insignificant. This led them to the creation of an Olympus of their African spirit invested not by the majestic personnages of their daily scene like the elephant and the lion, but also by an insect whose physical beginning was so minute as to be almost invisible, and who had for his companions such intangibles as the colours of the rainbow. So when faced with one of the greatest and most gracious temples in the midst of a learned discourse on its history and doctrine, what really impressed me was the information that it attached special importance to the preservation and display of some thirty different kinds of moss.

It seemed to me proof that this lofty exegesis of Buddhistic enlightenment to which we were exposed, would not have been possible had the continuity with the basic feeling for nature not remained unbroken. The mind was still taking instruction as though newly arrived in the kindergarten of Nature where the teaching of the alphabet of Life began with moss.

I felt this most of all when we came after many a temple to that of the 'Dragon's Repose'. A nephew of Mr Shirakawa who had temporarily taken over from Mr Tajima the duty of guide and interpreter, had been so long explaining to us the importance of The Temple of The Celestial Dragon and its garden that we had little time left for more. We were grateful to him for the temple and the garden onto which it looks because it was one of the most significant and beautiful of all. 'Beautiful' is not what a European gardener might have termed it. I fear the gardeners of my world would have dismissed it as a naïve, quaint, picturesque and even a distorted evocation of nature. Again there was no ostentation or rhetoric and exhibitionism of flowers, or ornamental shrubbery. The garden's inspiration came from the same area of the imagination out of which shrines, pagodas and temples arose. There was no preconceived pattern of the apparently beautiful and decorative imposed from without on this garden. Indeed, it was as remote as it could be from our obsession for imposing our own laws on, subduing and mastering nature for our own ends. It was the product of a profound and trusted partnership with the natural. Although that is a simplification, it would be true to say that where the continental garden was a creation of intellect, this temple garden, and a thousand and one others, was born of a love of nature in which each and everything had a shape, a validity and a right to maintain an inalienable dignity of its own. The gardener was the explorer of an inter-relationship, a secret sympathy, hidden in the profusion and diversity of nature, and so pledged to reproduce its underlying pattern on however small a scale so that the original design was there for the spirit to contemplate and follow.

It was another leap forward in a process of increasing affirmation not unlike the mapping of the new worlds discovered by the explorers in whose wake we were travelling; tentative, yet true enough to guide others until all was laid out as this garden was, complete and in truth like Mercator's infallible projection of the rounded earth. The garden, of course, was also part of a process of discovery of the world within and a poetic grasp at its elusive meaning. The result for me was that this garden appeared not apart from the temple so much as an integral part of the same indivisible whole; that it was, as a priest told us, brought constantly by itself out of the nature of its conception into the temple. I was not astonished so much, therefore, as excited when told that the spirit which presided over the conception of Temple and Garden in the thirteenth century was one of the most exceptional in Japanese history. It was so exceptional, I was told, because his life had been transformed by a dream. No-one could there and then tell me precisely what the dream was. All they could say was that the influence of the dream on his life was such that after his death another name was added to his family one. He was literally called 'Window Dream', meaning that he himself had been a 'Window on the Dream'. I seemed to know it all then: the temple was the window and the garden the dream.

We left then as we seemed to be doing all the time, in a brutal haste. But I myself was feeling that this last thought had brought me as far as someone of my upbringing could go into the language of the garden and that I had been taken to the frontier where poetry and prayer met. Yet there was more to come. We were now taken from the temple of the Celestial Dragon and its garden to that of the Dragon's Repose, presumably the same Dragon as the one honoured in the first, without a special reason. It too had a garden, which has remained for me ever since as the garden of gardens. I have been back to it several times, the last only recently, and it is today even more crowded with pilgrims and visitors than it was then; the numbers of pilgrims falling, those of visitors increasing. It is now world-famous, and the Zen Buddhism of which it is a product has a growing international following. But on that late afternoon in the autumn of 1926 there was only a priest at work in it, purifying it of the stain of the day for its journey into the night; and an old lady, in the sombre, unflowered kimono of a grandmother, enduring her going hence as she had her coming hither, holding a young girl in a kimono of the loveliest colours, by the hand; and we ourselves.

It was, I knew already, labelled as 'The Abstract Garden' but the label did nothing to protect me against the shock of the abstraction. At first glance there seemed nothing to justify calling it a garden at all. It consisted of a stretch of well-raked gravelly sand from a river bed, with some fifteen stones, large and small, carved by time into shapes of an indefinable but definite meaning. Moreover, they were placed in such a way that walls, gravel and all were related to one another in harmony. I observed them, moved beyond reason, as one might a devout and devoted family in an hour of hate and envy. But even so, was it a garden? The question raised itself immediately because at first glance nothing seemed to grow in it, and what was a garden if not for growth? Then suddenly they fell on a sleeve of green lichen clinging to a little stone and further on some moss round the base of the greatest in the gathering. The moment I saw the moss, the process of recognition began. It quickly swelled not into my rational perception of meaning so much as an emotion of sharing the intent, which produced a conclusion that it was a garden before and beyond all gardens. In its self-denial of all the means of eloquence at a gardener's disposal, and leaving so much unsaid, it evoked a whole far more poignantly than any complex statement, however inspired, might have done. I remember unbidden thoughts coming like spontaneous happenings over me: as, for instance, that rivers flow on and empty themselves in the sea; the earth may be torn and washed away; forests of great trees shattered and uprooted. But the beds to which this raked-up ground belonged, these stones that once supported the earth, and the moss which was the yeast that quickened cold clay to grow tall, remained, so that when the storms of wind and time passed, the earth would rally again around those stones; the rain would fall once more; the rivers flow and, through the moss, renew grass and forest. The face of things change and shatter; but these things remained. So this then, was it not a garden for all seasons in the life and mind of man? And. . . .

I was interrupted there by William. He had been moved to sombre thought of his own, although it led in a direction he normally either avoided or as a rule preferred not to discuss. The reference as often was oblique and came in a remark that pretended to be lighthearted.

"You know, Lorenzo," he remarked with a smile, "this garden makes me think of what Eliot said about Webster: 'Webster was much obsessed with death. He saw the skull, beneath the skin, eyes that dull and a lipless grin'. And if you ask me, that old lady there feels something of the same sort as well."

He referred, of course, to the old Japanese lady. She was standing still, her eyes on the garden, and her grip on the hand of the beautiful young girl tighter than ever. There were bright tears on her finely creased cheeks, although the face was a mask of composure without a hint of external expression. The girl was watching her intently but it seemed to me to be with a strange look of concern, which worried me until it occurred to me that perhaps the tears were not caused by distress but by release at finding in the garden a reassurance which life beyond its walls could not give.

I watched the pair as closely as I could without being offensive, and became confident that my interpretation was not wide of the mark. The pair stood there as if youth and age, beginning and end, were one. More, they left still silent and hand-in-hand; the way they bowed as the people of Japan bowed only to the Emperor whose title 'Honourable Gate' implies that he himself was meant to be a signpost to the gods, was to me as much affirmation as example.

Meanwhile our guide had talked to the young priest who at first did not like the interruption of his work of ablution, but in the end he came over to meet us. The look on his finely-boned, inwardly-turned features made me feel as if we had interrupted someone at prayer. We both accordingly asked our questions as politely and quickly as possible. Yes! The garden, he said, was conceived at the end of the fifteenth century, by the great Zen tea-master, and creator also of the garden of the famous Silver Pavilion. As for its purpose, it was there for each one who came to it, to read for himself by his own light and not by that of others which would wither, blind or dim his own. But if pressed he would tell us a story briefly and quickly because he had work to do and it was always later than men thought. . . .

There was once an abbot of a Buddhist monastery who was a truly holy man. Near the monastery was a small country village. One day a great scandal came to light in the village. A beautiful and unmarried girl was found to be pregnant. She refused to say who the father was until the child was born. Then, under great pressure, she said, falsely, that the father was the abbot. The village, enraged, confronted him and demanded that he took care of the child. His celibate monks, outraged, demanded that he expel himself. He made no effort to defend himself or deny the falsehood. He looked calmly at his accusers and just exclaimed: "Is that so?", took the child from the mother, settled it in his arms and went out into the world as a beggar. After years of bringing up the child on alms, begged in the harshest of ways, he found himself back at the village once more. Monks and villagers confronted him, all in tears as they prayed for his forgiveness. The mother had confessed. She had lied. He was not the father. They had sought him for years. The mother herself was there to beg him for pardon and to ask for the return of the child.

"Is that so?" the holy man exclaimed, as he urged the child towards the mother. Then without another word he turned his back on monastery and village and went calmly on his way.

This then was the garden of such acceptance. To ignore falsehood, injustice, the unreality of appearances, and to take on as one's proper task the need that was nearest and most immediate, that perhaps was a way of reading the garden. People call it abstract as if it were remote from the world. But it was 'here and now' because it was also 'before and after'. In fact it was the 'that' in 'is that so?', and with this remark the abbot turned away abruptly and resumed his raking.

On that note of life and not death, we went back to the city to meet Mr Tajima. The twilight was fast losing its colour, and the lights of paper lanterns of all shapes and colours were making a dawn of their own out of the darkening crepuscule. This dawn was most colourful in the ancient restaurant and Geisha quarters where we had our rendezvous. The light, the colours of the kimonos, the kara-koro of Geta, the conversation more in urgent whispers as of a world on its way into the night, created a mood in me as between dreaming and waking. My own self seemed in suspense and oddly provisional, so that our reunion with Mr Tajima was in a way a return to lost reality.

I had persuaded Mr Tajima to allow me to take us to the best restaurant that he could find, with one reservation which embarrassed me greatly: namely, could I borrow the money from his company on the understanding that I would pay it back the moment I returned home? The persuasion had been difficult not because of the money – that he made clear was of such unimportance as not to be worth mentioning – but as a matter of honour. I was certain then, I had said, how he would understand, how much my own honour was also involved. He had looked at me in a new way, quickly and intently, and unexpectedly for someone with so sophisticated a dress and appearance, bowed and, coming out of his bow, remarked softly that paying respect to the demands of honour of the guest was also an obligation of the host.

With that he became a man of the world again and said, smilingly, "Besides, Post-San, I would be very surprised if you get your bill within a year. In this restaurant we are going to, these things are still the matters of great delicacy they were when Kyoto was first founded. Money is never mentioned within them, even now, to gentlemen." And, he explained in detail, the question of a bill and its payment was so involved that it was left to emissaries who were specialists and knew precisely how not to violate either the sensitivities of the guest or the needs of the hosts for an appropriate settlement. I was only to appreciate fully what he meant some nine months later when I received the account with many apologies for the rude haste with which it was being presented and the trouble to which the unpardonable imposition of so trivial and unworthy an affair would put me! But by then I was not surprised.

The restaurant was one of the oldest and most honoured in Kyoto. One was learning fast, no matter where one went in city or town, there was no place or person, however humble and exacting might be the social priorities, that did not possess honour of their own. The result was that they ultimately honoured as they were honoured. And in restaurants such as these, it was not the guest who conferred honour on it, but very much the other way round. Indeed so much did this subtle rule apply that men of ambition struggled for admission into such restaurants because to be known that they had been there was a reference of their worth and of great pragmatic value in the furtherance of careers. As a result many came but few were admitted. The restaurant was so conscious of the obligations heaped on it by the honour in its keeping, that it had no hesitation in turning rich and powerful men away. Many a politician of note had been denied, while men without worldly power and little wealth were welcome because of the quality of their being. In this the family who owned it were supported by a highly selective tradition evolved all over the city in the four centuries since the restaurant was founded by their ancestors.

We ourselves were admitted only because the Foreign Secretary, Baron Shidehara, had asked the restaurant to receive us on behalf of the nation. I have deliberately elaborated because, small and insignificant a matter as dining out in Kyoto might appear, this was a clear illustration of the city's unique character. Although no longer the political capital of the country, poor by comparison with Tokyo and the bigger rapidly expanding cities of industrial wealth, like Osaka, it was still the capital in the heart and spirit of the land and as such it was a keeper of riches of a different kind. Its own native people were so conscious of this that more and more they were held by the rest to be proud and full of presumptions of superiority. They even spoke in accents and used a vocabulary too archaic and difficult for the emerging generations. And I was to meet exceptional men from elsewhere who worked in it as exiles in a strange land.

Yet again outwardly the restaurant had put out no obvious signs to proclaim its unique character. Like the houses in the residential quarters of the city, its back was turned on the world, being unremarkable and indifferent to appearances. One had to enter to discover the quality and expression of a singular and finely introverted beauty. Even our reception was different. There was nothing commercial or obsequious about it. From the moment we left the world, with our shoes behind us, and mounted the wooden step, shining like a mirror, the ladies of the house conducted us to our room as if they were royalty of a kind receiving royalty. Yet all was of the simplest. The room was only half the size of our hotel at Moji, and its beauty lay entirely in its cleanliness, proportions, texture of mats of finest rice-straw, quality of the wood, shaped and cared for with a love as if the material was possessed of spirit itself. As a result the devout attention was there as a patina of bronze on the surface of the wood and the grain brought out as fingerprints of the hand of time itself so that the attention of centuries was upon it. For decoration there was only, I remember in passing, a single sprig of scarlet maple in a black vase, at an asymmetrical angle which evoked in the Zen way vision of the symmetry to come. There was also a scroll of writing in a hand whose beauty seemed exceptional even to me. It was written, we were told, by a famous monk at a time when all this beauty and dedication to refinement, delicacy and a state of spirit, was belied by the brutality and disorder in between the wars which convulsed the country. This intrusion of history was the first of these portents of an as yet unresolved dichotomy on earth as in the spirit of man. Seen out of the context of its own day, one was profoundly disturbed that all this dedication to beauty and quality of spirit might still have failed to cancel out its dark and brutal opposite. But to return to the monk of the scroll. He came there often, we were told, and on his last visit wrote something to the effect that: "Dining here with a friend simply, I discover a tranquil mind and the shadow of the world outside is lifted: joy and harmony here, but no excess even of pleasure."

It was, as William remarked, characteristically tempering emotion with a joke, the first time we would have been in a restaurant where the guests were warned in advance not to over-eat!

Typically, too, there were no Geisha. We each had only our own waitress, at our side to feed and entertain us. Yet their clothes were of a beauty of great antiquity that came from afar. In between courses, two other young ladies came to play the samisen, sang and danced quietly for us. The words of the songs inevitably in the esoteric vocabulary used only by women, and moreover the women of Kyoto at its beginnings, were out of my reach, and even Mr Tajima's paraphrasing was unsure and brief. I remember mainly the mood and atmosphere created as if charged with nostalgia for things past. I remember in particular when our meal ended with a lacquered bowl not full but holding only a disciplined helping of steaming white rice like the rarest of dishes. Then came the tone and sound of the music that followed.

The samisen was given over to the oldest of the young ladies. Like Teruha, she had that ancient Yamato look with a purity of line yet charged with an intensity of spirit that her heavy make-up could not disguise, for it glowed in her delicately pencilled eyes, like a fire about to burst into flame. In her hands the endemic melancholy of the samisen became the tone of anguish of soul as to the voice that joined it. It came from far back, shadowed by this reality of a Kyoto, split between a love of beauty and denial of all it implied in a world of killing and disorder; so much so that the notes plucked from the samisen were hurt and dark with injury . . . and the young voice of the singer came not from mouth and throat as much as from the pit of her stomach, with a sound as old as time itself. It was indeed as if she were a ventriloquism of history so convincing that music and words were relevant not just for Kyoto but all ages and places. She sang, I gathered, of the unreality of the world, and life brief and brittle within it. She sang also of a secret flower in the human heart which, despite its fading too, did so without losing its colour. Somewhere in that was something of the spirit of Kyoto for me.

So not surprisingly we talked in such a place and atmosphere in a way that we had not talked before. My account of our meeting in the Abstract Garden and the priest's story, unlocked a flow of anecdotes and stories both from Mr Tajima and our student guide. The two of them were united in a preference for Zen before all other forms of Buddhism. They had an instinctive aversion to doctrine, dogma and metaphysics. The Zen that held their Shinto-rooted imaginations was non-rational, unconceptualized and a growth in the Japanese spirit, and it sprouted from the seeds of stories like parables regarding the conduct and example of centuries of Zen masters and their pupils. And it came over me that the story could be so important in that regard because, if truly told, it had the seeds of new being in it. None of us, it struck me then, could take such a story into our imaginations without being changed by it. That was the urge which forced the Masters of Zen to resist the written word and pass on in endless anecdotes and stories from lip to lips of living men, alive and dynamic in their search for enlightenment. For the first time I realized that this was precisely what Christ had done. Although he could confound Pharisees and scholars in the Temple of temples with his learning, he too avoided the written word in delivering the message which was the 'end unto which he had been born'. He left the writing to others, and conveyed to his disciples and the world the meaning of his coming only through example and the living word of the living God with whom it had been in the beginning. He refused to pen it down in unchanging sentences that would start the process of special pleading for absolutes that men call dogma and doctrine. How blind I had been not to notice the significance before of so obvious a fact in the life of the greatest source of enlightenment in Christendom! And why could not so great a community of seeking as evidenced there be pooled, and all made to enrich rather than divide one another?

Two days earlier I had bought Okakura's great work, *The Book of Tea*. It had been published in English I noticed, with a quickening of senses, in the year that I was born. Okakura referred there to the ancient Chinese maxim that when knowledge and intelligence appeared in the world, the 'Great Artifice began' and started to interpose iself, more and more, between man and his natural participation in the movement of the universe that was necessary for his attainment of true wisdom. It had become vital therefore to shock man out of his intellectual and ideological presumptions by paradox; by practical jokes that lowered his egotistical esteem; physical blows that inflicted pain which exposed the limitations of the flesh and so freed him for the inflow of Instinct, which itself could bring about the reconciliation between the self and the non-self so essential for the whole man, but which reason alone is unable to perceive, let alone achieve.

"So the sages spoke in paradoxes", *The Book of Tea* had told me, "for they were afraid of uttering half-truths. They began talking like fools and ended by making their hearers wise." Moreover they did not seek to escape from the world but fled towards it, fingers pointed at the heart of man, where dew, grass, the little snail upon it, trees, clouds, mountains and wind were more real and closer to him than the presumptions of world and empire. Adopting the utmost simplicity and avoiding excess, they attained a greater quality of being capable of 'action out of non-action.' It was an occasion therefore of such importance to me that it lasted far into the night. I discovered a part of myself that had been starved without my knowing that talk of this kind existed.

The stories that came out as examples were endless. The priest's story told impatiently in the Abstract Garden was followed here by another seminal one, that has since become so well-known that I have heard it exploited even in a political election. We were told of two Zen monks, sworn to celibacy, walking through a remote countryside after a flood. They came to a river where the only bridge had been washed away. On the far side stood a beautiful and distraught girl.

"Oh, you holy men, please help me," she called out. "I must cross this stream and hasten to my mother who is helpless and seriously ill."

One monk immediately lifted his habit, tied it round his waist, waded across the stream, put the girl on his shoulders, carried her across and put her down. Ignoring her thanks, he re-crossed the torrent and went striding on purposefully ahead, ignoring the fact that an unusually silent companion was lagging behind. Suddenly there was a flurry of garments at his back and his companion hurried to his side, exclaiming: "Brother, I am deeply troubled! We have both taken a vow of celibacy and pledged ourselves never to touch any woman. Yet we have done just that today and broken a sacred vow."

The first monk looked him straight in the eyes and said sternly: "Brother, all I saw was a human being in need and I did at once what was necessary. When I carried the girl across the stream and put her down, I had done with her. But you brother, who did not help, have carried the girl for five miles."

And there were many more set in train by this as if to follow a winding, moss-covered trail through the great forest of the influence of Zen on all aspects of Japanese life over some six centuries. Through all this ran, as a golden thread in the darkness of the original labyrinth, the profound passion of commitment of the totality of the person to his allotted task which is natural to the Japanese. There was, for instance, the story told that night of the great painter Sesshu. The abbot of the temple where he was the youngest novice thought his love of painting made him too slow in learning the sacred books. He reproved the painter over and over again without any effect except that Sesshu himself began to feel guilty. His sense of guilt became so extreme that he asked his superiors to tie him hands and feet and upright to a temple pillar.

"If I can break this bad habit for just one day," he told them, "I could perhaps free myself of it forever." But as the day went on, frustration turned to a feeling of doom, and tears flowed down his cheeks, inside his kimono on to the earth beside his feet. They did so for so long and so fast that earth became wet and soft. He looked down and saw his big toe had begun to move of its own volition. When his superiors came to him at the end of the day, they thought they saw a dead mouse at his feet but it was, of course, a mouse only drawn with his toe in the wet around him.

"It cannot be helped," they said as they untied him, "you must go and do what you have to do immediately."

Yet what was there when parable and paradox failed to produce the state of grace which was the dawn of enlightenment? There was always the world of nature that in its heart nourished the sympathy deep in all things for one another which 'The Great Artifice' denied. The meaning of that too lived on in a story of two close friends. They had not seen each other for ten years and felt so maimed by the separation that they decided they had to meet. They chose the most beautiful of places for the meeting: a clearing in the forest by the banks of a stream, large enough to allow the sun to weave its light in plaits of green-gold grass. Deeply moved as they met, they could only bow to each other because all the words they knew diminished what the occasion meant to them. For hours they sat side by side in silence in the clearing on the edge of the onflowing water. Then when the shadow of evening fell over them, they rose together, faced each other again, bowed deeply and exclaimed: "What a perfect day!" Resolved and, though separate, still at one, they turned about and went their ways.

There is also the story of the painter Chôdensu who painted the immense scene of a great Buddha dying with a vast gathering of grieving animals around him. Everyday while he was painting in a courtyard of the Tôfukuji Temple in Kyoto itself, a strange black cat appeared and sat from beginning to end of the day, beside the painter, watching without sound or movement. The painter became more and more mystified by the cat, until suddenly he realized there was no cat among the death-bed assembly, and indeed that no cat had ever been included in any other paintings of similar subjects. Contrite at such injustice, he immediately painted the cat beside him on the scroll of silk. The moment the painting of the cat was complete, its model vanished from his side and never reappeared.

This happened to be the last story on the long night of our long last day in Kyoto and seemed a singularly appropriate end. And so, as well prepared as we could be, we went straight on to Isé.

I do not think the pattern of our journey had so far been deliberately planned. I believe that our own natural interest revealed itself on our journey and was enough to suggest the next stage to our hosts. As a result its design was non-rational and intuitive, rather than calculated. But whatever the cause, this aspect of our encounter with Japan, which I have summarized so arbitrarily, needed the experience of Isé for its provisional completion. At Isé, though technically moved forward in time and space, we found ourselves back at perhaps the most remote and crucial expression of the impact of the search in the Japanese spirit for the source and meaning of creation. We arrived at the site of the two most sacred places in Japan: an outer shrine dedicated to the Goddess of Rich Harvests, the national holy of holies; and the inner shrine, dedicated to Amterasu Omikami, the Goddess of the Sun. The two shrines were set some miles apart but contained in the same landscape of forest-covered hills. Clear streams tinkled like glass as they flowed over shining stones to wind their way through a wood of an unmeasured antiquity. Imposing trees seemed deployed like a brigade of divine Household Guards to protect from worldly men the reality of which Isé was the elected centre. And this was not as fanciful as it may sound. On arrival we found that world represented in abundance, and ourselves in the thick of a great and eager crowd of all sections of Japanese life from peasants, aristocrats, shopkeepers, industrialists, prostitutes and Geisha, to children, babies, mothers, politicians, professors, civil servants, soldiers and priests. All were pressing towards the same goal.

Churches in my world had begun to empty fast. But here still was an attraction to religion that was as yet in no danger of decline. Far more impressive than the numbers was the atmosphere of singleness among such numbers and diversities of people, and the gentle manner in which they ploughed forward like a river over the immaculate gravel avenues lined by those impressive cypresses of Japan, called cryptomerias, which reach to a height no ordinary cypress ever achieves. They did so no less ardently than those candle-lit cypresses of Van Gogh, straining as in a mounting fever, towards a sun whirling as on a St Katherine's wheel of fire across the burning sky of Provence. One look up and along the ranks of those sentries of the forest was enough to command one to silence and a new humility. But when we arrived, what we saw was unlike anything for which the temples and cathedrals of Europe could possibly have prepared us.

The physically great obviously has as important a role in the architecture of religion as the small. In Kyoto already we had seen a crescendo of scale in temples without sacrifice of proportion or grace. But for something dedicated to the primordial moment of departure of the human spirit into exile in order to regain a more meaningful reconcilia-tion with all that expelled it in the first place, nothing could have been more appropriate than what we now saw: a small thatched building of cypress wood, in a style so old that no-one knew either its time or place of origin. Documentary recognition occurs first only in the sixth century. But all the circumstantial evidence suggests that the original design arises from the pre-historic Japanese from their unknown land of origin, and yet is preserved intact to this day. And perhaps the word 'preserved' is misleading, for the building is renewed every twenty years. Indeed every twenty years from as far back as can be remembered, this shrine has been dismantled and then reconstructed according to an unchanging plan with all the rigour of a supreme religious rite. In consequence the intermediate spirits of nature who inhabit it shall not suffer disturbance.

Perhaps one of the saddest things in life is the recurrent illusion of human beings that they can improve on the truth. The unchanging architecture of this modest building was without this or any other illusion. It was as simple and pure a building as I had ever seen. It could easily have been the ultimate illustration of the Confucian maxim that 'to be modest about modesty is the greatest of all virtues'. There was no decoration and no varnishing or secondary treatment of the original material.

The nobility and beauty came from the total absence of pomp and ostentation. Even the entrance to the inmost shrine itself was totally unadorned, with only a silk curtain between it and the world of men and its winds of change. We could not see beyond the curtain, indeed could not have attempted to do so without forfeiting our lives. Only the priests in charge, the Emperor and a few of his family, are allowed to pass beyond. And a Japanese Minister of Education who had not even tried to enter beyond but merely flicked the curtain aside with his walking stick, had roused such revulsion that he was killed and his assassin deified in the Shinto pantheon of great men. . . . Yet what lay beyond the curtain, we were assured, was simple and pure, and designed only to hold the sacred mirror which is the image of the ultimate Shinto-seeking.

Standing there to attention, as the ancient custom demanded, before bowing deeply to the veiled truth, and clapping my hands twice to alert the spirits of my presence which, I had been assured, since Moji was prayer enough, it was as if a burden had fallen from me. It was to take many years before I had an inkling of precisely what was the nature of the burden dropped that day. All I know is that the extraordinary gaiety of the immense and diverse crowd who flowed all around us on the way out, was most marked, and proof too that they had at least been lightened if not enlightened.

I should add only one ominous afterthought which struck me as we walked away. It seemed to me no accident that the first and only transgressor of the laws of the reverence due to the shrine of Amaterasu Omikami should have been a modern Minister of Education; another of those promoters of 'The Great Artifice' which made reason and intellect not the partners of feeling and intuition that nature intended them to be, but rather imposters and tyrants. It was he and his kind who had encouraged an intellectualism already recognised by me at my own school, which prevented a true balance between us and the so-called civilized, and despised primitive. This happened not only in my native continent but also in the invisible first world that we all have in our keeping within. That, however, was as far as I could take this aspect of life in Japan at the time. It had confirmed in me a sense of the relativity of religion and culture that was never to leave me again but to become an inexhaustible source of enrichment of life. I seemed to be free from there on to look around me with a clearer and younger eye, and to observe other significant and at times even deeply disturbing things.

It was in fact remarkable how from that moment our journey took us into another dimension of life in Japan and switched over, as of its own accord, to an extravert level from the introverted one on which it had engaged my imagination. Indeed the very next morning gave us our first experience of how the paradox of earth and spirit at which I have hinted, and which was so contradictory an element in the story of Kyoto, was still tunnelling in the national spirit.

It came upon us as we were making our way up Momo-Yama, the Peach Tree Hill of history, near Kyoto, which owed its name to the numbers of fruit trees grown around its base. It was a morning of uniquely autumnal beauty. The sky was blue; the air clear, true and precise with a valedictory delicacy as if it knew it was looking its last on the summer scene before the cold and storms of winter were let loose against it. There was no wind and a solemn stillness over all to match the quality of the air. All these elements in the day continued to make us unusually silent as we came to climb the flight of hundreds of wide stone steps leading up to the tomb of the Emperor Meiji which our guides insisted we should see. The steps, the gravel at the side, the trees, all were as clear and without stain as the day. As a result, after taking the stiff climb slowly – I counted two hundred steps before I gave up – we paused at the top to rest and enjoy the pure and transparent view.

Standing there relaxed, I did what I had come to enjoy at the end of long climbs: I celebrated it by taking a cigarette from the tin in my jacket and lighting it. I remember it well although I have not smoked since the War, because as I lit it, I thought that with its name 'Kensitas' and with a Victorian sketch in colour on the tin which included a butler, it could hardly have been more inappropriate in such a place.

The thought should have come as a forewarning, perhaps, instead of the mere absurdity that it seemed at that moment. We had hardly turned about and begun our approach to the outer walls of the tomb and its grandiose iron-gate which looked as uncharacteristically out of place, when it happened. Disaster struck and, as always, so swiftly that I was surprised and totally unprepared. Somewhere on my right the silence was broken by the loudest, strangest and most irreverent of noises. It was a sound I had not heard before and seemed hardly human because it came not shaped out of mouth and throat so much as violent and chaotic from the pit of some stomach specially designed for the onomatopoeia of anger. It was in human terms the equivalent of the rumble of the earthquake that destroys cities and lays temples low. Alarmed, because all my instincts suggested a danger of being laid low myself, I turned towards the sound with hands free, if needed for my protection. I turned so suddenly that the smoking delegate of Messrs Kensitas and the Imperial Tobacco Company was left on its own between my taut lips. I had time only to see a small man in uniform leaping at me. As he leapt, a sword, which must have been drawn the moment the first eruption of sound in his abdomen contorted him, flashed in the sun. It came streaking down hard by my forehead and just failed to deprive me of nose tip and lips, before it cut 'Kensitas' from my mouth. The feat would have caused rounds of applause in a circus but considering the speed with which it was done and the fact that it was performed on a moving target while tremors of outrage were still shaking the soul, hand and body of the man in uniform, it was miraculous. I was told later that William almost fainted. However, that was not the end of the affair.

For some ten minutes or more while the swordsman upbraided me, I had to stand silent and still as if facing a rhinoceros trying to decide whether it should charge and impale me on its horn. I knew there was no point in explaining that I would not have been so insensitive as to present myself smoking at the gates of the tomb. I was certain that any sound of protest or movement from me would be too much for the little authority which made this small man gigantic. I dismissed even the thought of apologizing in the highest degree of politeness as instructed by my teacher. I somehow knew that by revealing some knowledge of Japanese I would have convinced him that I could have not failed to know also what 'crucial neglect of the respect' due to an Emperor, above all an Emperor exalted in death, it was to smoke near a place sanctifying his memory. Even so, still and unprovocative as I tried to remain, there were moments in this storm of harangue when new gusts of rage would shake him and make him brandish his sword again in a manner that suggested it would take very little more to make him slash me as well. My companions were silent as I was. The fact that our guides made no effort to intervene and explain confirmed my own reading of the situation. I thought it wise also to avoid his eyes but as he continued to rumble and tremble like a human volcano in eruption, the odd glimpses of his face were like that of a mask of a Samurai of the Dark Ages at the climax in a melodrama of revenge for some ancient wrong in the oldest theatre in Japan. If ever one were to see outside a lunatic asylum an illustration of what happens to the human being when he moves out of his own body and lets in negations of collective and private history to take full possession, this little man was it. When he sheathed his sword at last and waved me away disdainfully and I could look at my friends again, their faces were pale and strained with fear for me.

We were about to go on to the tomb, still some distance away, when I was shouted at again but this time in a voice in which the animal had receded and the almost human appeared near enough to make me hear the words and make it plain that I was being commanded in the lowest degree of politeness to pick up the mutilated 'Kensitas' and restore it to its tin. Only then were we reluctantly allowed to approach the tomb. But the day had been spoilt for us all, so much that I felt almost affronted at the way its beauty had not been unimpaired.

As a result, our visit to the tomb was perfunctory and hurried, and we spoke little until we were off the mountain again, and then only very little, about the affair with the small man in uniform. But the most revealing part of the incident came much later from Mr Tajima. After characteristically taking the blame on himself and apologizing for not realizing sooner how strict a ritualistic etiquette had to be observed on Peach Tree Hill, it was as if he felt that the conduct of the guard of the tomb needed not apology, but rather explanation.

"You see, Post-San," he said with a note closer to approval than courtesy would allow him to admit, "it was not the man but the uniform."

And there, however epigrammatic the utterance, was enough evidence of the basic national paradox on which to hang the tale of horror which was about to unfold for Japan and to send it unerringly to its own catastrophic awakening at Hiroshima.

Of course, I did not ask Mr Tajima to elaborate on a matter I knew was hurtful to him, and I was as glad for him as for myself that, before the day ended, we had a happier illustration of what 'the uniform' represented at its best to the Japanese. That afternoon we travelled by train for the first time and met our first Japanese station-master. He appeared on the crowded country station platform some minutes before our train was due.

It was, of course, a time when the train was still increasing in the high standing it already possessed in the life and imagination of men everywhere. It was a moment, for instance, when no railway station in the British countryside would have risked the inevitable loss of self-respect if it did not present itself to the public with superb little gardens of shrubs, flowers and lawns to grace it. Nor would any British guard or ticket-inspector have appeared on duty without a flower in the buttonhole of an immaculate lapel, so that they could walk the lurching corridors of their train like sidesmen down the aisles of churches, examining mere tickets as if they were invitations to a wedding. One felt such men, with their look of 'servants with a mission', if ever faced with the ineluctable choice between two extremes, would have chosen to leave their trousers rather than their buttonholes behind at their beflowered railway base. Moreover, a new type of locomotive – another attack on the record between the great cities of the world and the building of a new railway line – was banner headline news, and involved the imagination of the peoples everywhere in a way neither automobile nor aeroplane were ever able to equal. But no matter how great the romance which, unseen, was bringing the latest nine-fifteen, its polished engine gleaming like jewels up to platforms all over the world, I doubt if it ever flowered as spectacularly in the popular spirit of any land as it did in Japan.

For the Japanese it was a matter of love at first sight. And from the day the first railway line was inaugurated, and by the very Emperor Meiji whose tomb was guarded with the archaic fanaticism that I had encountered that morning, a rare love-match was made and a union consummated between man and his transport unequalled in the story of communication. The evidence is on record and still makes touching reading in this day of the world-wide decline and fall of the great empire of rail and steam, except in Japan, where only the other day I saw the same uniform and same looks on the faces of station-masters as on that limpid afternoon in the deepening autumn of 1926.

The 'rapid course' of the train in which the Emperor travelled, one newspaper had it, was 'like the wind or the clouds' and 'No such event was recorded in history since the days of our first Emperor Jimmu' – which even for the Japanese was saying more than most as the observation spanned some 2,000 years! Even the official artist of the occasion in a country where modesty was among the greatest virtues, exceeded its exacting proportions. At the end of his sketching, he laid down his brush in despair and declared: "My power of observation fails me. I fear I have made many errors in my sketches."

But there was no despair and neither humility nor pride in the bearing of our first station-master as he walked on to his little wayside stage. There was only a natural and ancient dignity in his manner and a serene air of a man certain of his command because of the knowledge that in a land where all authority was one and indivisible, his too came from an Emperor, who was a Son of Heaven and derived his own by right of Royal Blood direct from a Divine Sun. His uniform was, therefore, as for the sworded fanatic of Peach Tree Hill, a badge and seal of Imperial Office; and he wore it in a way that made him a benign and benevolent symbol that hinted by comparison at a dichotomy in the national spirit. The colour of the uniform itself was a violet of rather a feminine tone and without a trace of male aggression. His cap was pillar-box red with gold lace worthy of the hat of an Admiral of a High-Seas' fleet. And indeed at times he walked the platform as an Admiral does his quarter-deck, where all those waiting on the platform made way for him and bowed low to him as he passed; the older persons hissing in the loudest and most polite way of their ancestors – through their teeth – while doing so. But in general he carried himself, as far as I was concerned, more as a priest in charge of a roadside chapel about to perform a mystical rite, and never more so than when he extracted his watch from his uniform pocket. It was an unusually substantial instrument of silver on a heavy chain, and he would uncover its face with a delicate hand to look at the time, concentrating as if he were the man who first 'consulted' rather than looked at a watch. For the country people waiting for the train with a rapt look on their faces, his watch might have been a vessel for splashing holy water on them, rather than a mere container of something with such a deplorable inbuilt obsolescence as time. When the train finally arrived, they gave him warm looks of gratitude, as if he were personally responsible for such a miracle of transport, and he responded with a regard which was an assurance that the source of such riches was inexhaustible and well within the ordained dispensation of the gift of the vicarious living to which he had been called. Accordingly, the train was boarded and dispatched in a high spirit of exaltation that released the passengers from social and personal inhibitions and made them all missionaries of travel. Everything that had been unpleasant in our day fell away like the leaves of autumn. The lightness of the gay, warm and happy spirit in the compartments and corridors all around us took over. I thought I had never travelled in better and happier company that made me capable of such high enjoyment myself. And that, without exception, was my experience in the days that followed.

From there on until I left we went everywhere by train but rarely on any main lines; instead almost always on trains in the country where neither time nor speed were obsessions. I had never imagined that just travelling could make human beings so happy. First-class compartments were rare and second-class ones few, I am certain not because they were hard to obtain or too expensive but because they would have fragmented that sense of one-ness and all-belonging in the consummate joy of travel. For as always the first sign of happiness in the human heart is a desire to share with one and all; share food, jokes, laughter, tears, conversation, news, space, sympathy, consideration and everything that could be imagined which was best accomplished when the journey was without measure of class and everyone travelled third, as we all did. The coaches' green velvet had come down in life and tended to be out at the elbows of the arms of their benches. If ever there were a people over whose spirits there flew a banner of conviction that it was "better to travel hopefully", as R. L. Stevenson had it, "than to arrive", our country companions were they. I thought that nothing would have made them happier than an assurance that this journey in a sanctified train would never end. Moreover, everyone from the highest to the lowest: rich man, poor man, beggar man, thief; marchers; virgins, prostitutes, children in uniformed droves or clinging in ones and twos to vivid kimonos; priests, soldiers, honeymoon couples, run-away lovers, bereaved and betrothed; all always at the drop of a fan, took to trains all the time, or so it seemed to me, and with a confidence of imperial blessing and an imperial station-master priest waiting on every station platform obliged to serve their journeying spirit.

The stops at sidings and stations also had the happy compensation that we could invariably buy souvenirs and food from colourful vendors there. These bamboo luncheon baskets and little wooden food baskets were unbelievable sources of pleasure to us. They were, however, modest and unadorned; always works of art and therefore obedient to its own disciplines of simplicity, respect for the nature of the material, precision, thoroughness and proportion. After eating the contents, good as any feast to me, it still seemed a waste to discard them because, even empty, they contained so much food for the eye.

There were two other experiences that I owe to these journeys by train. Without them I do not believe I would have discovered in so short a time how robust, natural, young and irrepressibly human the ordinary people of Japan remain despite the exhausting centuries of extremes of religious, aesthetic, moral and political disciplines imposed on them. Not only did I find their capacity for enjoyment, laughter and curiosity unimpaired, but also a tough sense of realism, a certain healthy vulgarity and robust disdain of bombast and presumption in their society. Far from being cowed by their despotic past, they were alive as any people I had ever known and so immensely stimulating companions.

However, I soon learnt that the particular aspect of the national character I enjoyed so much in trains, went far enough back into their past to ensure that such loss of reflexes would not be lasting, and invariably seems to have been unleashed whenever they have taken to the road. I read very little on these journeys except books by Japanese themselves because the experience was so precious and personal to me that I wanted no second-hand interference, however expert or sensitive, from other foreign observers, to come between me and it. There was, for instance, a satirical novel from the eighteenth century called the *Hizakurige* that taught me how even then the popular world of Japan was not all Murasaki and Blue Trousers; Shei-Shonugon and Pillow Books; Bashō and Hokku, Samurai and Hara-Kiri, but also one of an intense almost volcanic humanity, surviving despite the lofty pressures from above because of a certain creative irreverence and capacity for improvisation, and living by its wits in between the lines of great authoritarian writ. For this novel chooses as its heroes two attractive but outrageous delinquents, and describes their adventures on the Tokaido, the Eastern Seaway, the Great High Road of history built some six hundred years before to do for Japan then what its main railway line was doing for it now. And there was far more of the stuff of *Falstaff* and *The Merry Wives of Windsor* in those pages than Zen, Bushido and endless flower arrangements. There was also the gentle, elegiac painter Hiroshige, who was spurned into unpredictable satire by the great road and, in his Fifty-Three Stages on the Tokaido, who lifts a curtain on this vital, burning, unrepentant aspect of life in Japan as, for instance, where he mocks some great lord travelling with an attempt at mediaeval pomp on the road, by reducing his followers to a retinue of grasshoppers.

Finally, without these journeys by train, I could never have known what a beautiful land Japan was. Since it is so mountainous, almost every line had to be tunnelled and dramatically cut out of steep slopes and flung across wooded valleys with flashing streams and gleaming rice paddy waters. The constantly changing scene was so attractive that reading and conversation were spasmodic and irregular, and talk on the whole came mostly from our guides and companions, who were eager that we should not only look but listen to the story connected with what we were seeing. That was one of the most impressive features of all about these journeys: every acre of land seemed endowed and endeared with a story of its own. Among the many, two moments stand out in my recollections. One was of my first view of Fuji at dawn. I saw it from our compartment coming out of the night and though I could not fail to recognize it, since it was just about the most advertized mountain outline in the world, it was as if I were seeing it for the first time. For a moment I could not believe in it nor accept that it was real, but once it established itself firmly in my senses, it was as if I not only saw it but heard it as well. The dawn happened to be exceptionally clear and the mountain emerged out of the dark a profound purple. The purple was all the more alive by contrast with two startling white feathers of snow brushing lightly against the delicate curve of the arms which held the great volcano in so fast a sleep under a Hokusai-blue sky, where one lone star was sinking into the flooding tide of a sea of light. The colour was so overwhelming, and the shape of the mountain so elegiac and eloquent, that they continued to make a deep humming noise in my ears as if bees that made honey of the sky in the East were there singing a hymnal to their Queen as they went winging out to gather more of the last sweetness of light before winter shut them deep in their mountain hive. In that instant, I believe, I knew not only why but also how profound and subtle a service this mountain of perfection had rendered and must still be rendering to the spirit of its land. At the same time I knew that it was of an eventfulness I could not describe. One could only look at it with awe and let it happen in one's senses, as it was happening then to me, and leave the wonder it evoked to do the rest.

Verification of all this was there too in the look on the face of a little Japanese girl who shared the experience with me. She was Mr Shirakawa's daughter and was just about four years old. Her father and William were asleep in our compartment but there she was suddenly at my side. All the way from Osaka she had not spoken once but sat silent by her father's side; staring out of wide, dark, glowing, slanted eyes in the way Japanese children did, as if the life of their elders and the grown-up world were a façade through which they saw only too clearly the reality. She appeared at my side so unobtrusively that I did not notice she was there at the window next to me until she announced with a small but distinct and awe-stricken voice: "Lord Fuji".

She was pointing at the mountain and, when she caught my glance, became shy and apologetic at such boldness and hastened to add: "Please be so good as to excuse. Lord Fuji murmured and it could not be helped."

She finished bewildered by her boldness and looked away at our sleeping companions and again up at the mountain. Her shy confusion changed to amazement that they had not heard the purple bee-swarming mountain sound. Brief as that moment was, it and all connected with it were stamped indelibly on her mind so that, in a recent letter telling me of her father's death, she could recall in detail even the impression which my outlandish clothes had made on her at the time and which added to the strangeness of the experience. As for me, the young child's face has continued to stare back with undiminished clarity. From that moment of first light in the presence of Fuji it soon became like one of those Renoir renderings of the mystery of the feminine, already present on the uplifted face of the young. So that to this day the memory of Lord Fuji has provided me with an authorized version of the image of the child we are exhorted to re-discover if we are ever to enter the Kindgom of Heaven – a kingdom of which we had so imperative a glimpse on that lovely morning before the shadow of winter fell over it.

The other moment of our debt to the train was our journey to Nikko, where the Tokugawa Shoguns had built vast and grandiloquent mausoleums to promote their memory. It was perhaps the most beautiful of all settings for uplifting the imagination with intimations of immortality and inflaming it with longing of the infinite. Indeed its natural sanctity had been recognized by a great Priest already in the eighth century as one of the holiest of shrines built on the mountain, whose name was changed at once from Niko to Nikko, because Nikko means the 'radiant beams of the divine sun of Japan'. Indeed, so sacred was the shrine and intense the feelings induced by the scene that the inspired Poet-Priest Bashō, who climbed the mountain in the seventeenth century, beyond recording the historic fact, refused to elaborate because "to say more about the shrine would be to violate its holiness". Significantly too he ignored the mausoleums of the Shoguns who, with the energy, determination and sense of absolute power that characterized their dynasty, had begun to violate the scene. The builders of the Nara and Kyoto that we had just left would have understood a corner of Japan that is blessed with valleys, peaks, volcanoes, forests, waterfalls, lakes, with twist after twist of silk-black gorges and flower-embroidered meadows as in no other part of the country. And they had obeyed the awe they had always felt before such confrontations of nature with a selfless modesty and reverence. The Shoguns had no such inhibitions, and exploited without taste or shame almost every vantage point, confusing the grandiloquent and pompous, the magnificent and bombastic, with art, in a way that is typical of imaginations obsessed with power. I remembered an African saying: "Great birds of prey build bad nests and do not sing".

Happily we had hardly arrived in Nikko when the clouds came down from the mountains and the rain set in. We barely had time to observe the famous lacquer bridge which is an exhibitionist arch of red cross a crystal stream, bringing the scarlet make-up of a red-light district into surroundings that asked only for simple, unstained and cared-for wood.

Then the cloud covered the hills and a mist spread like muslin over the valley. We had to walk these long avenues of giant funereal cryptomerias leading to the great mausoleum, cold and wet and yet deeply impressed by the stature of the trees that were the best part of the excursion but no preparation for the vulgarity of display of the mausoleum itself. It was, it is true, a vulgarity of genius and the energy and abundance of bad taste raised to a level of perverted inspiration which was frightening. One such mausoleum on such a day was enough even for our guides. So at nightfall, hastened by storm and looming mountain shadow, we made a perilous, almost vertical journey up the mountain pass cut out of sheer rock faces by local taxi to Lake Chuzenji, which was a fashionable hill station not only for Europeans in Japan but for those thousands of their kinsmen who found the heat of the summer unendurable and sought relief there in great numbers. But at our hotel where the stormcloud and dangers of the narrow road had made us late, there was no sign of anyone. We had the large hotel to ourselves. We had the sound of rain which is always music to African ears, the stillness, the warmth, and food that were all there to enjoy without pressure. I went to sleep as if re-absorbed into nature: that was one of the great gifts of travel at the country heart of Japan.

I was woken up by the silence. The sound of the rain had gone and nothing had moved in to take its place. It was the sort of silence that used to be called 'absolute' and is seldom part of contemporary life. There was nothing frigthening about it but something so positive that I thought of it as a trumpet call to the day. I looked out of the window on one of the most beautiful and serene scenes that was ever to come my way. A silver mist was rising from a lake of unbelievable purity so that, as a result of the capacity of the water to reflect, truly it was a two-way mist: it not only moved upwards but also withdrew in the same measure downwards into the water; and as it withdrew, fires seemed to break out at different points on the slopes of hidden hilltops around us, and to send them up with flames so fiercely that soon the mist looked like smoke, and even the lake itself like the place where glass for the windows of another Chartres was being melted and coloured. It was my first experience of maple forests in the fall, which comes early among these clear-cut mountain tops. They were already, as the Japanese said, in full flower, and it is almost impossible to describe what the sight did for eyes trained in Africa where, though there is an autumn too, and uplands of extreme cold, there are not the trees, plants and leaves to express this seasonal transition as at Chuzenji. I felt that visually I had never known autumn before; so I abandoned myself there and then to a resolve to make as much as I could of the event.

William, for all his experience of autumn in England, was almost as affected as I was. We hardly spoke because it seemd all-important to listen to the silver voice of the silence and its instruments of mist, flame and stained-glass water. So, as soon as we could, we made for the water, and William, who loved rowing, took us out to the centre of the lake where the creaking of the oars in the row-locks and swish of water from their blades were like those sounds of respectful shuffling that accompany all marches to the grave, and were never more reverent than at this personal and private mountain funeral of summer. So clear was the water that I could see the shadow of our boat following us on the bottom, and when we returned finally after floating at the centre of all that maple flame, as if in the heart of the burning rose that took Dante transfigured to heaven, there was not a ripple to distort the reflection.

We came back prepared to meet our guides for a walk through the woods to one of the greatest and holiest of waterfalls in Japan: the Kegon Waterfall. It seemed right that so perfect a lake should have its outlet in so long and delicate a fall of water. We watched it all in silence until suddenly our guide left us to follow a precarious track to the precipitous top of the fall. I watched him until, to my relief, he stopped by a tree just short of the edge, looked all round, and came back to us, it seemed, oddly despondent. He explained that often one found farewell notes and pleas of pardon for the trouble caused, pinned to a tree, by persons who had killed themselves by throwing themselves over the edge with the water of the fall. It was a favourite place for suicide for couples frustrated in their love by the demands of family and duty to society. It was so common indeed that at one time notices were posted in the valley below requesting "honourable suicides" to leave their names and addresses with the police at Nikko before taking their lives as "it would greatly help the authorities in communicating with their relations afterwards".

But why so many suicides, William asked. Our guide seemed taken aback, I thought, perhaps by what he took to be an excursion into the obvious. He replied that it was in the nature of life to make people tired of it, and that one of the commonest reasons given was 'weariness of the world' or 'tiredness of life'. But in my own mind I was more concerned as to why young couples should make so exquisite a waterfall a popular place for taking their lives. I have already mentioned how Japan tended to romanticize death. Waterfalls with us may have notorious but happier attractions for honeymoon couples. Yet from the little I knew of the Japanese, there seemed to be a logic of the national disposition in this tendency to make so beautiful a scene the setting for honeymooning with death. The thought made the moment more autumnal, and the winter hastening nearer, while the natural reverence of all around us, as we walked away, found a voice in the sound of a temple bell that came from below to join the wind-rush music of the Kegon Fall.

I thought again of Bashō, whose feet had walked these hills three hundred years before, and his poem:

> *"The calling bell*
> *Travels the curling*
> *Mist-ways . . .*
> *Autumn morning."*

And time lost its distance.

So, not fully armed and yet better prepared than most foreigners, we came to our encounter with Tokyo. I cannot pretend that I liked it then as a city. But I thought how miraculously it had risen from the ashes of the fires and the levelling of the great earthquake of 1923. It was already sprawling over the cold Musashi plain without the least hangover of disaster: a teeming modern city of an abundant and dynamic activity. If one had any doubts about the capacity of Japan to acquire a modern technological dress for itself, a day in Tokyo was enough to disperse them. Nonetheless we did have doubts after what we had seen; we wondered, for instance, whether the technological power acquired would be used in a truly contemporary and not retrogressive way. Outwardly all made for a reassurance formidable enough to satisfy the Japanese 'establishment' that we met. But that did not relieve a thought that it was either more plausible than convincing, or at best a critical issue just capable of a positive outcome although as yet undecided. As we were conducted round the capital by Mori, who had joined us and taken command, we were shown by preference the modern rather than the ancient relics. We met and dined with Chambers of Commerce, and odd millionaires, and were asked questions which neither of us was qualified by experience or temperament to answer. Our questioners already wore the clothes and had acquired the figures that were generic in their species all over the world. We addressed outcrops of this world, thought essential for the image of their establishment and their country. A swarming Rotary Club gave a lunch in our honour. Again, as elsewhere in the world, the Rotary movement's motto 'Service before self' was displayed on a large banner of blue hung across the wall behind the Chairman's place at a table set for a terrifying European meal. The motto elsewhere had often produced cynical reactions that were unusual for me. But here, as I listened to the talk at table and was struck by something new in some most unusual and untypical big business faces, I heard things and caught expressions that made the motto seem somewhat less hypocritical than elsewhere. And I had my first intimations of a profound national mystique which made the Japanese go, as it were, religiously, into industry and commerce.

We were, of course, also taken to meet the Japanese Foreign Minister, Baron Shidehara. I had half-hoped, half-expected to see at his side the young emisssary who had met us at Moji in another age, sobered up and perhaps emancipated from plus-fours. To my regret, he was not there, and my enquiry after him seemed to puzzle and startle the Baron. That the Baron did not even know the name of one who had sought to serve to the extremes of absurdity beyond his own capacity for recognition or foresight, invested his memory with some of the intensity of the anonymous with which Chekhov makes the tragedy of so many of his characters universal.

Meanwhile, dressed like a pre-war inmate of the Temple in London and well-schooled in the manners and patter of the diplomacy of the day, the Baron arrived by a round-about way at the cause of his interest in us. He had just asked William where he came from and what his nationality was. William, who by now was well on his progression from his African self towards England and Englishness, had answered that he just happened to have been born in Africa but was really English. I felt a shift of focus in the Baron but still he persevered with William.

"I hear you are to become a second Lafcadio Hearn?" he observed, with a flattering look.

For a moment I thought William's sense of humour would desert him. He hesitated before he managed to answer, without too obvious a bite in his tone: "No! I shall be the first William Plomer."

The Baron thereupon turned to me and asked the same question of me. I said, of course, that I was born in the interior of southern Africa and added that I had been educated and grown up there. Whereupon the Baron wanted to know which city I had worked in and, when I said Port Natal, his round full face was transformed with something approaching pleasure and a recognizable interest: "Ah! Is that so?" he exclaimed with a faint hiss of polite Japanese, before adding in English: "Then you must know Abyssinia well!"

The Baron's ignorance of the thousands of miles between Durban and Addis Ababa was to me as naïve and ludicrous as his disregard of his devout envoy had seemed humanly regrettable. I dared not look at William. I concentrated instead on the questioning that followed without pause. I had never been to Abyssinia but fortunately read all I could about it and had made friends in my Hockey Club, the Nomads, with a remarkable Cornishman – called Trevelyan – who had prospected all over Abyssinia for precious metals. In any case it was not difficult to know more about the subject than the Baron; and from the manner in which he thanked Mori, I was certain he felt the encounter was not wasted. But what interested me most, finally, was the objective confirmation that the occasion offered of Japan's secret dream and ambitions of empire, rumours of which had troubled Mombasa and Nairobi as they had enraged Mussolini in Milan. They suggested that already Abyssinia had been tentatively singled out as a potential colony of Japan in Africa. Perhaps most disturbing of all was that so well-travelled and highly placed a person as the Baron could be so naïve and so out of touch with the forces of transition in the contemporary spirit. As one who had dreamt a great deal and had been troubled by them ever since he could remember, I already suspected that dreams could not be taken either for granted or at their face value, and that to be naïve about the dreaming process could be even more disastrous for nations than individuals. For a while afterwards I was preoccupied with the suspicion that even our presence there might be part of a nation-wide sleepwalking venture; and was only shaken out of this mood by another meal with Geisha in attendance.

This meal was presided over by a more exalted Foreign Office personage than our missing Moji official, and given in one of Tokyo's most prestigious restaurants. It had none of the complications or consequences of its memorable Moji prototype but led to one incident which was of personal significance to me through what it revealed of William and Mori. Towards the end of our meal, a lively discussion took place between our hosts, Mori and Mr Shirakawa, as to how the evening should be concluded. There seemed to be a majority feeling, opposed by Mori, that we should be taken to the Yoshiwara, the brothel city and capital of sin in Tokyo which was still a subject of scandalized public discussion, because it was the first institution to be rebuilt after the earthquake and fire three years before. Mori, I gathered, argued in vain that they had misunderstood us. He argued hotly enough to give me a new glimpse of the Puritan in him which, as an adolescent, had made him think seriously of castrating himself. We were not the normal run of foreigner, he emphasized, and would be as outraged as he would be. However, the majority feeling finally was that we should be offered the hospitality of this famous Japanese brothel. It was discussed in Japanese but William soon had the drift of it, and becoming, I thought, disproportionately agitated, he finally panicked and turned to me.

"My God, Lorenzo," he burst out, "they are determined to take us to a brothel. What *shall* we do?"

"I think I can stop them," I tried to reassure him, my confidence greater than normal thanks to all the drink I had taken. "But if we are forced to go out of politeness, I promise you, we shall only look and nothing else." I had hardly spoken when the invitation was addressed to William. I had to field it quickly and just managed to hold it as one might a tricky catch in the slips. I thanked them warmly but said that, as we had so little time in Tokyo, would they forgive us if we asked them to show us something less old-fashioned and more modern, say, a 'night club'. The words 'old-fashioned' and 'modern' were enough to produce an eager unanimity to prove that night clubs were as much the rage in Tokyo as London or Paris, and we were led off to an establishment called the Black Cat.

I do not know who was more relieved, William or Mori, but as far as I was concerned, it could not have had a happier outcome. I danced with a charming, cultured and beautiful Japanese hostess who told me how exalted a place the cat occupied in the Japanese pantheon of animals. She explained that there was even an animal cemetery where impressive services were held annually for the souls of dead animals, and even insects, but above all for the souls of cats. I should know, she stressed, that the famous Neko-Bashi bridge in Tokyo was so called to commemorate a great cat who had kept a sick and poor mistress alive by stealing gold from a nearby money-lender!

I was so captivated that I asked her for breakfast at our hotel, unwisely because of the inevitable headache that these late alcholic nights produced and which made me, I fear, not nearly as good a host as she had been a hostess. Yet I could have done no less as the only return of which I was capable for another valuable insight into the feeling of brotherhood still existing then between man, his animal and other natural neighbours – a natural all-belonging which I was certain had prevented the puritan in Mori from violating his own physical self.

The rest of the day was almost entirely spent buying things for myself. William, it was finally decided, to Mori's undisguised jubilation, would stay on in Japan. I was to go back to Africa on an even longer and more round-about voyage, and there were things I obviously needed. Much of the little money I could spare, once I had bought the presents that social obligations demanded, went on books. I found all I wanted at a remarkable department store called Maruzen, where one could buy anything from a safety pin to shares on the stock exchange, and inoculations against cholera and bubonic plague. The book department of the store was most impressive, and I had no difficulty in getting enough books for the months at sea ahead of me. Some, like Dostoievsky's *The Brothers Karamazov*, were perhaps not surprisingly available in numbers, but what was astonishing was the presence on the shelves of more esoteric volumes like Herman Hesse's critical essays, which I still think his best work and which included one on Dostoievsky, 'Ein Blick ins Chaos', that is even more relevant today than it was then since it reveals how we ourselves are still, in our mind-intoxicated manner, going the drunken way towards the inhuman world of Dostoievsky's prophesy and his grand inquisitor. There was also, in Dutch, Louis Couperus' remarkable novel of Java, *De Stille Kracht*, which, in its English translation under the title of *The Hidden Force*, had influenced William profoundly, as well as his remarkable *Eline Vere*, Europe's first purely psychological novel and alas not yet translated. There was D. H. Lawrence's *The White Peacock* that gave me far more than *The Plumed Serpent* and *Kangaroo* which outnumbered it on the shelves. There were also his early poems 'Amores', as well as, 'Look! We have come Through'. There was a collection of Burckhardt's essays on history in German that were to influence me greatly; T. S. Eliot's *The Waste Land* and *Prufrock* poems; Balzac's *La Cousine Bette;* Stendhal's *La Chartreuse de Parme*, and a remarkable translation of Homer into Dutch by Frederick van Eden whose *The Little Johannes*, a Dutch version of the eternal Peter Pan pattern, earlier and more profound than Barrie's version, had been fed to us as children by a formidable governess brought from the Hague by my parents to our unlikely corner of Africa. All these books were to serve me well, perhaps even better than was good for me, because there is no place to equal the effects of reading at sea, particularly when alone as I was about to be. The list illustrates how widely cast was the imagination of the intellectual Japan of the day: an historic gift destined for temporary eclipse in War, but irresistible and ultimately bound to reassert itself even more significantly in the future.

At the end of the day we were free for what was perhaps the most important event in my experience in Tokyo. We went to a Noh Theatre and saw among other pieces one called 'Sumida-Gawa'. It was composed at the end of the fourteenth century by the greatest and most prolific of Noh actor-writers, Ze-Ami, the son also of a great Noh composer and actor, Kan-Ami. I had no inkling then of the effect and consequences that this piece would have on William and myself. The story was simple enough. A bereaved mother sets out alone to brave the danger and tumult of the dark ages of Japan for news of her son, whom rumours declared dead, and, if true, she will bring him back across the fatal Black River. It was acted and danced to music of the simplest organic kind from a Japanese chorus of actors, as in a Greek play, with the principals in masks which made the faces timeless, and the grief and suffering on them recurrent in life to come. The result was another forever which went so deep that it became for me almost more like a happening in a dream than something externally observed on a mere stage.

Consequently, as we walked back in the company of Mr Shirakawa and Mori, I had no desire to talk. I could only smile feebly at Mr Shirakawa's pleasantry aimed at calming the storm raised in all of us by the play. Punning on the literal translation of his family name, he said: "Although I am only an insignificant little white river myself, I must show you the August Black River itself." So we went home by way of the great Sumida-Gawa that streams through the heart of Tokyo. It was not black but glittering and vividly coloured with reflections of electric light and illuminated advertisements. But as we looked into it, Mr Shirakawa rescued it from its modern make-up. If we had been there in August, he told us, we would have seen priests, monks and people of all classes and vocations come down the Sumida in boats, to pay their respects to the spirits of the drowned and say prayers for their souls, and also beg forgiveness of the fish for taking their lives so that they, the people, could live. The association of the river with the frontier between life and death obviously was older by far even than was our play. It explained why no other river could have served the drama's symbolism half as well, and had inevitably drawn a stricken mother in a mask of immoveable and unchanging grief to sway on its banks, as she had done in the play, in a slow trance-like dance, a twig of new green in her hand to represent a son too young to have so brutally lopped off from the tree of life. I seemed to know at that moment how the capacity to grieve, which we seemed to be losing fast in my world, was as important to life as that of laughter. Yet the river went even blacker when Mr Shirakawa added: "You know, Post-San, the evening after the earthquake, this river wound through the glowing ashes of a thousand fires covered thick with the grease of the flesh of human beings burnt in the flames that followed the disaster."

Our last morning in Tokyo was taken up in a visit to the Yasakun shrine where Mori wanted to present himself and us to the nation's spirits for protection, and to pray for us all. The next day already found us back at Osaka and visiting its largest cotton factory. We arrived just in time to see some thousands of young girls, employed in the factory, the managerial staff and directors of the company, preparing to say goodbye to the old looms discarded for the modern types which Britain ironically could supply but still thought too expensive for a declining Lancashire. The girls were neatly dressed, well-nourished and looked far too happy for the slave-labour their competitors in Britain and Europe accused them of being. They welcomed us with smiles, giggles and shy delight, to become serious only when the moment of farewell to their old machines arrived. They bowed deeply to their looms as though they were people; thanked them for their loyal service and wished them well in their future industrial reincarnations. The mystique of commerce and religious fervour that went into industry of which I had a hint in Tokyo was made manifest in a manner there, formidable with implications for the future which are only now becoming apparent to the world.

By the afternoon we were back in Kobe harbour and on board the *Canada Maru* again. Only William and Mr Tajima boarded the ship with us. I do not remember the details of our last conversation; only the feelings that the occasion provoked and which remain deep and fresh and still incapable of being spoken. I hated leaving Japan as much as I was worried about William. I was worried because he had no definite employment as yet and hardly any money left. I relieved my anxieties somewhat by giving him my warmest clothes for the coming winter; the last of my pocket money, and a promise to cable more when I reached Africa. This added together was not much to fill the yawning afternoon and yet it seemed over-full and went by with the speed of a catspaw of wind darkening a plain of grass before a storm. One moment William had been there on deck, bravely laughing that deep uninhibited laugh of his with Mori and myself; and then suddenly he was in the Company's launch with Mr Tajima by his side and receding fast towards the quayside in Kobe harbour. The sun, as it seems to have a propensity for doing on these occasions, was beginning to go down and spread a pathway of yellow light on the dark blue water for the Company's launch to follow back into the harbour. Moreover, its light fell on William and Mr Tajima at right-angles so that, flood-lit as in a *son et lumière* of their own, they were visible far longer than usual.

William was waving his black hat which he wore, as a badge of the artist that he was, high above his head and with great vigour. In the process my best suit of clothes on him, too small for his taller person, looked inadequate to the point of caricature with which fate so often converts the sad into the tragic. The sleeves of his jacket rode up his arms almost to the elbows, and the hems of the trousers, though worn with braces extended to the utmost, rose almost to his calves and gave him a disturbingly Dickensian appearance as of an orphan seeking food and asylum in the slums of a great city. This element of the absurd heightened the impression of his vulnerability, so that it redeemed the occasion from all that was ordinary and raised the final valedictory appearance to an heroic scale which I found almost unbearably poignant.

William had never lacked courage, but he had never possessed it in greater measure than in that autumnal sunset moment of farewell. What made the parting for me even keener was a premonition that I fought and would not accept for years, namely that it was a double farewell: we were not only saying goodbye to each other but also to a William whom neither of us would ever see again, and whom the England to which he was committing himself with such conscious determination would never know. He vanished from view to a farewell blast on the *Canada Maru*'s horn with that bass note of the irrevocable which only ships' sirens command. Then came the grating sound of chains on the forward winches as they began to weigh their anchor. With the sun going down, we headed straight into an after-glow of an Odyssian red towards Moji and the open sea.

I thought Mori would be devastated without William. I am certain he missed him greatly but he was really overjoyed that William had decided to stay on in Japan. At dinner that night he could talk of little else to me than William's 'genius' and how William would be able now to understand and appreciate the 'true Japan' and interpret it to the world.

As happy as Mori was Teruha when I saw her again the next day at Moji. Though I was not invited ashore, since our call was only for the purpose of picking up two first-class passengers and some light express cargo, she came on board in the Company's launch to see Mori. She had lost none of her attraction for me, nor had she acquired any greater interest in me than before, and regarded my role as a bearer of news of William as almost my only importance. When I told her that he was to stay on indefinitely in Japan, she could disguise neither her pleasure nor some fresh expectation and subtle design for the future. She became more animated and talked to me unprofessionally and openly about a remarkable variety of things with an insight of her own and total freedom of convention and cliché.

I thought I saw in the mainstream of our conversation a religious undertone that at times produced the look of a vestal virgin on her fine-boned face; this was incongruous in that dress and foreign to her profession but did not surprise me at all. It added to her beauty and confirmed the quality I detected in her. When Mori came to send her ashore, like a favourite daughter back to school, I was acutely saddened, but this time there was no hint of caricature to deepen the sadness of yet another farewell. Before she sat down in the launch, she looked up at me and Mori on deck beside the ladder above her and then bowed slowly, deeply and formally, her dark head moving forward and over like a black tulip caught in a wind of change. And as a flower the colours of her sash, Kimono and head-dress drifted out of sight, not vanishing so much as wilting and fading with distance. She might have been some bright petal under an extreme of sun, drooping until, deprived of light, it sank in the sea of garden shadow around it, for she went on to lose colour and definition and to fade at last, launch and all, into the green blue water. I watched until the end, hoping perhaps that she would look back, but she did not. Her spirit clearly turned its pages naturally and firmly in the approved mythological manner, so that whatever my own feelings about the content of the moment, it was clear-cut and accorded the proportions and dignity of a timeless shape. I went into my cabin and began reading *The White Peacock*, as if, in turning its pages, I would turn my own as well, despite, my far less deft and unwilling hand.

CHAPTER SIX

The Shadow in Between

THE voyage was a good deal longer than its predecessor because we went by way of Hong Kong, the mainland of China, Singapore, Ceylon, Zanzibar, Kilindini, Dar-es-Salaam and Lourenço Marques, now Maputo. It produced new material but on so remote and external a perimeter of the world in which I had become engaged that it had no claim to be considered. The sea, the ship and my life within it were happily a continuation of all that had gone before, and continued for me in an even more significant manner. Though now alone, I was never lonely. Once Moji and the great South Island of Japan had fallen away in the dark behind us, I looked forward, as far as the exteriors were concerned, only to such events as those in the world of Conrad: its mirror of sea, its surround of jungled and volcanic earth, and cloud and sun, which had seemed so much a part of an inner-eventfulness. Indeed, we broached this territorial imperative of Conrad once again by just slipping across the frontier between two of the storms that patrol its northern marches.

On the mainland of China we picked up some hundreds of Chinese deck passengers who camped all over the ship from bow to stern so that only the tips of winches, ventilators and donkey engines were visible. Yet they did so with a genius for making much out of the small and insignificant which was even greater than that of the Japanese. They were never a shapeless and mindless crowd, nor ever untidy or obtrusive. Despite the press, families somehow remained distinct and self-contained patterns, and one of my regular evening attractions was watching each family grouped around a mat as if it were a mandarin table, eating, talking and enjoying a pipe or two of opium before settling down for the night. It would all look so innocent and inviting that I would go down from the boat-deck to be among them, and was touched how these poorest of poor people would invite me to join them and in particular offer me a pipe of their precious opium, until, out of fear of giving offence, I overcame all my horrific associations and deeply engraved scruples, and tried a pipe. Its only effect was to make me so sick that I have never been tempted to do so again. I was helped in this perhaps also by an observation of our ship's doctor. Opium, he told me in a voice I thought full of implications of reproof and warning, was the greatest and most remorseless of drugs. Hashish for instance, which he declared was a formidable rival, gave those who smoked it a conviction of power that they were capable of doing anything, no matter how difficult. Opium went much further. It gave a man the ultimate of convictions: whatever it was that needed doing one had only to inhale one's opium and one had accomplished it. But perhaps the most important consequence of travelling with so many Chinese on board was some new self-knowledge derived from my contact with them.

As the only so-called European among about a thousand people of the Far East, I might have been expected to feel more self-conscious of my Europeanism and so more of a foreigner. But, on the contrary, I felt at home to an extraordinary degree. And in so far as the obvious and real differences of race and culture were concerned, they seemed to me as valid and valuable or irrelevant as my own, and so a potential source of belonging rather than alienation. What was more, I surprised myself by finding, when comparing differences, that I tended to do so from a Japanese rather than a European point of departure. It was my first glimmering of an awareness that was to prevent me from ever feeling, I believe without arrogance and presumption that I was a foreigner no matter where I went in the world. Indeed, only such an awareness, already a force in being from birth, could have made me as at home in the *Canada Maru* and Japan as I had been among the primitive peoples of Africa or indeed on my mother's farm. I could not foresee then how my life was to involve me deeply and even fatefully with a greater variety of races and cultures than perhaps anyone else of my generation. As a result of this seed of new awareness nourished in the *Canada Maru*, I was beginning to think of myself as a native of the world, and my love of Africa and the Britain to which I was increasingly committed became a universal provincialism destined to transcend and render archaic the imposing array of powerful nationalisms which still dominated the emotions of my time. It is true I would have to start out as a stranger, but the strangeness was relative and never greater or more exacting than the feeling of being a stranger in my own beloved province, and indeed alienated where I had taken belonging and understanding for granted. For all my unknowing, there was awakening then something which was about to make the world my home, a home of many mansions where my closest neighbours were not next door but in the far off wings. Something of this awakening of a new sense of family made this part of the voyage even more rewarding than before.

It also had other great and immediate compensations. I now saw much of Mori and came to know him well. The more I knew him the more I liked him and appreciated the elements of greatness in him. He was by far the greatest sea captain I was ever to meet. For him sailing the seas was not just a physical and technical venture but the fulfilment of a mission to which he felt himself born by life itself. He knew his task as a ship's master by heart and performed it superbly; and with an authority that never lost touch with humanity. But ultimately there was far more than that in his conception of his vocation. Love of his country and people, love of duty and, above all, love of a life of meaning and a search for greater being were all united in it. This enabled him to contain extremes of paradox and tensions that daily tested him. As we talked at the heart of his special world, reflected once more in mirrors of his special sea and uplifted in temples of cloud, I often thought that Conrad would have loved Mori too and put him in a story where through the heightening of perception, which is the role of fiction, the essential quality and truth of the child-like man would shine out. In this way all could see what was hidden or blurred and enigmatic in the complexities of character and behaviour of the captain in total command on the bridge of his ship. We became close friends, and though I am certain I never measured up to William in Mori's estimation, that was enough for me. It is a friendship that will, I feel, outlast time.

Meanwhile, my relationship with my teacher also deepened and my study of Japanese broadened so that it moved into that most exciting of all dimensions where the facts, grammar and bones of the process daily acquired flesh, atmosphere and nuance. We talked more and more about the great Japanese intangibles. One of these was the all-pervasive 'momo-aware' of their spirit which is beyond explanation or definition and has to be experienced before it produces a fallout in the heart and mind that passes for understanding of sorts. We discussed it most as a sense of sympathy built deeply in all things seen and unseen for one another, even as they are made manifest in the same all-inclusive moment of the unfolding dimensions of time. My teacher managed to make me grasp an elementary axiom of its presence in the mind and life of men only by giving me, not without pain to himself, an example of its working in his own life. If it would help, and if I promised to 'pardon the presumption', he would tell me of his own experience of momo-aware.

Many years ago he was suddenly informed that the girl he loved had taken her life. She was an unusual and beautiful girl, and had died in an equally unusual manner. She had taken the trouble to climb to the summit of a volcano and thrown herself into the crater but, of course, not without leaving a letter of apology behind. He stopped there for a moment. I could only suppress my own horror because of the evident support he found from the observance of this propriety at the manner of her going; equally because of the unimportance of my own feelings in comparison with what he must have suffered and still was suffering. After a while, he went on. It was his mother who told him what had happened and, after the telling, she bowed deeply to him and to his sorrow before he turned quickly into the little garden at the back of his home. He sat there in a trance of grief under a cherry tree in full flower. It was a beautiful spring day, and he was crushed by his failure to understand how a person so lovely and loved could turn to death when around them the whole of nature was erupting with life. Soon, his mother brought him tea for comfort. He put out his hand for his cup and was withdrawing it from the tray when a petal of a cherry blossom fell into the cup and floated there, alive with colour on the pale green tea. That, he said gravely, was momo-aware – a sign from the great togetherness in which nothing, however small or abhorred, was excluded, and the beginning of a slow growth into understanding and acceptance of the fall of his own love from life.

He had hardly finished when the thunder spoke behind us and ship and sea trembled with the power and glory of its voice. I heard myself say, as from far away: "And is not that momo-aware too?"

He looked at me closely and long before saying: "I think, Post-san, you have begun to understand. Thank you for helping me as I have all along wanted to help you."

The order of merit his remark conferred ranks among the highest I have received, and very nearly brought me to tears which neither he nor Mori would have despised because they both already had impressed on me how the Japanese concept of 'a knight without reproach' was composed of three imperatives: one always spoke the truth; one never lacked courage; but these were as nothing if one did not also weep easily.

I also practised my Japanese on our two first-class passengers who appeared regularly at table and sometimes came to join me on the little boat-deck where I had my cabin. The senior of the two was a retired businessman called Mr Fujita. He was small and slight even by Japanese standards. At first glance he looked the homeliest of men: a somewhat crooked shape in drab clothes; a wizened and incredibly creased face, eyes obscured behind thick lenses and irregular, protruding and discoloured teeth. He seldom looked one in the eye, not, I discovered, because of anything to hide but because he was exceptionally sensitive and shy. He talked in a low voice that tended both to hiss and utter in rushes. He rarely laughed but on the whole preferred to listen in silence, which he appeared to enjoy. At times he would remove his glasses and wipe his eyes as if in pain. Then for a moment one would just catch in them a look of a strange certainty and calm of resolution that made the face beautiful.

His companion, Mr Nakamura, was a rather solid and much younger man of a cheerful but unremarkable appearance. But his energy and capacity to laugh loudly were both exceptional. He was chosen obviously for his extrovert qualities that Mr Fujita either lacked or had forsaken. And one never saw them except as a pair. They were on their way to Singapore to receive a rare bird from Brazil. It had a beautiful voice and was for Mr Fujita.

"Yes!" they stressed, their mission was of unique importance "because no such bird has ever been seen or heard in Japan."

Such was their high regard for their mission that we rarely talked of anything except the bird, until one day, feeling I had to show my appreciation of its importance which was clearly declining under the wear and tear of reiteration, I asked if I could be allowed to see the bird when I called to say goodbye to them at Raffles Hotel before we sailed from Singapore. I thought my request was granted, but perhaps not without some misgiving. This was so slight that I overlooked it until the next day when neither Mr Fujita nor Mr Nakamura appeared at breakfast or lunch.

"Were they perhaps unwell?" I asked Mori.

"No", he said, somewhat embarrassed. "Not unwell, but worried in case they might offend you."

Mr Fujita, I learnt, had in fact been sleepless all night for fear of giving offence. Astonished, I asked why. It was the little bird, Mori explained. Mr Fujita was worried about its future well-being. He was already afraid that his own and Mr Nakamura's faces might be too much for it. Another strange face and presence around a bird so rare and sensitive could easily be disastrous. Could I possibly understand?

I could not only understand but was somehow touched and humbled, and promised to set it right if Mori would only see that Mr Fujita came to light again. He did so for cocktails on the boat-deck at sundown. I immediately went and begged him to forgive me for having been so forward as to ask to see the honourable bird he was going to meet. It had worried me ever since because I should have known better. After all, I came from an Africa where I had grown up with many kinds of birds and should have remembered how sensitive and easily upset by strangers the best of them could be. I felt that my interest had got the better of me and made me ask for an introduction that was selfish and potentially harmful to a wonderful little bird. Would he forgive me if I withdrew my request?

Mr Fujita's relief made him bow low to me with a hiss of emotion like a human equivalent of a volcano about to relieve itself of an intolerable charge of lava. Our acquaintanceship from then on flourished once more; and at Singapore I gave Raffles a wide berth in case a passing glimpse of me would renew alarm and despondency in Mr Fujita's inflammable heart.

On my last night on the *Canada Maru* in the eastern approaches of Port Natal, we sat down to a dinner of many courses, specially prepared in my honour. That familiar menu-card which had greeted me at my first table in the saloon, with a small boat and Fuji-san, graced my last, and had in large type 'Sayonara Diner' written across a list of courses that exhausted the space below – and the misprint, the fatal touch of the absurd I have mentioned, was there again to increase all the urgent emotions of the 'if it must be' implied in the 'Sayonara' which was the national idiom of farewell. Before going to bed, I stood at the boat-rail outside my cabin staring across water that I had first sailed with Thor Kaspersen, until the light on the Bluff overlooking Port Natal began to sweep the horizon. For the first time I almost hated it for reminding me of what awaited me. I was startled by how great – even greater than I had expected – was my reluctance to leave the *Canada Maru*. I had learnt a good deal of the Japanese spirit but obviously not enough to see, as they did, an inevitable logic of sequence from the 'if it must be' to the high style of the 'it cannot be helped' with which they fulfilled their Sayonaras.

Therefore, I went ashore the next day with a heavy heart, but comforted for the moment that Mori and the ship would be in harbour for another fortnight, and we had arranged that I would see them all every day. But I should have been warned perhaps that the misprint on the menu might represent also a misprint of chance. I woke that night with an alarming fever, and my doctor, summoned in haste, ordered me to keep to my bed. By the next evening my temperature was hovering on 104° and he was thoroughly alarmed, the more so because he could find no explanation for my condition. He and the specialists consulted could only drug me against the extreme pain I felt and try to keep my temperature down. For a week I was in a critical condition and then, as suddenly as they had come, the fever and the pain left me. So, despite the doctor's disapproval, I insisted on seeing Mori and the *Canada Maru* out of harbour.

This time I was allowed to stand by Mori's side on the bridge as the pilot, Captain Robinson, an old friend, took us safely through the narrow entrance out to sea. I was still so weak that following him down the rope-ladder of the waiting tug, heaving even more than the ship in a swaying sea, made me sweat. I stood beside him as the tug's cheeky whistle said a cheerful sort of 'bye-bye' to the *Canada Maru*. When Captain Robinson returned Mori's salute from the bridge, I waved my own hand in farewell. Then I saw Mori had lined up the whole of the starboard rail with the noon-day watch, and all his officers, including my teacher, were in their best uniforms. As I waved and the *Canada Maru*'s siren pronounced its own deep-sea form of sayonara, they all, as one, bowed formally and imperially to us. I stopped waving, came to attention and tried to bow likewise. When I came out of my bow I could hardly see the ship, my eyes were so blurred.

Captain Robinson, seeing my distress, put his hand on my shoulder and said: "Never you mind, young fellow me-lad, I have seen ships come and go most of my life and I am still not proof against it."

I could not say to him that my emotion was no ordinary one because it was black with the belief that I would never see the *Canada Maru* and all those who sailed in her ever again.

The years that followed seemed to support the belief without qualifications. I stayed only another year in Port Natal. Roy, Mary, little Tessa and Anna had moved on to England as, unbeknown to them, I grappled with my strange fever. Also the voyage to Japan and back had worked such a deep-sea change in me that I felt unreal and a stranger in an harbour city, which had once given me so much. I would have left sooner, but I thought that a year was the least I owed my editor who had so generously helped to make the voyage possible. Yet even with him, I was for the moment no longer at ease – I say, for the moment, because the years, which are far more forgiving than their seconds, put that right. We had disagreements over various issues of life and art, particularly art as relevant to the work of Campbell and Plomer, which seemed to me so important that I debated them with him in the letters-to-the-editor columns on his revered Leader Page. It says something about his quality that he would publish them in full and in a sense dignify them by full-length leaders in a reply expressed in sonorous Ruskinian phrases and generalizations that made my sense of our differences more acute.

I could barely serve out the sentence of the twelve months I had imposed on myself, but did so in an incoherent and inwardly rebellious manner. At the end, I hastened up-country to say goodbye to my family and was shaken to discover how detribalized and displaced a person I had already become and how only those ancient hills, vast plains and immense distances in a gloaming of blue illuminated by the candle-flicker of flame of antelope and gazelle going down the long corridors of their miles until even they too were snuffed out by their opaque horizons. I thought of a saying of my father's French mother: *"La tristesse n'est jamais dans le paysage, mais en soi"*, translated from her native tongue at the beginning for my benefit, as: "The sadness is never in the landscape but in oneself".

In the New Year I was back in Port Natal to sail for England in a P. & O. emigrant ship, *Balranald*, on its way back from Fremantle in Australia to Southampton. What with paying back the money I had borrowed to make our voyage to Japan possible, settling things like our mediaeval bill for a meal in Kyoto which seemed more and more to have been an experience in a dream rather than so-called real life, and sending William the money I had promised him, I had little left for starting a new life in London – so little that I changed all I possessed into gold sovereigns and gold ten-shilling pieces to save the rates of exchange I would otherwise have had to pay. However, as this is not an autobiography but a story of a special relationship with the sea, the detail of the long years which followed belongs elsewhere. What matters is that from then onwards I travelled a great deal in ships mostly between Africa and England, until the War.

Of this part of my sea experience, the first harsh voyage in *Balranald*, one of the bleakest and most clinical of ships in which I ever travelled, was perhaps the most important. It did not matter that the food was spare, tasteless and at times inedible, and, after the fastidiousness of the *Canada Maru* saloon, served without grace or style. The captain may, like Mori, have sung in his bath; but I could not know because I never even saw him, let alone came near enough to overhear. If he did, I am certain it was not the equal of the abdominal classical music that Mori loved, but something more like the song which I heard night and day on board and which became the signature tune of the voyage: the 'Charmain' out of the musical play of that name which was drawing crowded houses all the year round in London. I heard it so often that, when it is repeated in the programmes of songs of remembrance which radio and television render from time to time, I am, by the intractable reflex it conditions in me, back in the *Balranald* and hearing not only the song but the rush of wind and waves, and seeing again all that was on my own on the voyage.

My first experience was of the ambivalent Cape sinking into the sea: the patrol of gulls, a flash of white seen under their wings and breast, piping us out of the shining roadstead to where the dreadnought albatross on their hulls and spans of ectoplasmic snow could close in and relieve them. My first view of the Atlantic main with all its fetches and stretches from Antarctic to the western approaches of Europe: its dawns, sunsets and sunrises achieving a cosmic flamboyance under tropical skies until the Southern Cross, which had for so long sanctified the night, itself went down, and the ship drove to its home under a moonless heaven packed with dancing stars over a dancing deck on dancing water. It is true that, central to the memory, was an acute longing for a return to a time when the only musical overture to night was the pure and lucid Shakohachi and the feeling of the absolute and irrevocable, imparted thereby through its absence; but that never failed to add to the immediacy and violence of the reflex. Nor did it matter that I made no special friends among crews or passengers, as I unfailingly did ever afterwards. I was utterly contained in a deep-sea pursuit of my own: I was completing, although in reverse, the other half of the voyage in history which had made a halfway house of the country in which I had been born, and to whose fulfilment into far more ways and dimensions of life than I could possibly foresee I was nevertheless, by my own nature almost in the classical sense, demonically contracted.

Out of that sense of contract, I read ever more widely than in the *Canada Maru* because I no longer, alas, had my studies to do. Hard as I had worked, I had mastered only some eleven hundred Japanese and Chinese ideograms because of a basic incapacity of my hand to combine with the eye and reproduce what it sees. And that was not enough to read a Japanese 'tabloid', let alone the books which were the only ones that would have interested me. So I read all day long in English, German and occasionally Dutch; but mostly in English which I had only been able to do before when alone on our broad farm in the Interior of Africa, and herding sheep under a blue of sky which, whenever the bleat of an ewe for a lamb recalled me to duty and I looked about me, made me feel as if my biblical flock, I and all that vast land, were at the bottom of an unfathomed and fathomless sea. And on that sea the clouds had all sails set and filled with the wind of the world as they passed by overhead on some lawful occasion of the universe.

I dwell on this aspect of the voyage because it was an elemental factor of my experience of the sea which increased my love of it, and determined that most of the adult reading that was to influence me at the most profound levels of imagination was done in ships. As a result, I had constant company of my own within; and almost at once came an upsurge of excitement with the discovery of how timeless and close a neighbourhood the world of art and literature and the spirit were to me. They were, it seemed, a reality outside space and time, and their seasons, without separation and death of created and creator or, if subjected to death, then freed again through catharsis for resurrection. Even when resolved in statistics to determine relative distances between one neighbour's door and the next, measurements were insignificant and ultimately established an enduring propinquity rather than segregation. For who could find Homer and his Helen and Hector, Odysseus and Penelope, to be far away when one came to realise, as I did when I read again of the battles on the great plain of Troy and the return to Ithaca, with the smoke still coming down among its housetops as if no dynasties had passed in between, that they were contemporary events in this world of imagination outside time, and that two thousand four hundred years or so which separated me and them, numbered in begetters of children, were at the very most eighty, which was half the company of infantry wherein national conscription had drafted me for my own military preparation? Remembering also that I had a vivid description, many times repeated, from my grandfather of a battle in which he had fought against the British in 1848, the exchequer of my memory found the expense of such manpower between Homer and myself to be spendthrift, and had no hesitation in reducing its allowance to accommodate a mere platoon of the memories of twenty.

In this manner, inwardly more prepared than I had ever been but outwardly more vulnerable even than William had been in Japan, I arrived in England early in 1928 and still in the first of one of its coldest winters. The winter was all the colder in my imagination because Britain and the world were sliding rapidly into one of the greatest economic depressions within living memory, and my own preparation and experience of Africa and Japan seemed unwanted and irrelevant. As far as Japan was concerned, no-one was interested in a measure which was not quickly filled and overflowing. And had it not been for a chance meeting with Arthur Waley, I would have had no-one with whom I could positively share my experience and love. Again, the detail of my friendship with Waley and a remarkable woman in his life, the Beryl de Zoete who translated *The Confessions of Zeno*, belongs elsewhere. What matters here is that the same hand of chance and inner sense of occasion which had taken me to Japan, directed me infallibly to them and, through them, to a chain of other acquaintanceships that took me into the heart of Bloomsbury where I met among many others: the Bells, Duncan Grant, Roger Fry, Lopokova, Maynard Keynes and, most important of all for me, Virginia and Leonard Woolf. These two I can never thank enough for being the only publishers in the world to take on my first English book of fiction, *In a Province*, and publishing it under the imprint of The Hogarth Press.

For some years I would dine with Arthur and Beryl once a week in a restaurant called 'Le Dîner Français'. Often it was my only substantial meal for days; and invariably Arthur Waley paid. In addition to his love of the civilizations of China and Japan, we shared a love of mountains and, in a precious little world of singularly cloistered and over-refined values of art and mind, full of contempt for the physical, his robust indulgence in the skiing he did superbly, was an example of a natural corrective of excess and another close bond. But as far as Japan was concerned, the singularity of this aspect of my relationship with Arthur Waley and Beryl stressed, so it seemed consciously, the probability that it was a once-only experience; and that, apart from the widening of the range of imagination and sense of the law of relativity on which a being in search of truth and wholeness is as bound as the universe in Einstein's time-space continuum, and which it had conferred on me, was the end of the matter. And yet a kind of non-rational something which was neither a hope nor a gambler's notion of the nature of the throw of fate to come, smouldered on underneath the growing pile of ashes of the dying fire of the experience behind me. It smouldered on in a feeling that all around me, and everywhere else in the world, unseen and unbeknown, something even more important to man than the discovery of fire was in preparation, and that I had a special responsibility of my own, however minute, in preparing myself and others for proper participation in a universal act.

The more I saw of Britain and Europe, the more I realized how all was changing within men, between their societies and between nations in a manner that has been so well documented and analysed that it is unnecessary to add to it. I have only to speak of how often this feeling brought a shiver of spirit and flesh into my life, despite my own battle for survival in a London dark with increasing economic depression, with unemployed and unwanted lives and abysmal aimlessness around me, and I would remember another saying reported of the French grandmother I never knew. Sitting by the fire one cold winter's night in our home in the Interior and hearing the jackals yanking and the hyaenas howling with the melancholy only the carrion of night know, she remarked to my father: "*Écoute! On se prépare. Ici auprès des chandelles tout est clair et constant et on est très bien et calme, mais dehors au delà et dessus de la nuit, l'univers se prépare, et moi j'ai peur.*"

I too had an increasing fear that the world was walking in its sleep towards unparalleled disaster because of its failure to become aware and join freely in whatever was in preparation for life. Unless we did so freely, I suspected it was all to be forced on us by catastrophic suffering. Unless we let it in by the front door of all we had of reason, feeling and intuition, it would break in by force through some back door of our spirit in a dark blacker than the night which had filled my gallant grandmother with fear.

In this way, I felt compelled to march with the unemployed in Hyde Park and join in passionately with the debate which shook my generation over the nature of the radical reform and renewal that the societies and nations and their institutions urgently needed if life itself were not to be imperilled. But I cannot pretend that I had any clear vision of precisely how we had to prepare. After the many night-long debates with friends, I always came back to a kind of Zen-like conviction that for me the only true preparation was to do always what was closest and more urgent; never to say 'no' to what life deposited on my own doorstep, however enigmatic, but to take it in and give it a home in heart and mind.

"If I mind my own step as best I can", I told myself again and again, "the miles will take care of themselves. As for the horizons, watch them as clearly and steadily as your nature allows but watch them all around and in the whole, as you did with Thor Kaspersen on the Indian Ocean."

Almost at once I was at odds with my friends because I could not join in their game and the 'taking sides' in life which they called commitment. They, all in one voice, accused me of neutrality and fence-sitting and reminded me of Lloyd George's description of Ramsay MacDonald as "a man who sat on the fence for so long that the iron entered his soul". But already I had a glimmering that theirs was an archaic sense of commitment, if not an evasion of it; that life was greater than fundamentalist concepts of action and reaction, cause and effect, which dominated science, theology and political philosophy. A truly contemporary commitment, I was beginning to suspect, would bind one to a search and service of wholeness that transcended and resolved this parochial and tactical demand for 'taking sides', and so contained rather than opposed. As a result I was at the time the only one in my group of many friends who became famous as poets, artists, actors, politicians and revolutionaries, who rejected Communism.

Indeed, my first book of fiction was, among other things, a rejection of the lure and blasphemy of Communism and as a result, so I was told by a host of generous friends and critics like Herbert Read, Stephen Spender, C. Day Lewis, René Janin, Lilian Bowes-Lyon, Ernst Robert Curtius, and so on, it had a transformative effect on them all. I had come across something in Goethe which, despite a certain allergy I always had to his conscious cult of greatness, said what I instinctively felt: "I hate every revolution because as much good as bad is destroyed thereby; I hate those who cause revolutions as much as those who make them." But on this occasion, my only quarrel with that statement was the resort to 'hate'.

As a result I tended to say 'yes' rather than 'no', perhaps too often, to what chance and circumstance brought to my desperate door. Yet I never numbered among these, the instant 'Yes' to a cable from Desmond Young which reached me in the summer of 1929.

"Have just taken on Day editorship Cape Times", it read. "Please come and help me clear up mess in this country."

Apart from the fact that Desmond had never said 'No' to any plea for help of mine, as in getting William to Japan, and I was indebted to him on many other scores, his cable seemed to correspond also to a summons of my own from within. I cabled 'Yes' without a moment's hesitation, despite the realisation that it would interrupt work on a full-length story that I had in mind. It demonstrated a dilemma which I was never to resolve. I had always wanted to be a writer more than anything else. I had, in fact, written my own form of Haiku in High Dutch at the age of six; and remember the time, the place where the puddle in the arid earth that was a sea to me evoked the verse; and the words of the beginning come as clearly to me as the day on which I wrote them:

> *Stille Wateren,*
> *Groene Bladeren,*
> *Donkere Schaduws,*
> *Bonte Swaluws*
> *Vliegen Over al.*

which, translated in prose, reads: "Still waters, green leaves, dark shadows, motley swallows are flying everywhere."

But I had never felt my longing privileged enough, nor my gift sufficiently great, to allow me to be only a writer. There seemed to me moments in a desperate time when one had also to do and act on the ordinary everyday human scene. Art and writing, it seemed, ultimately demanded not only expression in their own idiom but also translation into behaviour and action on the part of their begetters. Being and doing, doing and being, for me were profoundly interdependent, particularly in a world where increasingly it seemed to me the 'doers' did not think and the thinkers did not 'do'. But if I had to lay an emphasis on any one of these considerations which faced me as a result of Desmond's cable, it would have been on the 'being'. In the Western World to which I belonged, all the stress was on the 'doing' without awareness of the importance to it of the 'being'. Somewhere in this over-balance of contemporary spirit, there appeared to be an increasing loss of meaning through the growing failure to realise how 'being' was in itself primal action, and that at the core of 'being' was a dynamic element of 'becoming' which gave life its quality and from which it derived its values and overall sense of direction. Because of a lack of such 'being', we were constantly in danger of becoming too busy to live. I longed to 'do' as a consequence of my 'being', but feared the busy-ness of my day like a plague.

None of these simplifications of a protracted process of reflection and reappraisal are in the least products of hindsight. They are a mere summary of what became clear to me on the voyage that followed from Southampton to the Cape, and are elementary illustrations of the thoughts that came to me at sea, leading to a belief that thinking was not an egotistical process alone. That much quoted Descartes' "I think, therefore I am", and the whole trend of rationalism in all the disciples of my day, had the makings of a dangerous hubris for me because ultimately in the decisive and larger issues of life and death I knew now that there was something objective and infinite thinking through me.

Again, I was certain that, as far as I was concerned, these things could not have occurred half as meaningfully anywhere else than at sea. Indeed, the long voyage to the Cape established for good a pattern begun with Thor Kaspersen and Mori which was to recur like an indivisible decimal in my reckoning with life, to reveal and confirm stages of evolution and elucidation that were a constant, until traffic at sea abandoned me half a century later. On this particular occasion, there was also a singular felicity in that in response to Desmond's cable I was travelling in a ship which continued the discovery of new and even closer neighbours in this timeless dimension of art, this forever 'now' which contained and sustained me when alone in the *Canada Maru* and established far more personal and immediate links with the special aspect of history to which my own life seemed so singularly and inevitably connected. It did this because I was sailing the High Seas for the first time in a Royal Mail Ship.

The Union Castle Line to which it belonged had brought to a glittering summit the traditions and service of the centuries of traffic and travel by sea which had gone into the making of our country. The impact of this particular voyage could not be properly understood wihout some illustrations of the story I carried as an important part of my baggage on board at Southampton. The Line was the work of one man more than any other, a Scot called Donald Currie. He was not only a great 'owner' in the classical British sense but a powerful, complex and far-seeing man of many dynamic parts – a great friend of Gladstone, a Liberal member of Parliament, and so trusted by the new Boer republics in nineteenth-century Africa that they turned to him again and again for help in their worsening relations with Britain. He was the possessor, too, of a flair, rare in his day, for keeping public imagination involved in his affairs. As he built one ship after another, each faster and more luxurious than its predecessor, he gave parties which, in comparison with even the most elaborate public relations exercise of today, seem now to have been unbelievably lush and pretentious.

The initiation ceremony in one of his favourite ships, the *Pembroke Castle*, which took him to sea in 1893, tells it all. The distinguished company invited to try her for their private delight before the ship became a servant of the world, included not only the Prime Minister, Mr Gladstone, complete with bags, side-whiskers and all – except permission from his Queen to leave the shores of England – but also the aged Poet Laureate, Alfred Tennyson, and his son. The poet would often lean over the side of the ship, look deep into the waves and, as his grandson would put it, "the broad silver path of the moon on the rolling sea" and compare it to "a great river rushing to the City of God".

In the evenings he would read aloud the 'Northern Cobbler', 'Rizpaz', 'The Children's Hospital', 'Ask me no more', 'The Bugle Song', 'The Two Voices', 'The Promise of May' and 'The Grandmother', over which he broke down on one occasion.

The voyage was such a success that it lasted longer than intended and took the *Pembroke Castle* to anchor at Copenhagen, where it lay for several self-indulgent days. The ship's company were invited to the Royal Palace, and by way of reciprocation one of the most remarkable gatherings ever held in any ship took place on this liner, destined merely for service to the Cape. It gave a party for which, no matter how hard one tries to side-step it, only the archaic epithet of 'banquet' will do. Apart from sixteen distinguished service officers and Ambassadors, it included twenty-nine royal personages. Among these were the King and Queen of Denmark, the Emperor and Empress of Russia, the King and Queen of Greece, the Princess of Wales (Queen Alexandra) and the Royal Family of Hanover. When Gladstone proposed the toast of the Czar, the news was signalled to a Russian warship nearby and the man-of-war immediately joined in with a broadside from all the guns. At a late luncheon that was more like dinner, the Queen of Denmark rose and drank to the poet.

The Czarina said to him: "What a kind sympathetic man Mr Gladstone is! How he stood by little Montenegro."

The Czar, on the other hand, observed: "I should like to be King of Denmark."

When the ladies left the gentlemen to their drink and cigars, Captain Harrison, the Commodore of the Currie Fleet, extracted Tennyson from the group and took him to the little smoking-room where the royal ladies were gathered and made them listen, overcome with unaccustomed emotion, to the Poet Laureate, reciting in a suitably modified form of his natural Balaclava voice 'The Bugle Song' and 'The Grandmother'. He did so seated between the Princess of Wales and the Czarina. A breeze came through the porthole, and a solicitous Princess whispered to him to put on his hat. But he refused, explaining that he ought, if possible, to make himself even balder before so many royalty. The Czarina paid him a graceful compliment and he, being short-sighted, took her for a maid of honour and patted her affectionately on the shoulder, much to the consternation of her entourage.

Outside, in the yellowing light of a September afternoon, ships of war from the many countries manned their yards, the national anthems of the royal visitors were played and everyone was as touched as they were over-awed, with the exception of Queen Victoria. She was not at all amused at what she regarded as irresponsible behaviour on the part of a minister. The ensuing correspondence between her and Gladstone, who responded endearingly rather like a scoutmaster caught unprepared, has to be read to believed. 'Escapade' was one of the least of the royal labels attached to his conduct. Ultimately he was forgiven, but at a price: the peerage he had requested for Tennyson was granted only on condition that it was postponed a year. It may all sound petty today, but in an age when royal occasions had marked political consequences, Gladstone – in the eyes of a Sovereign known to our African selves as the great White Queen – had put British Foreign Policy at risk.

Yet an unrepentant Mr Gladstone was to sail once more, some twelve years later, in a Currie liner, the *Tantallon Castle*, again under the command of 'Holy Joe' Robinson and Sir Ronald Currie, as ever the grateful and generous host, with a hundred other distinguished guests and representatives of "statesmanship, politics, art, literature, science and international commerce", as the log recorded it. It was Gladstone's fourth and final voyage in a Castle ship, and he was on his way to witness the last of the Kaisers open the Kiel Canal. Already there was a touch of the winter of history in what appeared to be such a warm yellow harvest air of a tranquil summer. And as I read of it in the ample library of my own mailship, it seemed all part of this sense of transition and preparation that I had brought on board with me.

I read how both Alexander of Russia, who had been with Gladstone in the *Pembroke*, and his successor, had been assassinated. The Russian boy (and cousin of both the Kaiser and the Prince of Wales), who had romped around the decks while his parents dined below, was on the Russian throne. Gladstone had resigned from politics, Lord Rosebury's government was about to fall, and Gladstone referred to himself as the survivor of a bygone age.

Yet he did not behave like one. He committed what he himself called 'perjury' by going back on his statement to Parliament that he would never speak in public again, and rose to his feet on several occasions to address royalty and other great dignatories who rushed to see him wherever the *Tantallon Castle* put into port. Moreover, the same intellectual drive which made him capable of addressing the first Greek parliament at Corfu in his university Greek, made him study Danish so that he could address the King and Queen of Denmark in their own language. But despite the signs of tragedy and storms of change, what seemed to concern Gladstone most was the fact that Parliament had passed a bill to erect a monument to Cromwell. He confessed that if he had still been in Parliament he might not have voted in favour of the monument even though his own son Herbert had charge of the project. He admitted the greatness of Cromwell, but he said he could not forgive him the Irish massacres, and drank a toast to the Irish members who had voted against the proposition in Parliament.

"Carlyle's Cromwell", he said in a voice sonorous with passion (everyone noticed how alive he became when on his feet with an audience to address), "was a piece of pure fetishism". He then went on a rampage against all autocrats. Charles II "stuck in his gizzard"; if he had his way "Louis Napoleon would have been struck off the roll of the monarchs of France for having waded to power through the blood of a coup d'état". Louis XIV, the Sun King, and Louis XV were "the most abominable Kings ever" who "made monarchy in France impossible". But in the evenings he would always relent to the music after dinner when his daughter-in-law played the violin and a certain Mr Nicholl sang the most dreamy of ballads. One night, listening to a rendering of 'Auld Robin Gray' there was, the log book says, "brought over the wonderfully green field of his memory, a vision of a little boy baffled by his first problem in political economy".

"I remember", Mr Gladstone declared, "when I was a child hearing that sound and being struck by the lines:

> To mak' the crown a pound, my Jamie went to sea,
> And the crown and pound were baith for me.

"How, I asked," Gladstone exclaimed, "could they baith be for the girl when the crown was an integral part of the pound, not two distinct acquisitions but the larger including the less?"

Somehow, that one slight incident tells nothing about political economy, which is unreal and a non-idea to me. But it does tell more of what was right and wrong with Gladstone than any more portentous incident recorded in the biographies that I know. What was of special concern to me in an Union Castle cabin in the time was one of those coincidences that are never meaningless for me. 'Auld Robin Gray' was composed by Lady Anne Barnard. Lady Anne Barnard was both a Lindsay and a Crawford and, in the time that she spent at the Cape during its occupation by the British as the wife of a certain Captain Barnard, she wrote those amazing letters home to Britain. This was the beginning of English literature in the South Africa which the *Tantallon Castle* had been built to serve.

With such traditions and owners, it was not surprising that the *Walmer Castle* had its own testimony of continuity in living history moving on to the future. Like the founders, who built their *Walmer Castle*, its sister ships, and proliferating intermediaries that were drawing the whole of Africa and the Mediterranean into their orbit, the officers and crews who manned them all possessed to an extraordinary degree, either by instinct, education or a combination of both, the quintessentials of that unfailingly over-criticized and under-rated Victorian age.

I remembered with great discomfort in this regard an argument with Virginia Woolf over Charles Dickens, into which I fell as some insect into a spider's trap. The argument took place just before I sailed and alarmed me with what it showed of cold steel, and the ability to wield it coldly in someone held to be too frail and vulnerable for 'real' life. She despised Dickens. She said he was the kind of person she would instinctively avoid by darting down an alley if she saw him coming down the street towards her. She said that there were no gentlemen in Dickens unless they were gentlemen who had come down and were seedy and soiled in life. Though young and very much in awe of a lady of such gifts, such beauty, such an extreme sensitivity and refinement of temperament, I was almost overcome by the paradox of ruthless judgement and uncompromising vehemence displayed. As a result I only ventured to suggest that she might, perhaps, be too harsh because, for me, one of Dickens' outstanding qualities was his extraordinary perception that the concept of what was called 'the gentleman', together with the chivalry that went with it, had penetrated so deeply into the British character that it was not the possession of any particular class but was to be found everywhere in all the subtle nuances of the highly-nuanced class system of Great Britain. I added that the greatest gentlemen, the knight errants, as it were, of Dickens' novels, appeared to me to be drawn from the working classes, and even the criminals of society. I myself thought perhaps the greatest gentleman of all was the convict hero of *Great Expectations*. Virginia was neither amused nor converted by this hesitant observation; but that did not prevent me from feeling that something of this generalized truth appeared to be invested in the way of life in all these ships.

If this were not so, how did they fail to go unhesitatingly to the rescue of one another, even when their most hated rivals were in trouble? What else could have made a ship of the Union Castle Line be as imaginative, compassionate and tender in the solicitude with which they conducted the tragic Empress Eugénie by sea to Port Natal on her pilgrimage to the scene in Zululand of the death of her only son, the Prince Imperial? They behaved similarly to the King of the Zulus, Cetewayo, whose warriors had done the killing and had been responsible for the disaster and slaughter of Isandhlawana, when they took Cetewayo into exile in the Cape. They even diverted a ship, at much expense, to St Helena so that the Empress could pay her respects to Napoleon's tomb. As for the Zulu King Cetewayo, when they heard that he would be heart-broken if he could not sail with his three favourite wives, two chief indunas and two children, they built an imitation Zulu kraal for him on their immaculate decks so that the King and his wives could live there in the station and in the manner to which they were accustomed. And what else could explain the reaction of J. C. Robinson, 'Holy Joe' as he was affectionately called (Currie's favourite master with whom I once sailed from Durban to the Cape), when he heard that his flag-ship, the *Kildonan Castle*, had to receive 2,500 of my countrymen made prisoners in the Anglo-Boer War. He mustered all hands on deck to remind them that their prisoners were not to be regarded as "anything but honourable foes who had fought freedom", and that by "consideration, kindness and courtesy, the ship's company might in no small way contribute to a better understanding in mutual relations when the war was over".

He then gave orders to holy-stone, scrub and wash the decks of the ship so that it was as spotless for their reception as for that of any king. So much were my imprisoned countrymen affected by this treatment that some days later, when the news of Queen Victoria's death reached the ship and its company broke down and wept, the prisoners sent two of their leading men to express the sympathy of all their ranks to the Captain, his crew and the people of their enemy, Britain, in general. And, when after six weeks the *Kildonan Castle* was able to hand over its prisoners to two other transports, these men of the bush and veld climbed up the rigging and to every other available point of vantage in their new prison-ships and first cheered the *Kildonan Castle* as they sailed past it on their way to imprisonment and exile, before breaking into their great old Dutch psalm and battle hymn: "Rough storms may rage, around us all is night, but God, our God, shall not forsake us". They sang till their voices faded into the sunset of a placid summer evening, and so on and over the gleaming rim of Table Bay. That last story, in particular, made this feeling of propinquity of history all the keener because I heard it also from a cousin of my mother's who was one of this choir of prisoners.

The *Walmer Castle* was renowned for the intensity of the vibrations induced in steel and timbers by her powerful engines. But she was for me something almost legendary because of the beauty of her line, the wooden panelling of the elegant public rooms, and the orchestra gallery of gold-fluted pillars and graceful latticed balconies in the dining-saloon. She was like something lifted straight from the best of a nineteenth-century Viennese operetta. Just being at sea in her was, for me, a full-time occupation. No sooner was I awake than I would rush on deck in dressing-gown and slippers as if to reassure myself, as in my days with Thor Kaspersen, that there was no land yet in sight to compromise the quality of the view. My fellow passengers were interesting and varied, including the new Governor-General of a colony in Africa; other distinguished colonial servants and their families; statesmen continuing a special mission begun by the unfairly forgotten Leo Amery; some Johannesburg millionaires, an actor-manager who after a successful London season toured the Empire with his company, as was the habit then; the young Moiseivitch; professors, soldiers, students, sailors and prospectors with sun-dark faces. But I soon discovered that my own nature drew me to the officers, engineers and crew, above all the crew. Through them one had not only a sort of infra-red vision into what was concealed behind the social masks of fellow passengers, but also much of what had been significant in the humanity of those who had travelled in the ship in the past.

There was an old steward who had resolutely refused, with rare exceptions, to serve in any other ship, despite offers of advancement. A young man in the early years of the century, he was already a waiter in the first-class saloon at the table at which Kipling and his family sat, as well as Dr 'Jim' Jameson of the infamous Raid which was the catalyst of the totally unnecessary Anglo-Boer War whose consequences are with us still. This steward was known as 'In-a-manner-of-speaking' Jack, because almost every sentence he spoke incorporated that phrase. When I asked him what Kipling was like, he drew himself up the full height of his six feet and then looked away as if seeing through the porthole the smoke of a funnel of a legendary ship just vanishing over the far horizon: "In a manner of speaking, Kipling was not half-weird but, if you ask me, in a manner of speaking, he was a blinking weirdie all over."

When I asked him why, he said that first of all, if I knew what he meant, "A fellow would never know whether Kipling himself, and his mind, were, in a manner of speaking, always in one and the same place at one and the same time. But when they were," and at this point he would bend down and almost whisper with the enormity of the weight of the memory, "one would know all about it because he would then become, all of a sudden like, the nosiest passenger I have ever met." Apparently he wanted to know about everything: one's life, quarters, pay, the nature of one's enlistment, one's work and so on. In this mood, Kipling would, 'in a mannner of speaking', cross-examine everyone from the Captain and deck apprentices, chief engineer and bosun, right down to able seamen and stokers. What was stranger still, unlike most passengers who thought deck officers straight out of the top drawer of the ship, Kipling in this mood had a marked preference for engineers and engine-room staff. It was almost, 'in a manner of speaking', as if he loved the engines better than any other part of the ship. And then, weirdest of all, when Kipling got a fellow talking despite himself, he could listen as no-one he had ever met could listen, almost as if somewhere in his mind he was writing down every word a fellow said. Yes, one had to give him that, he was a good listener but, 'in a manner of speaking', too good almost because Kipling 'made a bloke feel he would round on him like a blooming prosecutor at a murder trial if he strayed an inch from his own experience and trimmed his sail to some fancy of his own'.

There was confirmation of this from the fact that Kipling himself, as I read in the library, often spoke of how much he and his family loved travelling in this and other Castle ships; how fascinating he found his multitudinous contacts with so great a variety of men and women in these ships and how, talking and listening, he would extract significant patterns in what he heard, and make stories of what he was told. There was a story he called 'Little Foxes', overheard originally in the *Walmer Castle* from an army officer and rendered so truthfully that, according to Kipling, after seeing the story in print, an ordinary superintendent of police once wrote to him from Port Sudan asking how he had come to know the names of the hounds in the very pack to which he had been whip in his youth. Kipling had to write back and tell him that he had had it all from the lips of the man's Master of Hounds himself.

As for Dr Jameson, according to my old steward, he was a rougher sort; fond of his food and wine, to the extent that he was always complaining about the ship's food. I hasten to add that, for some reason I have never fully understood, this is a habit that many people cultivated at sea, perhaps as a defence against the threat to their 'land identities' from the subversive power of ships and sea. One day when the lady at his table complained of the quality of the food, calling it 'prison fare', Jameson replied: "Speaking as one of the original classes, I assure you that it is worse". Jameson, of course, had been in a Boer gaol under sentence of death for his infamous Raid. But, unlike the Boers, the lady at Kipling's table showed no mercy and, because of the suspicions aroused by this remark, she saw to it that he was removed by higher authority as unsuitable for the Kipling table.

And then there were Kitchener and his Lord Roberts rushing out to South Africa to rescue the British from the disasters inflicted on them, observed in the *Dunnottar Castle* with the unconditioned eyes of the young man that the old steward was then: "That Lord Roberts, the one so-called 'Bobs'," he would tell me, "was the kind of gent, in a manner of speaking, that everyone liked. But that Kitchener, with his big black moustache and eyes bulging like great marbles, he was another one. He frightened me proper, he did! I would not like to speak ill of the dead, in a manner of speaking, particularly one who died a sailor's death by drowning as he did. But he was the talk of the ship because he was afraid of the ship's cat! Now I ask you, in a manner of speaking, how could a man beat off the whirling fuzzy-wuzzies at one minute, as he was said to have done, and the next run away from a cat which, if he had allowed everyone to stroke him as everyone wanted to, he would have had no hair on his back. God help poor Tommy Atkins, we all thought, with a skipper like that. But thank God for Tommy Atkins' sake for the one so-called Bobs.''

Finally I learnt for good how, with a totally unqualified removal of responsibility for one's own life on a voyage of this kind, and with nothing around one but the sea out of which life is supposed to come, most people are briefly restored, as my steward would have had it, in a manner of speaking, to a more innocent and natural version of themselves. One moved into a dimension of which the increasingly metropolitan context of life deprives most human relationships, even those that pass in town for the closest of friendships. They deny people the experience of doings things together and leave them to base all, as a rule, on eating, talking and going to places of entertainment together. But at sea, as they travelled then, they were still finding the human condition enlarged by being actively in one community for days on end, sharing and contributing the same forward-moving experience.

What greatly helped this return to natural grace was the fact that ships were built so that one was still incredibly close to the sea. One of my main objections to travelling in the 'Queens' and the great greyhounds of the United States Line was the difficulty one had in finding a view of the sea! But in the *Walmer Castle* one was still very near to the sea itself and thus able to observe the things that came out of it in a profusion which seems almost fabulous. Coming back to my own cabin one night whilst in the tropics, I found I had left the light on, and so close was the open porthole to the water that a flying fish drawn towards the bunk-side lamp was there, all silver and asleep, on my pillow.

I realized for the first time how good a guide to what the sea was like then is *The Flaming Terapin* by Roy Campbell. It remains as fine a record of those seas as it is of the wide wild world of Africa with which his own naturally wild temperament was so closely identified. He describes how the wind of the speed of his archaic ship would unwind one smoking iris after another in the white spray raised by its bows, and flying fish in coats of silver would rise up like stars and spatter down like hail, not spluttering around as I had seen them do in the Indian Ocean from the deck of the *Canada Maru*. Even for me there was hardly a day without such irises and flying fish, as well as schools of dolphins leaping, lacquered with light, from wave to wave to surprise and delight, and never failing to raise a glow of pleasure in the dullest heart; porpoises and dolphins, too, leaping phosphorescent out of the black water at midnight to shake the Bible-dark night into what D. H. Lawrence in 'Amores' called 'a dazzle of living'; and sometimes even some lonely Old Testament leviathan suddenly sending up the fountain of his breath to stand briefly like a palm of silver in a mirage on a shining horizon. Yet all these served to resurrect a dead, deeply buried Merlinesque aspect of oneself and made one feel oddly harmonized again as it had done to me when I had sailed with Kaspersen. Indeed the berths in the *Walmer Castle* were still so closely clutched to the sea that it was not uncommon to meet old sailors who claimed to have seen sea serpents of a St. George's design; whales the size of dreadnoughts; and the kraken itself like a large baroque *cathédrale engloutie*.

This was, I felt, a poet's ship which, together with the deep poetic waters, reminded me not only of Roy Campbell but of Camões and many contemporary poets and writers, from Macaulay to Baudelaire and Olive Schreiner. Perhaps Baudelaire could never have been half the poet he became had his parents not committed him to a voyage round the Cape and back in order to get him away from acquaintances they found undesirable. One has only to read again his great 'Albatross' and poems like the 'Voyage' which I once translated in full and in which the "real travellers are those who go purely for the going, hearts light as balloons of air", and so towards that moving reflection straight from his travel-stained heart:

> *To the child in love with his maps and his stamps,*
> *The world is big enough for his appetite vast,*
> *But to eyes looking back in the light of lamps,*
> *How small it is at the last!*

Towards the end of the seventeen and a half days, which seemed almost too much at the beginning and now had come to feel so little, the experience had become so unique and precious that one instinctively feared the land ahead, no matter how much one loved the people waiting on the quay. I knew now, after the *Canada Maru*, how life as lived on land would instantly pounce like a tiger on the experience behind and drag it away to devour it instantly in the civilian jungle. Indeed, I came across many a retired person on this and other pre-war voyages, when time had a steadier pulse than now, who valued this aspect of life in ships enough to leave one ship only to board another as speedily as possible – vanishing into the future and pledged to voyaging without end. So much did I feel this that, on the last evening before arrival, the long cabin trunks and suitcases stacked on deck ready for a quick landing in the morning looked to me like coffins in which the dead voyage was about to be buried.

And yet something imperishable of these things remained in safe-keeping in some bank of human memory where, strung like beads, they form a rare kind of necklace ready for wear when the right moment and properly initiated company are summoned again for another reunion at sea. So from that trance-like *Walmer Castle* moment of the late twenties to this day, I have, with rare exceptions, travelled in nothing but Union Castle ships. I was in a sense lucky also to have been just in time to see how they still reflected the clear shape and nicely, if not ruthlessly, defined social order which still governed the lives of men. Everything and everybody still appeared to have a definite place and space of their own in life. The feeling of being displaced, detribalised or disorientated beings caught up with no appreciated or appreciable identity in an irresistible process of change without form, name or direction that appears to be one of the universals of our time, was still a massing of tensions deep down below the surfaces of all except the most unusual of minds. Accordingly, the *Walmer Castle*, like all the best ships, was divided into the three estates of man, the third with a nuance almost marked enough to constitute a fourth.

There were in each mail ship first, second and third class, and a rougher degree of the third called 'steerage'. There were firm barriers across the decks to mark one from another; and it was noticeable how these barriers, which were so rigidly respected at the beginning of a voyage, seemed to become more and more flexible as it went on, until they became physical rather than social or psychological impediments. It was always a source of amusement to see how, within days, acquaintanceships that grew into friendships were formed across these 'barriers'. But when the classes were not over-crowded, as now in the *Walmer Castle*, first and second, third and steerage, combined for a fancy-dress ball; a Derby night – where ladies served as horses and raced one another with scissors to cut equal lengths of tape in two in the fastest time, while the gentlemen, dressed as Epsom bookmakers, called out the odds. There were also the inevitable ship's concerts. The great European carnivals I have seen, like the Pan-like Bock Bierfest of Hitler's Munich, or the elaborate tribal upsurge of a pre-dawn European spirit which annually, to the tap of thousand-year-old drums, invades the respectable citizens of the circumspect Basle, exceed the fancy-dress occasions on the main of the *Walmer Castle* pattern only in size. Anyone who was foolish enough to think that Dionysus was dead would only have had to see these ocean spectacles or even read about them, to know how alive and potent he remained and how great was his power still to overcome our resistances and transform external pretensions. I think it is proof enough to say that in the *Walmer Castle* I saw a distinguished colonial Governor (who seemed to speak to no-one but the captain, occasionally nodding like Homer at a not inconsiderable fellow passenger or two) now appear as a Roman Empress with a talent and zest for the role that would have easily earned him a principal part today in the 'I Claudius' of the television screen.

The Derby nights roused passions so great that they produced the only occasion on which I saw two gentlemen, both puce-faced manufacturers from the Midlands, exchange Victorian fisticuffs among startled vicarious gypsies, strawberry vendors, debutantes and grey, top-hatted onlookers. Finally, there were the ship's concerts, and although one was forewarned to dread them in those days before films were shown in ships and when, of course, everybody had to travel by sea and invent their own amusements, they could be unusual fun. But imagine what a Castle liner's concert must have been like when Dame Clara Butt was the star of the occasion, and sang 'Abide With Me'. My steward would remind me in a voice which 'in a manner of speaking' would, he feared, if I knew what he meant, sink the ship. But on this occasion I only heard one singer at sea out of the great English order of hilarious folk straight from the Edwardian music hall: the tall, angular Nellie Wallace, with red-hennaed hair, a long face and rush of teeth to the front, singing that she was a "bit of lost property left on the shelf" and "what old Pringle of her had left, she wanted for herself". She was sheer Toulouse Lautrec to look at but a warm-hearted and endearing sea companion. And there was Franklyn Dyall who had played the hero in the first performance of *The Importance of Being Earnest* but now had grown into actor-manager, taking off Beerbohm Tree and recounting diamond-pointed witticisms of the Oscar Wilde that he had brushed against. There was also music by the famous Polish Czerniavsky Trio, who played what remains for me an unequalled performance of Beethoven's 'Archduke'.

But much as they gave me, it did not stop me from an envy that was like a kind of hunger within me, when the senior purser told me how he had travelled out to South Africa in 1912 with Paderewski in the *Armadale Castle*, and listened to him practising day after day, and playing Chopin in the beautiful domed, wooden-panelled saloon of that graceful vessel. The purser would always comment how this great little man would play as if that moment at the piano were for him, not butterfly-brief but infinite, untroubled by the improbable intuition that one day he would have to renounce the piano in order to become Prime Minister of a then newly liberated Poland. And what inevitably added to the impact of this recollection was that it reminded me again what an intimate family affair European history in southern Africa had been; how always one had only to stretch out one's hand to touch either a direct ancestor, a family relation, or someone whom a member of one's family had known intimately and so discover them at the heart of the cataclysmic events of the three hundred years and more of our past.

For although I myself was not to hear Paderewski play either at sea or on land, a relation by marriage heard him play. Paderewski's tour of southern Africa had been a unique success, except in Port Natal. There, only a humiliating handful of people had troubled to turn out to hear him. According to my relation, who was as musical as she was beautiful, he played as if he were performing to all the angels in Heaven. For me it showed the quality of the spirit of Poland and the courage that sustains them and makes them one of the greatest and most gallant of people. The experience in Port Natal left in Paderewski a deep scar. Immediately after the war, when the heads of the allied nations were assembled at Nice to consider the future of a Poland which Paderewski had been chosen to represent, this relation of mine, sitting on a bench in the Promenade des Anglais, saw Paderewski walking slowly by her. Suddenly he stopped, turned fast about, looked intently at her, shook his head as if in unbelief, and began to walk on again for a pace or two, when he had second thoughts. He turned round and came back. Looking directly at her, he walked up and, raising his hat, bowed and said:

"Madame, forgive this intrusion, but I have to thank you for coming to hear me play in Port Natal in 1912."

She had, of course, not met or spoken to him before.

This, and so much else, struck all the deeper because it was unfailingly connected with our remote beginnings. Our Captain told me after one final concert that already in the year sixteen hundred and two an English captain, in one of the first ships bound for the East by way of the Cape, brought a virginal in order to provide music for his crew and officers. More impressive was his account of the Captain Keelinge who commanded the third English East India Fleet which called in at the Cape on December 17, 1607. He had aready made his ship's company perform Shakespeare's *Henry IV* while off Sierra Leone and, in the approaches to Table Bay itself, he had staged *Hamlet*, which an entry in his journal calls 'the newe and popular play in London', and establishes it as the first performance of Shakespeare's tragedy outside England, I felt that nothing could have been a more appropriate theme for so ambivalent a Cape; and nothing cast so much doubt on our notions of progress when I compared Captain Keelinge's seamen, who could be addicted to Shakespeare, with those in the *Walmer Castle*, estimable as they were, whose sense of the theatre did not stretch beyond Nellie Wallace and all that.

And so we came at last to the Cape. I was called an hour before dawn by Jack because the Captain had invited me to the bridge to see the dawn raise the Cape from the sea. The vision, with almost more of great music than sight in it for me, stays as one of the most moving experiences of personal recognition for me. It made me eager to go ashore despite the ritualized lament for the end of the voyage, which we had celebrated the night before. Moreover, the act of landing did not disappoint my eagerness because our reception left no doubt that the first harbour city fully shared my own feelings of historical occasion.

So important were the arrival and departures of these ships that my countrymen had evolved a national ritual which to me was more impressive, colourful and moving than many occasions of State. The East India Company had, for centuries, mounted a picket of soldiers on the peak of a high mountain some fifty miles north of the harbour, with orders to fire their cannon at the first glimpse of a sail pecking for crumbs on a blue horizon like a white homing-pigeon bound for its loft. At the Cape it had long ceased to be necessary to summon crowds of people by gunfire to witness these events.

The Castle ships were so regular and loyal that everyone knew the precise timing well in advance. This instinct of history and inborn reverence for our filial link with the sea and its ships had, in fact, become so strong that the arrival and departure of the mail ships were enough to draw far more onlookers to the relevant dock than those who were there to say hail or farewell to friends and relations. In the course of time, people had made it a part of their imperative national mores, which demanded that everyone who was someone should wear their best suits and dresses for the occasion. There was not a lady of fashion who would not have as a matter of course a new 'mail-boat dress' and special hat of incredible fantasy, in her wardrobe. I can remember homecomings when from the ship it looked as if we were sailing into the midst of the Cape's version of the Chelsea Flower Show.

Having already participated at Port Natal in far more of these occasions than it is possible to recall, I could vouch for the fact that these well-dressed gatherings continued against a backdrop of the long walls of soaring warehouses and gigantic cranes that loomed high above their grimy roofs like question marks of destiny. The effect, striking as it was, would often at a superficial glance seem highly incongruous. If one went by the dress and colour alone, one could easily have thought that the Governor-General and his Lady had themselves gone crazy and were holding their annual garden party for the King's birthday on the waterfront. Yet in some strange way there was something historically so appropriate about this approach to the ships that, however bewildered at first sight, one soon came to feel it was as fitting as it was uplifting. To explain the impact of my arrival by the *Walmer Castle* I should add, therefore, that countless repetitions and sustained familiarity never dulled arrival or dimmed the memory.

At the heart of my response, I believe, is some realization that only the sea and ships in their own unrivalled day, as a means of long-distance communication, could ever inspire such behaviour. Both sea and ships are in themselves natural symbols of royal and ancient standing in the mind of man. Inevitably, on their own account they provoke the most intangible nostaligia. One has only to compare the coming and going of ships with the arrival and departure of the latest aircraft to see how deprived our new ways of travel are in that regard. People who travel by air generally look as if their planes had ruthlessly bundled them out with ruffled feathers like birds for the first time pushed too soon out of rough nests. As for the crowds that wait to greet them, judging by their appearances, they might just have come from a football match at which their favourite team had been defeated.

But the sea and ships between them saw to it that the human imagination ordered these things better. As on all occasions of natural ritual and symbolism, they predetermined that there was music both on arrival and departure. I remember how, when the sound of the ship's temple gongs and the final long Olympian blast from its siren had sent all the last hardened visitors scurrying ashore, the passengers would man the decks and look down on the garden party spectacle to watch the gangways removed. Then the beautifully woven hawsers of yellow manilla hemp were cast off and they would listen to the sound of the winches rapidly winding them aboard and finally dying away. A strange sense of the irrevocable would possess their senses, and the ship's band or orchestra, as it was still called in the *Walmer Castle*, would hasten to take charge as if both to contain and express the unaccustomed emotion for them all. They would begin by playing popular music of the day. Some of the crowd would join in, singing or whistling at the receding ships like shepherds controlling a pet collie which had strayed too far, or calling out a message from their hearts in voices that could be heard nowhere else save in the midst of ascending music-hall sounds. But as both quayside interjections and individual outlines became less distinct, and the moment came when the special mark of tugs that only the Cape produced had turned the ship about to face the gateway to the sea, then the most telling of all sounds rose high above the music and voices. I always looked on those tugs as the Cockneys of the seas; so indestructible, irrepressible, nimble, and as quick in remarks and sound as in movement. Always when preparing to cast off the great liner and abandon it to the sea, they would release three quick, bright, impudent whistles of farewell. As the mailship boomed back its own bass blast of acknowledgement, the ship's band would play four tunes in this order: 'Life on the Ocean Wave', 'Home Sweet Home', 'Abide with Me', and 'God Save the King'. Instantly, even at that inexorable distance, one would see the handkerchiefs that had fluttered over that vast crowd of many colours like white admirals over a garden in full flower, vanish – one knew from experience – to do their duty around the eyes and faces of the spectators. But on this *Walmer Castle* morning the flutter of handkerchiefs, and waving of hats and hands, was in the warmest of welcomes, and of a finality of arrival, which uplifted and increased the quality of one's own subjective portion of the emotion.

I stayed, thereafter, in the Cape for some eighteen months, working harder than I had ever worked before, as special writer, leader-writer and a sort of junior assistant editor, committing myself utterly to what Desmond Young's cable had described as 'cleaning up the mess in the country'. He had summoned a friend we had known well in Natal, Leonard Barnes, who was to write *Caliban in Africa* as a result of this experience and, as his chief sub-editor, the Crichton-Mandel who had worked with Orage and Belloc on the *New Age*. I also tried to involve Plomer and Campbell in this venture as I saw in it a continuation of what we had done earlier as an isolated, suspect threesome, to overcome colour and racial prejudice in the land. They also both needed the money. William had arrived in England from Japan some months before I left, so I persuaded Desmond to commission him to write a weekly London column for the paper, and Roy Campbell to contribute a poem a week from Martigues in the Bouches du Rhône where his wild spirit was full of wounds inflicted on it by others and indeed by himself. He lived like someone in a fortress of wilderness, besieged by civilization and culture and almost without any means of self-support.

Desmond imported an extraordinary quality to the *Cape Times*, making it so great a newspaper that some years ago when we turned to it for supporting evidence in the trial of Laurie Gandar, the gallant editor of the *Rand Daily Mail*, Sidney Kentridge exclaimed: "I never knew newspapers here or elsewhere could rise to such heights."

His exclamation was sparked off by a glimpse of a succession of leader-page specials written in honour of the two-thousandth centenary of Virgil's birth. Sidney is one of the finest and bravest legal minds of the day; he is also a person with a wide and acute appreciation of music and literature. The purpose of our search were the pieces I had written about the abuse by police and state authority, of black and coloured people in their custody, so that we could illustrate the point that what was regarded as culpable in an editor's conduct in the sixties was normal editorial practice for us in 1929.

In particular we were looking for a leader called 'White Justice', first written by me, then revised against libel by Desmond, and finally re-revised by the Editor, who was so outraged that, against our advice, he deliberately inserted criticisms of his own. It all ended in one of the most expensive libel actions ever instituted in South Africa and fought through the lowest to the highest courts, where it was lost. By then the business management of the newspaper had had enough of us, as we had of them. In fact had I not had to wait to give evidence in the trial for libel

I would have left months before, knowing I could do no more in that dimension of life. As it happened, Desmond and I left almost simultaneously: he for India to take over the *Pioneer* on which the young Kipling had served. I was not to meet him again until many years later in the dark of night in the heart of Abyssinia where, behind the Italian lines, my patrol was to rescue him from some Shifta because, characteristically, he was spying out the land far in advance of his Indian Division, fighting its way towards us.

I myself was bound back to England, believing that at last the self-employed person I felt myself born to be, but one who had hitherto not yet possessed the skills to employ even himself let alone be employed for long by others, was now capable of doing so; all the while among a host of duties that had exhausted me so much, and in a sense involved me in such painful and powerful conflicts with my own history and people, that for many years I could not talk about it to anyone. I had never lost a sort of 'Golden Ariadne's thread' in that labyrinth of soul and labour which those long months had constituted for me, and which re-connected me with my *Canada Maru* self and all that preceded it.

As an example there had been a night when the chief sub-editor's stand-in, Bill Parsons, came to me at midnight. It happened to be my turn to see the paper to bed. He handed me a cable just received, with the words: "I thought, young Laurie, you would want to handle this yourself."

The cable read: "D. H. Lawrence – controversial writer artist died Bandol southern France this evening."

The same sort of instinct which made me intervene between the infuriated lady and Messrs Shirakawa and Hisatomi in a coffee ship in Pretoria, sent me hurrying down to the works section of the newspaper. The editorial page was already on the stone about to be closed and cast for the press. I asked the Father of the Chapel to hold it and hurried back to my own little office. The news had roused such a sense of loss in me that I felt compelled to express it. D. H. Lawrence meant so much to my generation, and he had been a great breakthrough for us. The ways in which he widened the range of contemporary imagination, shattered paralytic taboos and brought new considerations and outlawed themes back into the artist's legitimate field, are today all taken for granted and are perhaps only too obvious. The tendency is to see only the excess, the extremes, the exaggeration and the loudness which often accompany the spirit that feels itself to have gone so far ahead and alone in his time as to be almost out of touch and hearing. But for us, D. H. Lawrence changed the climate of the literature of his day, for good; and for me was an intimate part of a self, discovered on the voyage back from Japan to Africa. The fact that I had met him and not really been drawn to him as a person in no sense diminished my feeling of indebtedness.

Accordingly, I wrote a leading article about him. I went back to the stone and substituted it for a long leader by the Editor-in-Chief on an important political issue. I heard the building shake with the sound of the presses going into action (which always made me think of the paper as a ship setting out in the dark to sea) and I was certain I was sailing into dismissal the next day. When the Editor-in-Chief sent for me the following afternoon, therefore, I was fully prepared to be axed. I found him (his name was B. K. Long) at his desk smoking his pipe, as always, and reading Jane Austen's *Emma*.

"Ah, there you are," he said in a tone that brought a sharp sense of relief: "I just wanted you to know how nice it was this morning to read a really civilized leader on a civilized subject and to thank you."

Of course, this illustrates how he too was another vital factor in the raising of his paper to such unaccustomed heights and, incidentally, made it the only newspaper in the world to pay a special trubute to Lawrence. *The Times* in London did not even publish an official obituary! However, on a personal level, the importance of the incident was its re-direction of all my attention to the course on which I remained inwardly bound, and increasingly demanded a return to my own natural preoccupations.

All this became clear on the journey by sea back to England, for, like all my voyages, it was a necessity for crossing another frontier of myself and of a nature of circumstances as if bespoke in all its detail for my own evolution. Oddly, and significantly to me, this particular voyage too could not have come about without another intervention of chance. A burst appendix of one of the engine-room crew, and last, minute inside information from a friend, were its instruments. This friend in the Union Castle office in Cape Town knew I was trying to save the £19 necessary for a steerage fare to Southampton, and sent me panting to the docks. The chief engineer began the interview by ordering me to show him the palms of my hands. I feel he would have rejected me had there been time to look elsewhere, and had he not been, as I later discovered, a hockey addict. However, he remembered how the local papers a few days before had commented rather favourably on my performance in an inter-provincial hockey final. He signed me on there and then for the twenty-four day voyage, and set me to work as a trimmer in the hold of his ship.

Born in 1911, the *Gloucester Castle* at that moment was in her prime; tested in war and peace, on long voyages round Africa, and in many a gale in the Bay of Biscay. She had been built so well that a German torpedo in World War One, though killing three members of her crew, had been unable to weaken the bulkhead, and the ship's engineers kept her afloat until she could be safely towed to port. It was something of which the second engineer in charge of my watch was undeniably proud and commented on almost daily. My own watch was from eight to twelve noon and eight in the evening to midnight. My job consisted of bringing coal by wheelbarrow far down in the hold to the stokers who had to feed the fires in the ship's boilers. It was a matter of permanent engine-room debate which was the hardest work, that of trimmer or stoker. I myself thought that there was little to choose between them except that on this voyage I never saw a trimmer with damaged hands; but cannot recollect a stoker who had not lost a joint or two, sometimes even a whole finger of a hand, from the sweat poisoning so often induced by the heat of their labours in the tropics through which the ship steamed month in, year out, for days on end through every round voyage. The air-cooled ventilation, common in the cooler oil-burning ships of today, was still unknown. The heat below was far worse than that of the many deserts I have known; and the work demanded much harder than that asked of the modern sailor. Fortunately I had above-average physical strength and stamina, and just managed to accomplish what was asked of me. Even so, I would not have managed to do half as well had it not been for the engineer in charge of my watch.

He was a Scot from Glasgow and went by the un-Scottish name of Mr Smith. He was a little man with bandy legs, a physique and face which betrayed the malnutrition experienced from birth in a slum from which he was justly proud to have emancipated himself. He was a paradoxical man: intensely romantic but without illusion, and almost a martinet. One knew there was no pretence or device for slacking that he would not instantly spot. Yet he was eminently fair at heart, and so much of a poet that he was liked and respected by all. When he was not inspecting one of the bewildering number of brightly polished gauges in the engine-room, he would take up position between two boilers by the telegraph which connected him to the bridge, one hand ready on the wheel to carry out immediately whatever unpredictable order his captain might give him from above. Unlike the stokers who had to feed their hungry boilers a shovelful at a time, I found that if I hurried and deposited a heaped-up barrow of coal at my stoker's feet, I would just have time to exchange a few words with Mr Smith, for whom I had conceived an immense respect and an even greater liking than he had for me. He would let fall the kind of remark which only the Scots – natural philosophers that they are – can produce, until with the long voyage drawing towards its end, I understood why only a ship's engineer and a Scot could have evoked Kipling's 'MacAndrew's Hymn'.

Often he would treat me to a preliminary wash – he even photographed me at it – a bottle or two of cold beer and bacon and eggs in his quarters, and would talk to me about his philosophy of a sea he loved. One of his favourite assertions was that a life at sea was the greatest begetter of self-knowledge, and that there was no secret deep enough, no error or falsehood in the character of those who sailed, particularly those 'overfed passengers in second or third', which, sooner or later, it would not expose. It was a place where either hard and continuous work or total honesty with oneself was in the end one's only protection. Therefore those who had dedicated their lives to purely material needs, whether they knew it or not, if unemployed at sea were putting themselves at great risk. If this were not so, he would ask me in that Scottish intonation which seems to me always so unfairly endowed with accents of sincerity and truth that other voices lack, why did the great Barney Barnato, having made his millions out of Kimberley diamonds, suddenly, at the peak of his success, commit suicide by jumping overboard on the high seas from that most beautiful ship, the *Scot*? Why did I think the over-rich, super-merchant Gundelfinger of Port Natal, contemplating the wake of another ship, suddenly one night jumped likewise overboard? This last example shook me even more than the Barnato incident because I had known Karl Gundelfinger so well that he came to inspire a character in one of my books.

Nonetheless, Mr Smith would add, those who perservered to the true end, whatever their call to the sea, would find it had the power, unequalled by any other natural phenomenon, to transcend all and make mere man more than himself. Not surprisingly, therefore, his ship was for him far more than an assembly of inanimate matter, driven by man-contrived power, and his engines were the heart of potentially transfigurative matter, and made him kin to my *Canada Maru* teacher and his interpretation of the 'Maru' of ships. I could, and perhaps one day will, fill a book with what he evoked for me. But I must confess to one clear and startling coincidence, if not truth, as illustration of what he said about the power of the sea to expose hidden aspects of human nature.

All the stokers and trimmers on my watch had quarters aft in the stern, right over the propellers. And on our way to the engine-room and back we had to cross the space reserved for third and steerage passengers, four times every twenty-four hours. In the course of this coming and going, we had come to know by sight many of the passengers there, especially an attractive Portuguese girl. We soon learned that she had mde her fortune as a prostitute in Lourenço Marques and was on her way to retire in luxury in her native Portugal. Every morning between half past seven and eight, when the passengers paraded their cramped and crowded deck for a breath of air before breakfast, this girl, with her provocative clothes, incongruously aristocratic appearance, and marked physical attractions, stood out among the correct dames and dutiful daughters who tried very hard to behave as if she were invisible. But it was quite clear that the masses of male students, naval-ratings, young artisans and contract men returning to England, were hypersensitive to her presence, and that the mothers regarded their interest in her as a totally unmerited and incomprehensibe slight on their daughters' attractions.

When the *Gloucester Castle* sailed into the heat of the tropics, and the soaring temperature drove many a passenger to doss down on the decks at night, some of the young men borrowed a ship's hammock and slung it between the lowered arms of the after-derricks so that this young lady could sleep above them in comparative coolness. All might have been overloooked and contained if some conditioned reflex of her trade, and awareness of her natural powers, had not made her delay her getting out of her hammock until the deck was as full as it could possibly be of male witnesses to the event. She would then languidly lower herself to the deck like some Toulouse-Lautrec 'Venus' in her translucent silk pyjamas, and bend leisurely and dangerously forward to put on satin slippers before donning the thinnest and most beflowered of crêpe-de-chine negligees. It was this in the end that the ladies of third and fourth estates could not stomach.

One morning, in the name of outraged morality, instinctively and I believe without premeditation, they leapt into action. I was just coming up the ladder to the relevant deck when a terrifying scream of many female voices joined as one shattered the tranquil silence. The bright emerald tropical morning suddenly went black and cold around me, and it sounded as if all the harpies of legendary antiquity were at war there on the deck. My feet found the floor of the deck just in time to see masses of ladies with demented, distorted faces, stabbing at the hammock with the long gleaming hat-pins designed to hold their Sunday head-gear in place. The whole of my watch had to rush to the aid of the many male admirers of the Portuguese lady to beat off this attack. But the unfortunate lady nearly died all the same, and spent the rest of the voyage in the ship's hospital.

Apart from what the incident taught me of the underworld of the respectable mind, the strangest thing of all was that, once this unpleasant incident was over, the ship as a whole appeared happier than before and a better time was had by all. This did not surprise Mr Smith when I told him about it because he commented, Carlyle-like, that he had never completed a voyage yet which had not, one way or another, although never as drastically as on this occasion, demanded some sacrifice to the dark gods of life.

But of all the *Gloucester Castle* experience, perhaps the most precious to me, I think because I truly earned it alongside them, was the glimpse it gave me of the British working-classes whom the men I worked with represented. I cannot remember a single member of the crew who had any fat on him; just as now I rarely see a manual worker, miner, factory hand or even merchant sailor, who is thin; they have been putting on fat while the middle classes have been losing it.

Then their faces tended to be fine-drawn and ennobled by the great though inscrutable and uncompromising artist that sheer necessity is in the lives of those who are ruled by it. Most striking of all was the fact that, although all the social odds had been against them, they had a dignity, self-respect, assurance, lack of bitterness and almost innate sense of superiority over those raised above them that is mostly absent from the social scene today. It was all summed up by the last vision I had of them when they each received a meagre nine pounds, as I did, for the hardest of labour; and then I watched them go through the dock gates with duffle-bags and little papier-mâché suitcases. Each of them was dressed in a spotless suit of navy blue serge, exactly as I remembered some twenty-six thousand of them were, without exception, when they lounged about the waterfront of Port Natal during the great seamen's strike some years before. There was not one who did not pause when the customs guard at the gate had finished with him, to turn around and look long at the ship they had just brought home. It was an absorbed and reverent look which moved me all the more because there was nothing sour, disgruntled, uncertain and complaining about it, which alas are the hallmarks on so many faces today that have not half the cause they had then to become bitter.

The years that followed were perhaps the most unhappy of my life, in spite of the fact that I was writing at last on my own, singularly helped by recognition from critics, public and my peers. That was just about enough to help me contain a feeling as of a potential doom deep within the nature of my time, and the management of affairs and events charged with potentials of irresistible change like a thundercloud with forked lightning not only in Europe but the world. William, for instance, had brought back a sombre account of developments in Japan. He reminded me of an image I had thrown at him one day when we were caught in a crowd proliferating with terrifying speed and suffocating pressure before the Imperial Palace in Tokyo, that the Mikado was not unlike the queen of a hive and the Japanese nation a swarm of bees around him.

"And they are getting ready to swarm, Lorenzo," he said. "And they will swarm in anger and sting anything that moves in their way."

I was barely back in England when news came of the Japanese march into Manchuria. Despite my love of the country, all the alarm bells sounded within me. I had a vision of the taking over of a nation by millions of the fanatic kind of swordsman of our little Peach Tree Hill drama. But far worse was the dismay induced by the failure of France and Britain to respond to an opportunity for joining America in some form of intervention which might have prevented a foreseeable catastrophe. In particular, I resented the British Foreign Secretary – John Simon – who was then the highest in my scale of the feeble, the blind and uncaring, ever to occupy the office, so that I came to resent him as if he were doing me a personal mortal injury. Hard on this came Japan's defiance, which led within a few years to its walk-out at Geneva and the cocking of a snook at the League of Nations. This was as literal as it was metaphorical, and still no response of consequence came from a body in which my generation had invested great hope.

From that moment the ordered change and evolutionary dykes of history seemed to be breached, one after another, and a flood-tide of a sea of dark and angry forces of negation released in Europe, Asia and Africa. Hitler invaded the Rhineland; Mussolini marched into the Abyssinia after which Baron Shidehara and his band had secretly lusted; the Emperor of Abyssinia was driven out of his country and turned to plead in person, but in vain, to the League of Nations – a plea which was followed by the Hoare-Laval betrayal that wounded my own historical self and sense of the future even more deeply than John Simon had done. I was convinced from then on that a Second World War was inevitable. The whole Abyssinian scandal, far more than the Spanish Civil War which dominated the emotions of my contemporaries, was for me the crucial international issue and betrayal. The first death-watch beetle of the day had started a process of irreversible and accelerating rot behind the wainscots of the blind spirit of what passed for statesmanship.

I felt all this so acutely that the imagination and feeling which are the writer's only instruments were taken away from me. I could not write and hopefully improve on *In a Province* with one of the many stories knocking at my door. I have never been much of a group person, least of all among groups with artistic and aesthetic inclinations, and I thought I could recover my imagination for my inner commitments by withdrawing to the country, which was a natural love. So I bought a farm in the West Country and left London in the hope that I could farm with one hand and write with the other. Far from helping, it made the problem worse. I quickly discovered how naïve it was to assume that one could own property without being owned in return, according to the measure in which one was an owner. I had merely added to the complexity and weight of distraction and deprived myself of what little 'concentration' I possessed for the main theme of my life. All this led to a turmoil which convinced me I had somehow to pull out and find a centre within the storm blowing up, not only in myself but in the entire world.

As before, I turned to the sea for help in the essential reappraisal, and did three voyages to Africa and back as fast as the ship could turn about at its terminals. In between I visited my mother on her own favourite and special farm in the Interior. She was one of the greatest women I have known, and as a result also of experiencing her brand of 'being' again after many years, I felt compelled for the moment to change direction. As the chance which seemed to be pressing closer in order to take an even tighter hand in writing the script of my life would have it, I had taken a

German ship – the *Watussi*, of the Hamburg-Afrika line – to accomplish these voyages. The choice was all the more incomprehensible because already I abhorred and feared the rise of Hitler even more than the activities of Japan and Italy which had dismayed and unsettled me so. Nothing, I concluded, would stop the march of those countries to catastrophe except a greater catastrophe but, nonetheless, still a disaster which a West at bay could ultimately contain. I even had a nightmare about Mori which was a sort of Peach Tree Hill affair in what there was of prophecy in my own dreaming unconscious. I saw him in dream-light doing the sword dance which he had apparently done for fun by moonlight in the *Canada Maru*. His face was distorted with undisguised rage as he danced and screamed: "You Europeans have enjoyed your colour prejudices for centuries; we will show you now what it feels like to experience our white prejudices."

Most significant of all was his ambivalence in the dream. Although his face was recognizable as that of Mori, and so authentically Japanese, the shape of his head, particularly the back, was like that of a caricature of a Prussian militarist. I had, of course, told William, and he wrote a poem about it called 'Captain Maru' which should be read in its original and not the expurgated versions in which some of his writing about Japan appears today. All this and more went to strengthen a private conviction that events in Japan, Italy and Germany were erupting from the same volcanic subterranean area of the contemporary spirit, and connected with one another by an as yet undiscovered subliminal system of their own. So, on a literal level, my distaste of the German scene should have kept me from choosing a German ship. But if, below the manifest surface of meaning, the truth left in me was seeking objective confirmation of its own intuition that an all-powerful and uncontrollable Teutonic Germany would constitute the forces of evil which could extinguish civilization, my choice of ship and voyage could not have been bettered.

The ship was superbly run in the same Fascist spirit that built new autobahns; ensured that men did not question authority; convinced them that all their troubles were caused by villains in the external world – particularly the world beyond their frontiers; and rewarded them with trains which ran on time. It was clean and comfortable, and the food was of the best. Moreover, instead of an orchestra, it had a quartet of fine musicians who, once they had played the ship out to sea and fulfilled their duty of providing dance music for the passengers with perfunctory renderings of Strauss waltzes and extracts from *The Merry Widow* (which I knew already was Hitler's favourite music), gave chamber concerts of a high order. They usually did so at nights in mid-ocean under a great universal stillness and a sky full of stars on tip-toes, as if straining with expectancy for the first whisper of a new call from creation, before they invariably sent the ship to bed with Brahms' lullaby, 'Guten Abend, gute Nacht'. Outwardly all might have been order and harmony as of a world going about the business of classical Europe as usual. But the more I saw of the ship, the uneasier I became, and the wider appeared a dichotomy in this microcosm of the upsurging Germany between all that its love of music symbolized on the one hand, and what one saw of the passengers, crew and Captain on the other.

I shared a cabin with a young German who had a Jewish grandmother. Already he and his family had thought it wise to leave home and prosperity in Germany. He spoke of how Germany looked from within, which not only confirmed but increased my dismay. Through him I met many others in the ship, each an early warning system of horror to come. The case for immediate and decisive preventive action, their example suggested, could not have been more overwhelming; and yet the pretence that all was going to be well hardened without, as the unconscious doubt within increased in the mood of the great establishments of the world.

Apart from myself, the few non-Germans travelling in the ship did so only because it was the only way they could use money invested in pre-Hitler Germany. One of them, a naturalized American Englishman, Jim Williamson and his wife Elsie, became intimate companions. He was exceptionally knowledgeable about the new evil in Germany and relentless in his criticism of and distaste for the upsurge of Nazism. I was warned by my new friends among the refugees that he had been reported to the real controllers of this ship and should beware. He did not heed any warnings and, as a result, on a journey to Australia immediately after in a similar German ship, where his voice was raised even higher against the proliferating Nazis, his wife (who went to bed early) one night woke up somewhere in mid-ocean and found that he had not yet joined her. Alarmed, she went to search for him and roused the ship in vain. He had vanished on the calmest of nights, and she and I had no doubt that the 'real controllers' of the ship had dumped him in the sea.

I had no doubt because I learned on this voyage that the 'real controllers' were now Nazi cells in all ships. In the *Watussi* they were composed not of officers, but ratings whose leader was the Captain's personal steward. I was told that in matters of ship policy this steward ordered his Captain about, which explained perhaps why the ageing man looked so sad and seemed increasingly so over the spread of all three voyages; he concentrated more and more on the pedigree pigs that he bred on rich first-class swill on the highest boat-deck from which passengers were excluded, and on chamber music after dinner. I often wondered how he and his officers could endure the humiliation and did not resign. I had yet to learn that on the whole the spirit of honourable resignation and exception either had vanished already or was vanishing fast in Germany. This ship's phenomenon, that those nominally in control carried on as if control had not already been taken away from them and with the uncontrollable happening all around them, was a disturbing example of events in the world at large.

Even the lack of vision and authority of the Church was exemplified in what seemed more and more a voyage to the end of night. It was illustrated best by a German cardinal on board, with purple beneath his black suiting and large black hat. He was a tall, thin man, of a severe, sour, thin-lipped look, pale blue eyes, bad breath, and apparently friendless in that outrageous ship. He looked so isolated that I tried to talk to him from time to time, and once dared to ask him about the 'new' Germany. He became as aloof as he was severe, and said reprovingly, in a clipped voice: "The Church is above such political things. Its concerns are with matters of life and death, and you should think more of death, young man! Do you realize how uncertain, even here, life is? Only last night in the mist we were very nearly sunk by a large freighter, so the Captain's steward has just told me."

There was also a curious little incident in the Canary Islands which has stayed with me and may or may not one day fall into place in some future research into history. It took place at Las Palmas, where I had already been twice before, and so was thought qualified to act as guide to my friends in the ship. On one of our tours round the island we came back by the palace of the Spanish Governor-General. The Governor at that moment was an unknown general called Franco. It was already twilight, and hurrying away from the sentries at the gates, so it seemed to me, was a strangely familiar figure in a neat civilian suit. He was walking fast towards the harbour, face set and looking straight ahead, and as we came abreast in our car, I recognized the Captain's steward, the reputed Controller of our ship. Some days afterwards in the newsletter brought to me with my morning tea, I read that General Franco had been flown secretly by a small plane from Las Palmas and had landed in Spanish Morocco to begin the Civil War which so convulsed my own generation. Then on the two occasions when we called at Walfish Bay (the main port to the old German South-West Africa, for whose return to a new Imperial Teuton fold Hitler himself was hinting and some subservient followers already clamouring) all sorts of young Germans one had not noticed before appeared in ordered formations on deck. They were of all sizes and shapes but had the same masks of predetermination on their faces and the same uncomfortable light as of a secret source of excitement and exaltation within themselves. They went ashore and regrouped on the quay for their transport, more as soldiers than civilians.

Of course, there was much else of great personal significance to these voyages, not least the human relationships which were to transform my life and which continue to this day; but they are not relevant to the meaning I am after here. What mattered was that without these voyages I would not have seen so clearly into the nature of my own confusions, a culpable lack of self-knowledge which injured myself and others and added, however minutely, I was certain, to the general alienation and desperation of the time in which I live. Somehow I had to start with myself, disband and re-group in a new way. These moments at sea, and in my native Interior, determined that I would have to begin all over again and start with what was on the doorstep of my own mind, with what was nearest, and where I saw clearest, and not where some ideal collective vision called like a siren on the horizons of imagination. I was acutely aware of the massing of the forces of doom, and though I had an inkling of their true nature, this apparently was not shared by others. I would have, therefore, to begin by warning in the best way I could. Thereby, I would at the very least recover some continuity of purpose.

Ever since the age of sixteen I had done precisely that in my own country; I had continued it as a newspaperman and on through *Voorslag* with Campbell and Plomer; and had orchestrated all in my book, *In a Province*. I would try, therefore, to do something of the same for Europe and in particular the Britain I had come to love and whose story and great historical values corresponded most to my own seeking. I remember the calm which came over me, like moonrise above the rim of the Kalahari Desert in a dark night, on that last night of the last voyage. I prayed ardently that I could have had just one more day and night at sea to consolidate this calm before the land, with all the new uncertainties and the change which such a redirection of way implied, took over again. As chance would have it, a thick fog enveloped the ship, the sea and land that night, and we were forced to anchor in the Solent for another thirty-eight hours before we could berth at Southampton. Without the sea and that dispensation of Providence, I might not have had the courage to try and do what I had to do.

As soon as possible I began a book of warning. I had, ever since I could remember, had a passion for myth and legend; above all the myth and legend of Greece. I owed my father a debt I could never repay and am grateful that during the brief and enigmatic years I knew him, he was aware of it. The first book he gave me, as it happened on my fifth Christmas, was a simplified account of the myths and legends of Greece. He followed it up a year later with Cervantes' *Don Quixote*. From that day on to this, the mythology of Greece and Rome has been a kind of chart by which I have navigated my imagination; and the elongated knight of La Mancha, and his peasant corrective on his donkey, ride still as a shadow in a sunset hour by my side in the way that Daumier painted them. My father's third and last gift to me, the Christmas before he died, far too young and unachieved for death, was Mallory's *Morte d'Arthur*, which remains for me a contemporary testament of the universal quest for truth.

Through a quickening of awareness produced by the pursuit of this interest, I had become increasingly impatient with history as I was taught it. It seemed to me more and more a superficial catalogue of external events and a mobilization of statistics of time and change which ignored a deeper something seeking to accomplish itself and increase the meaning of life through men and their societies. Written history, in fact, was not the whole story that it should be, but a slanted and partial rendering of a transcendental totality – a process which I labelled as not history but 'historicism', to myself. The history of my own country and how it came about, and the role of exploration and discovery by the sea in our making, was no mere marshalling of happenings in the exterior. Rather it was an expression of a deep and dark universal intent seeking to surface in the light of the world and the spirit and lives of living men. This intent in the first instance stirred the spirit of men and moved them, whether they liked it or not, or knew it or not, as 'mythology', and made of them heraldic and legendary material. Knowing the history of my own country almost as a family story, and that of my own black and coloured countrymen through their legends and stories, I was made acutely aware of how the rationlizations produced as 'historical explanations' of their causes and effects were merely the excuses and catalysts of mythological stirring, and had raised the ordinary men who were held to have made our history, to extraordinary and legendary performances.

With such a start, it was not surprising, to me at least, that what was happening in Japan, Italy and above all Germany made sense ultimately only when seen as a misunderstood and misinterpreted mythological process. That is where all my own fear for the future began and remained concentrated. Of all the mythology I had come to know by then, German mythology seemed to me the darkest, the most undifferentiated, archaic, turgid and dangerous in a fashion that by comparison made daylight of the humblest mythology of Africa, above all that of stone-age Africa on which my old Bushman nurse had instructed me. German mythology was the only one I knew where the forces of darkness defeated the gods themselves. All that was called the history of Germany, from the rise of the Electorate of Brandenburg into Prussia and the steady takeover by Prussia of the classical federation of the southern and least Teutonic states which followed, their moulding into a tight-fisted Brandenburg whole, made sense to me only as the prelude to an eruption of volcanic forces of a negative mythological kind in the German soul. The Arabs had a saying: "Tell me what a man dreams, and I shall tell you what he is." It seemed to me as applicable to nations and complete only if rounded with the addition: "Tell me what gods rule a nation's soul and I shall tell you not only what they are but what they will become." Fate, all mythologies had taught me, or so I believed, always intervened with disaster when confronted with a personal and collective hubris on the scale with which the rise of Prussianism had confronted classical Europe. I was strengthened in my belief that this was not a subjective rendering, but something objectively confirmed, by my reading of Burckhardt's essays in the *Canada Maru*, as well as his great stories of the Renaissance and Greece and Rome in the vast library with which my father had equipped us in Africa. Two illustrations are still vivid among my *Canada Maru* memories. When the news of the crowning of the first Kaiser Wilhelm, at Versailles, reached Burckhardt in Basle, he exclaimed, instantly recognizing the fatal, mythological hubris implicit in the deed: "This is the doom of Germany." Then there was his observation that Faust was essentially a German myth drawing attention to an unresolved dichotomy in the German spirit, and Goethe's attempt at a symbolic presentation of its challenge and resolution. That had truly tolled like a bell in my imagination because I had already thought of Hamlet as a personification of an essentially English mythological pattern. His philosophical malingering, and distaste of the imperative action demanded of him, expressed the trend I saw orchestrated in British statesmanship of the day which, as in *Hamlet*, would lead to greater disaster than an immediate and active national response to the obligation which preceded it would have produced.

On lines such as these, I began writing my book of warning. Progress on the book was slow, however, because in order to keep my hold on the farm, and myself alive, I had to return to journalism and become a Fleet Street correspondent. I was charged, in the main by choice, with diplomatic affairs, so that in a day-to-day fashion I could continue my own small warning system and preparation of Britain and its Commonwealth for what was to come unless they and France acted resolutely and swiftly against Hitler.

I was morally greatly helped and sustained in this by two senior French correspondents who worked in London and became great friends. In their own lucid and rational French way they saw the future as I did. The senior was Pierre Maillaud, destined to become Vichy's greatest adversary on radio and de Gaulle's first Minister of Culture and Information. A nephew and friend of a painter of distinction of that name, he was urged to meet me by Roy Campbell with whom they had stayed at Martigues in the South of France, which he loved and where he was to meet, after all he had endured, his death by drowning, much too soon. The other was Siriex, who was to become a first-rate Governor-General of French Somaliland, and whom I was to meet later with a group of Free French Spahis behind Italian lines in Abyssinia.

What we three learned of British and French statesmanship in the course of our duties only served to raise dismay into near dispair, and fear into horror. One episode is enough to speak for our loss of faith and hope in the British and French establishments of the day. Late one Friday afternoon Maillaud and I were summoned to an urgent press conference at the Foreign Office. Daladier, the French Prime Minister, at his request had just had an emergency meeting with Lord Halifax that day to discuss the German threat, and the news that the Japanese, whom everyone thought committed beyond their strength to full-scale invasion of Manchuria and Northern China, had suddenly occupied an archipelago called the Sprattley Islands, of which none of us knew and which we could not find on any of the office atlases. The conference was taken by Halifax himself and, as he sat at his desk, the light of the evening sun falling through a rather dusty window full on his head and face, made a rather noble, sensitive and fine portrait of the man. I marvelled how, as the brush of sunlight made a swift impressionist rendering of his aristocratic features, he continued to stare unblinking out of blue eyes straight into it. He had nothing to say about Germany except that France and Britain would continue, united as always, to watch developments closely, would continue intensive diplomatic action to bring reason and moderation into their relationship with the Nazi regime, and were hopeful of success. As for the Sprattley Islands, held to be French, both countries were immediately ordering a *joint-de-marche* in Tokyo.

When our time for questions came, I asked: "Could you please tell us, Sir, exactly where the Sprattley Islands are?"

For a moment he looked utterly taken aback, then smiled gently, as if reprovingly at himself, and said: "I am afraid I cannot do that. You must ask the head of the South-East Asia desk for that."

Neither Maillaud nor I could see what value discussion between statesmen with such lack of precision could have had, and hastened to telephone the man in charge of the relevant desk. He had just left for the weekend and it was much regretted that no one still left in his office knew where the Sprattley Islands were. We telephoned the French and Japanese embassies. Neither of them could help, blaming the late hour and absence of staff. Maillaud telephoned Paris in vain, and we had to wait until the next morning before we found the answer ourselves in a library in Fleet Street where an old map clearly located them in the ocean below the southernmost tip of the French Indo-China that is Vietnam today. Their importance both as potential aircraft bases of great strategic value, as well as yet another portent of future aggression, were only too obvious. The final irony, however, was that this map showed them clearly to be British, as, of course, their name implied.

All this and so much else, like the Nuremberg *Walpurgis Nacht* Festival of horror I attended, revealed an indifference and neurotic refusal to accept self-evident reality, joined to a Hamlet-wise reluctance to take the action which the increasing responsibility laid upon us as more and more knowledge of what was portended became confirmed and reconfirmed. I could only work the harder at my book in what felt by the day like a swiftly darkening *Götterdämmerung*, or a 'twilight of the gods' moment. Finally, there were two utterances which greatly increased the objective incentives to my personal endeavour, and the weight and obviousness of prophetic warning implicit in events themselves which went so unheeded by authority. One came from Hitler himself in his perversely arrogant declaration: "I will go the way fate has pointed me like a man walking in his sleep."

Nothing, I thought, could be plainer, more accurate and alarming than that. Yet the general response seemed to be summed up in an illustration in, I think, *Punch*, which continued its uninterrupted Victorian preference for using charladies to raise a laugh. Everywhere I went, the illustration was thrown at me with the accusation: "If only you newspaper people would concentrate on the good and not inflate the bad in your newsgathering, all would be well." The drawing was of two charladies sharing a pot of tea and one saying to the other: "That Hitler, dear! He does fidget so!"

The other came from a remarkable American correspondent called Knickerbocker whom I met in London, and who told me something which latched on of its own accord to the significance of Hitler's remark and placed it, with unassailable authority and precision, in the dimension where I was darkly groping. Knickerbocker had just interviewed C. G. Jung, whom I had resolutely rejected, at his home in Zürich. In tones full of foreboding, Jung had told him: "In a way you can say that Mussolini rules Italy but Hitler does not rule Germany. He *is* Germany. He is more of a myth than a man. He is the loudspeaker that makes audible all the inaudible murmurings of the German soul."

Again it struck me like a blow in the pit of my stomach, for what could be clearer or more conclusive? Yet it had no general impact whatsoever. Indeed after reporting it for my papers and repeating it to friends with whom I went to make up a cricket eleven playing in a Surrey village over the weekend, I could not keep it all to myself. It was at a moment when Jung was still inaccurately bracketed with Freud, and one member of our team, at a vast Dickensian Sunday breakfast, raised a great laugh by greeting me with a parody of the words and song from one of the Disney Silly Symphonies of the day which had such a high intellectual following, including E. M. Forster: "Whose a-Freud of the big black Wolf, big bad Wolf."

I could have said I was indeed as afraid as I could be for Britain and Europe. I had married ten years before, and already, at the beginning of 1938, I had sent my wife and our two children to South Africa for safe-keeping, for I had no family in England and knew I would have to go to war when it came. I thought and almost hoped it would come in the autumn of that year, since it seemed the only way of stopping Hitler and his potential collaborators practising for their real war, as it were, in the wings of the main theatre. But it came, of course, technically only at 11 a.m. on 3 September 1939.

The process of war, however, started for me the day before I was told at the Foreign Office of the reluctant Neville Chamberlain's ultimatum to Hitler over Poland. It started in fact almost to the hour, thirteen years after I had sailed out of Port Natal for Japan in the *Canada Maru*.

Thirteen, I believe, has always been my lucky number. I was a thirteenth child, born on the thirteenth of December, certain that if there had been a thirteenth month I would have been born in that. I lived at Number 13 Cadogan Street, and as the coincidence registered I hoped that somewhere it concealed a solemn promise of luck to us all, which, in my reading of the word, served a meaning greater than personal and national happiness or unhappiness. Yet my unfortunate book was only three-quarters finished under a provisional title of 'The Rainbow Bridge' which, of course, was so vital an element in the extinction of the gods and light in Valhalla. It was what I called a 'seed' title to myself, but on this occasion a seed destined not to grow unto fulfilment. Ironically it was to be destroyed in one of the fires started by Hitler's bombers. All that remains of it today is a recapitulation of its essentials in my book *The Dark Eye in Africa*, written in the fifties. I saw this neglect of the mythological pattern, the unrecognized 'dominants of history', as I called them, and the irresistible energies at their disposal, at work as hard as ever, sapping the subterranean levels of the human spirit everywhere, because we still deny it the light of our lives and expression in living behaviour. And if I emphasize this it is because it seems to me even more relevant to the contemporary scene in Great Britain and Europe where day by day, through a proliferation of a totalitarianism of the left, reason is being invaded by the unreason of negative collective myths, and their pocket Hitlerian loudspeakers are applauded as they peddle, in terms of our day, the same collective little lies, nonsenses and contradictions that Hitler did on his Wagnerian scale. If I did not believe in not speaking out on the same theme more than once, I would rewrite and expand it under the title *The Dark Eye in the World*.

But to return to the declaration of war. On the Friday night of 1 September I had attended the last peace-time session of Parliament. A great thunderstorm raged outside. It was so noisy that the voices, barely audible normally, were inaudible most of the time. Yet the Sunday evening of the first day of World War Two was immaculate and of incomparable lucidity, a calm and tender valediction of light that only autumn has at its disposal. As I drove towards London, my first view of the city was like an illustration from H. G. Wells' *War of the Worlds*. Wherever I looked, the silver balloons sent up so naïvely to safeguard London from air attack were floating and turning yellow in the last light of the sun setting on summer and the start of the harvest of all our long-neglected yesterdays. And yet there seemed to be meaning and reassurance for me then as from on high, that the light over England should be so lucid at the setting of the sun on its first day of another world war. Perhaps we had at last seen the light, and perhaps it might not be too late? As the phrase 'not too late' occurred to me, I remembered how disturbing a connection it had for me with the First World War. I remembered how in the remote Interior of Africa, all sorts of conditions of men who did not even speak English and had never seen the sea, let alone Britain and Europe, were seized with a strange fear that unless they hurried to enlist the war would be over and they would be denied a chance to take part in it. Droves of them set off to our little railway siding to catch the first train to 'join up' (that was the phrase on Boer lips then). And as they went off, their weeping women and distressed kinsfolk sang not the hymns my grandfather, my father, and their begetters had sung as with contrite and awesome spirits they too had gone to war, but 'It's a long, long way to Tipperary'. It was all the more lugubrious and of tragic absurdity because they did not understand the words, and knew them only in part; but they sang the song with a terrifying fervour nonetheless. The remembrance put into stark perspective, all the starker in the lucid imagery of autumnal light, the difference between that war and the one ahead. That first war was the last of the great 'romantic' wars for both us and the French. War would still stay romantic for the Germans once more, but for us it was a reluctant sort of police action; not even good but, as so often in life, the ineluctable 'least bad course' into which our evasions had forced us. The comparison had seeds of a new hope that the European spirit at last might have had enough illusions over the nature of war. And in that hope I hastened to join up myself.

CHAPTER SEVEN

The Sword and the Flower

THROUGH all those thirteen years, so crowded and eventful that they seem even longer in hindsight than they were in the rush-hour of history, what happened seemed predetermined. By the year, the experience of my youth in Japan, both on sea and land, had seemed to become more privileged and autonomous. It led a kind of legendary life of its own in my imagination. But whatever its importance to me, it had no consequence that my friends and their world would have considered to be of 'real and practical value'. Yet I never ceased to cling to the experience as one might to the memory and meaning within oneself of a dead but dearly loved friend who somehow had changed one's way of life.

However, by the time war came, that it could still assume an even greater significance than ever before in my life did not occur even to my natural intuitions. Suddenly, undetected, it came rushing up from behind, overtook me and opened my eyes. It happened on my birthday, 13 December 1941, when I was serving with the English Army in North Africa. News had come to us belatedly the evening before of Pearl Harbour, Japan's entry into the war, and the sinking of the *Repulse*, one of the most beautiful ships ever built, and of the *Prince of Wales*. With it came a signal for me to report as soon as possible to Terence Airey in Cairo. He was the principal Staff Officer at Army Headquarters in charge of Irregular Operations and had been, as it were, Wavell's overseer of our activities behind the Italian lines in Abyssinia. It was there that Orde Wingate had joined us in order to coordinate the patriot forces to which I was attached and which was officially labelled '101 Mission'. Airey told me in Cairo that there had been a request from Wavell himself that I should be released and sent to report to him in South-East Asia where he had just been appointed Commander-in-Chief of the allied forces; British, American, Dutch and French, hastily collected to stem the Japanese invasion of South-East Asia. At once it was as if all the missing pieces necessary to complete the jigsaw of my past were magically at hand, recognizable and about to slot into their preordained pattern. It was almost as if the speed of their appearance and clarification of meaning was so swift that it made a noise in my ears like the imperative beat of the wings of swans driving towards shelter against a rising gale. Nor was the missing piece of Japan the only important one. It might well have been the main and most immediate cause of this summons from a general whom I had admired more than any other. But my experience in Abyssinia, in a mission stamped like a signet ring with the crest of his military vision, obviously must also have contributed, and, whether known to him and his staff or not, there were other factors. Indeed that night, after Terence Airey's briefing, I became aware how not just my Japanese experience but my whole life hitherto, which had so dismayed me by its inadequacy and confusion, had begun to look more and more to be part of an act of preparation for a role of my own, however small, in the war.

In Abyssinia, my love of Africa which had made me travel it whenever I had money and leisure, and to study its myths, legends and stories, now seemed singularly pointed because it had enabled me to give some service already that I could not otherwise have rendered, and that certainly was beyond the capacity of the inexperienced officers and warrant officers, fresh from England, who had accompanied me. Had it not been for such preparation, I would not have known how to look after our camels as I did. My mission was to take much-needed supplies and ammunition from Um Idla on the borders of the Sudan, through the sleeping-sickness bushveld to the steep Ethiopian escarpment. There I was to meet mules to take over the heavy loads from the camels and climb the mountains of Gojjam to the highlands beyond, where I had already left patriot forces fighting the Italians.

My sense of contact with history and life became more pronounced when I joined the men and officers who were to accompany me and my camels at a little place called Shaba on the banks of the Nile, not far from Khartoum. The officer in command of the small base for those of us designated to go behind the Italian lines and stir up the long, simmering revolt against Mussolini's Viceroy was a gentle person called Hallam, a direct relation of Tennyson's closest friend whose tragic death inspired *In Memoriam*. He told me that Shaba was the birthplace of the queen who was the Helen of Africa and whose beauty Solomon celebrated in song and through whom he fathered the kings, queens and emperors who had ruled Ethiopia ever since.

That our point of departure should have such significance in history meant a great deal to me and it was increased by the unexpected arrival several nights later of Emperor Haile Selassie. He came to wish us well before we entrained for Sennar on the Blue Nile. He had that morning reached Khartoum after an exhausting journey and, hearing of our departure, typically insisted on visiting us. I was so unprepared for such a visit that I hardly knew how to present my little group to him suitably. I had given the men twenty-four hours' pre-entrainment leave. Miraculously, they all reported back at Shaba in time, the last man just before sundown, with our train due to leave at 9.00. All were in high spirits, almost everyone somewhat intoxicated, quite a number unsteady on their feet, and all barely capable of setting out their equipment properly for a final inspection.

At that moment a messenger arrived from Khartoum with the news that the Emperor would be at the railhead at 8.30 for a personal hail and farewell. With an extraordinary effort we managed to be on parade in time and would not have done so without embarrassment had it not been for the dark and the precaution we had taken of ensuring that the least sober were in the rear ranks where the dark was deepest. It was my first meeting with the Emperor and the beginning of a relationship which continued on and off until just before his brutal and totally unjustified murder by ungrateful and power-obsessed men.

Taken aback for a moment that he was even smaller than his official descriptions, I was at the same time impressed by the quality and unusual stature of spirit that this brief first contact conveyed to me. Fortunately I spoke French, which he preferred to English, and could interpret for him on what must have been one of the briefest of royal inspections. Yet the few words exchanged confirmed a feeling of the extraordinary degree of natural authority, dignity, grace and courage embodied in that small, slight person. Even the more disordered of spirits on parade discovered this and, as we passed three tall, slightly intoxicated sergeants culled from the Brigade of Guards as expert machine-gunners, and whom I had optimistically concealed in the dark, I burst into conversation in French to prevent the Emperor from overhearing their exuberant whispers. One comment was particularly hard to drown. "D'you see that, Bill?" the tallest of the three exclaimed. "Calls himself the Lion of Judah! More like Tiger Tim to me."

My efforts fortunately succeeded with the Emperor, but everyone else overheard the remark, and from then on he was Tiger Tim to them all; and myth and legend were, as so often, introduced through the side doors of our modern reckoning as clowning. But to me he was always the 'Janboy' his Amharic title proclaimed, for there seemed even more of history and myth and royal mystery in that.

The impact of such events, slight as they were in themselves, was all the keener for me because of an acute awareness of the perilous situation of the British world, abandoned by France, without allies, alone with yet another enemy in the shape of Mussolini's Italy to face and the shadow overall of the inevitability of defeat which so many took for granted. Only a few days later when I left our main forward base on the east bank of the Blue Nile, an Italian Division advancing on the opposite bank was reported only some forty miles away. The only forces we had to spare to meet it consisted of some fifty militarized Sudanese police, to whom I said goodbye in the late afternoon as they crossed the legendary river under command of their young District Commissioner. As a parting present I gave him a box of hand grenades which he returned with a gift of two bottles of brandy and the remark, uttered with a light-hearted, schoolboy smile so natural and uncontrived that it moved me greatly, "Strictly for medicinal purposes, old boy."

Transcending all sense of peril for Britain and world expectation of her defeat, however, was the heightening of perception and quickening of heart that came many miles and days later at our last base on the theoretical border of Abyssinia at Um Idla on the River Dinder. Somewhere between Roseires and Um Idla in dense bush we had left the inhabited world of the Sudan behind and re-entered primordial Africa. The Africa I knew from my youth in the Interior was still close to nature, and the original still magnetic in its design, and far from being irretrievably spoilt as it appears today. But it was not remotely like the vast sleeping-sickness country and the infected bush breathing heavily in the sun that stretched for some hundreds of miles between us and the great Abyssinian escarpment. It was the real Africa of which my grandfather, and grandmother, who was one of four child survivors of a Matabele massacre, had spoken as some great miracle of nature.

When time allowed, I would sit on the banks of the river and watch the deep pools turned a midnight blue, with row upon row of armour-plated crocodiles, disturbed by our presence, drawn away from the banks to watchful positions. Always a fish eagle, one of the great birds of the world, with the most exciting hunting call of all, sat with its breast on fire on the top of a dead tree, which threw a shadow that was rarely more than a paler form of sunlight. It would look as still and inanimate as an Egyptian hieroglyph, when suddenly it would swoop with the speed of a desperate sword thrust, pierce the shining breast-plate of water, and rise with a silvery fish between its claws, all so expertly and swiftly done that not a crocodile in formation blinked an eye. Towards evening the yellow river-bed was dark with baboons, apes, antelope and gazelle, walking to the sunset-stained water with the delicacy and sensitivity that only fear and the extreme need of vigilance in a world of hungry enemies have at their command. Watching the ritual, it was as if I recognized for the first time the totality of all the life that had come within me at birth. I felt almost ashamed of what I seemed to have overlooked in my race through the external world.

Somewhere between the ages of three and five I became aware of a frontier between the immense past, which was the hidden source of the meaning I sought, and the short term of life I had in which to live it and which would involve me in another act of painful and conscious birth of an individual self within. As a result, the boy that grasped that revelation remained ageless within me and became the tutor as well as examiner of the active knowledge I acquired of such a charge and potentiality of being and becoming, and still the sternest judge of how well I had discharged so sacred a trust. He still remained my severest critic and mentor, and in those days at Um Idla it was as if I were meeting him again after a sojourn of many years in a strange land and he took me without rebuke warmly by the hand. Implicit in this meeting between my grown-up self and this someone other, was an absence of any feeling of defeat, and it sent me with a strange peace and confidence at heart to take my camel train into the sleeping-sickness bush.

I was in command, in theory, of five operational centres, each consisting of one officer and three warrant officers, capable of training the Abyssinian patriots we had rallied to our cause, in the use of the World War One weapons we were taking to them by camel. These weapons, like the Lewis and Hotchkiss machine, were completely out of date. Rifles, such as the American Springfield, were packed like anchovies in coffin-like boxes by American ordinance for safe-keeping; ammunition and explosives, food and, most important, cases full of Maria Theresa dollars. These dollars were the only currency acceptable to the Abyssinians, despite some years of Italian rule and, although minted in Bombay under the direction of an official of the old Imperial Bank of Ethiopia, were the only counterfeit of which it could be said that it was better than the original. To carry these heavy supplies, each operational centre had some hundred camels in the care of their Sudanese civilian owners. I had the largest group, some hundred and fifty-seven in all, and as I was the first to go and the rest of my command was to follow at staggered intervals, my departure was made as much of an occasion as Um Idla, lost in the bush, could make it. All the officers, warrant officers and signalling staff came to send us off, surprised because I was going in the first fall of afternoon shadow and breath of evening cool.

This was done at the behest of my rediscovered African self and with great deliberation. Ian Gillespie, the head of veterinary services in the Sudan, almost an honorary camel himself because he knew and loved them much, had presented me with a mule. He said I would find it most helpful to lead camels who were fastidious creatures and stubborn in refusing to move at unpredictable moments, and were certainly bound, at first, to refuse to cross rivers. Water was sacred to them and inviolate to any part of their bodies except their long, sensitive, aristocratic and prehensile lips. I had christened the mule Prester John and put him now at the disposal of the one unattached officer in my group. He was the W. D. H. Allen who ran away from Eton, I believe, to take part in the Civil War in Georgia after the Russian Revolution. He wrote the best history of Georgia ever written in English. He was a paradoxical character: a friend of Professor Toynbee, but also a Mosley member for Belfast in a vital pre-war Parliament. On the outbreak of war, he joined the Household Cavalry. When the Cavalry, to his disgust, was being mechanized at Sarafand in Palestine, the call for volunteers for special duty in Abyssinia reminded him that it was great horseman country and he instantly made himself available along with two other countrymen, Bill Maclean of the Greys and Mark Pilkington of the Blues, who alas was to be killed in action later. The two last were to follow us, each with a camel train of his own, but I gladly and gratefully appropriated Bill Allen for my own train.

Some days before leaving Um Idla, I had sent him alone to scout forward on an old poachers' and slavers' track through the bush to select the best camp site he could find within four miles of base. I knew that camel trains as well as new engines and ships needed trial runs before they could be declared fit for their destined occasions. This he had done, and as the long level light lengthened and darkened the shadows in bush and yellow river, he and Prester John led off with a sense of historic adventure that I continue to remember. His tall frame was made for a horse rather than a mule, and the elongated shadow travelling by their side made it look as if the Knight of La Mancha and Sancho Panza had exchanged their mounts. I was the last to leave on foot, a rifle slung across my shoulder and a long staff in my hand. Again to some surprise, I had chosen to lead from behind. Africa had taught me that it was the point where trouble tended to collect. The result of our timing was that we could bed down in a carefully chosen camp site well before dark and put to admirable use the experience of this trial shuffle and make essential corrections in loads and knots and adjustments for men and beasts made obvious on a three-mile journey.

From there on we travelled only in the early morning and cool hours of the afternoon. On making camp we saw that every animal had been inspected for bruises, scars and scratches, and we made arrangements for them to feed and browse as a first priority. Each day when the camels and their owners rested, I would take two or three of the fittest camelmen out searching for game; I did this not only because I had been taught that one never hunted in strange country alone, but also so that the fanatical clansmen could see that the animals' throats were cut the moment I shot them. Otherwise they were not allowed to eat the meat. It was astonishing to witness the speed with which they achieved this after running like Olympian hurdlers over the tall buffalo and elephant grass and thorn, their knives flashing like mirrors, to where the antelope had gone down head foremost like a stricken ship at sea in some calm trough between the swell of that ocean of bush. I rarely came back with less than two buck for men who had been undernourished from birth, and that, in the end, made a difference for the good.

Meanwhile Bill Allen had already scouted ahead and invariably, once a camp site was selected, came to join me at the rear for a while to talk history and, above all, Russian literature. These discussions were often resumed round the camp fire, and on one occasion when the noise of lion, hyena, jackal, owl, night beetle, cricket and tree frog was so loud that we could only just hear ourselves speak, we had an argument over the book of prophesy that I took *The Brothers Karamazov* to be. The argument went on at white heat late into the night. This strange element, like the voice of a lost and abandoned calling, precipitated a strong dreamlike element for me in the heart of the experience, and all the re-remembering and recognition of lost quintessentials folded as the fire of the rose within its bud in my imagination. By day too this dream-like atmosphere, taut over the tranced and sleeping bush, was heightened by the silver voices of millions of sun beetles and, above all, the incessant calling of doves. From first light to the last dip of a scarlet feather of day in the inkwell night, the doves with throats on fire would call to one another, so that there was no expanse of silence or murmur or other insect noise too vast or great that was not dominated by them, until it was as if we were not at the heart of one of the widest of bush-covered plains in Africa but in the forecourt of an Olympian abode of a goddess of love.

Both day and night became stranger still when we moved through an area where the Italians, it was rumoured, suspected our presence but did not know precisely where we were. They had, it was thought, dropped incendiaries from the air and set great tracts of bush on fire to burn us out. Whether this was true or not, it was a grim fact that pillars of smoke began to appear wherever one looked by day and made us change and alter our course frequently to avoid the racing fires. Sometimes at night, camping in the wake of a vanished fire, it was as if we had happened on a place of antique carnival and all that was dry in trees still glowed red and even flickered like torches of the swift flame that a bacchanal of fire had made of them, and which their branches still held aloft in the night.

Once we were caught in a giant bamboo forest of fire. The fire was coming at us, roaring and exploding and crackling with the speed of an express train. Bill Allen came hurrying back with the news that he had found a broad river nearby and safety for us all beyond if we raced for it. Unfortunately the river ran deep below a high rocky west bank and the track down to it quickly narrowed between steep stone flanks with room for barely one camel at a time. In the hurry, and smarting under the lashing of their owners, the leading camel stumbled, fell and threw three more just behind it down, where they lay on their knees moaning in their most stubborn tone and refusing to budge. Although the head sheik belaboured the lead camel with his stick as only a whirling and enraged Dervish can, while his son screwed its tail in knots so that he could chew and bite it more painfully and effectively, the camel would not move. The noise, heat, smoke and flame came nearer and denser by the second, and I remember thinking of the camels loaded with ammunition and other explosives and finding no reassurance in my instructor's dictum that fire burnt but did not detonate explosives. Worse still was the terrible thought that we might have to abandon the camels and run for the river, which came to me just as Bill Allen, watching the scene with sardonic calm, said quietly, "Do you think, sir, it would help if I spoke French to it?"

The remark drew a silent retort from my African self that French would probably be useless but an African tongue might help. Ordering the sheik and his son to leave the animal alone, I went forward, took its head, tugged it gently and said in Zulu, "In the name of Umkulunkulu, the First Spirit, get up and follow me!"

At once it rose and followed. The others did likewise, and we all followed Prester John into the river and across the wide water just in time. No river had ever looked more beautiful. Once rested and before moving on in the afternoon, we christened it 'Lady Precious Stream', after the play which had enchanted London.

The real blow came when we arrived at the escarpment. The despair and frustration we experienced seemed for the moment beyond resolution when we realised that the mule trains that were to take over from the camels had not appeared, and were never likely to do so, however long we waited. This left me with no choice except the apparently impossible one of inducing camels and men from the soft, level desert earth to carry their heavy loads on and up the mountains and so into the interior towards our embattled patriot forces.

I remember looking at the grey stone walls of hills in front of me, the silence of the greater inner Africa like the murmur of a far-off sea in my ears. Only a narrow track, indeed at best a mule- and in parts only a goat-hoof path, with jack-knife corners and acute angles, led to the forbidding rim of rock above. I turned about, and from the rise of earth at the base of the escarpment saw nothing but the sea without end which the bush had become, and which had closed over the footpath of elephant, slave traders and poachers that we had followed from the Sudan. It was as if the path had never been. I surveyed my camels bedded down hard by their heavy loads. Their masters from Kordofan were close by them, keeping together in their various clans, each with a sheik of its own who had their fanatical loyalty. They were shivering in their cotton dresses designed for Sahara heat and not this upland air, cold in the first light of day. Thank heaven, the camels still looked well, but their owners were clearly worn and strained and over-full of feelings that they had done what they had promised and much more in circumstances of unforeseen hardship. It would obviously take little more to convince them they had had enough, and could shed their loads without fear of reproach and turn about as their contract with the military entitled them to do.

Only I knew how urgently our supplies were needed in the embattled world beyond the lofty summit with its impassive face proclaiming an ancient and instinctive African 'No!' to the presumptuous intruders ambitious to climb the wall of stone. Yet I knew I could not unload my camels and sit there among piles of badly needed war supplies, waiting for mules that I was certain would never come now, while our camels plodded back home to Kordofan. But what was I to do?

It was almost as an audible voice that a memory came to me of a passage in the army's Bible of King's Regulations. It was to the effect that when an officer carries out an order knowing that by carrying out the order he would actually defeat the purpose for which it was originally issued, he would be judged as not having complied with the order. This paradoxical pronouncement that I was to use again and again in my future career emancipated me immediately from a blind adherence to instructions, however imperative and solemn their first commandment, that I was on no account to take my camels beyond the escarpment, and above all not to expose the camel men, who were only civilians on contract, to danger in any field of action. But I could not breach my instructions without their consent if I breached them at all. Accordingly, I asked my liaison officer and interpreter to collect the men together in an open space round a large termite mound. He was a young, able, energetic and typically independent Sudanese, Ambasha, called Abu-Bakir, and he did so with despatch.

He soon had a reluctant group looking utterly incapable of further effort standing in their torn and soiled frocks like unwashed and outworn nightdresses around the ant-heap, as the sun was about to rise.

What could I possibly say to them to give them the will to do more than they had already done? It seemed a hopeless prospect and yet the voice urged me on and reminded me that all military leaders, from Alexander the Great and Julius Caesar to Frederick the Great and Napoleon, would have failed if they had not been inspired to use words to evoke in their men a resolve to persevere even though defeat seemed inevitable. I thought above all of Frederick's remarkable use of irony at various critical moments in his Seven Years' War, particularly at the battle of Rossbach when his troops were wavering and about to run. He rallied them to face once again the superior Austrian artillery by chiding them gently as he rode up calmly over a ridge into the great barrage himself, "What, lads? Do you mean to tell me you have no wish to die?"

I thought also of a lesser order of men, including my own grandfather's and father's experience of war against Bantu and British. Do not persuade or attempt any rhetoric, was the conclusion that emerged out of this lightning re-remembrance of the past. Speak the truth, you cannot improve on it, and let it do the rest. So I told them the truth, beginning with my acknowledgement of the terms of their contract and their right to turn back. I told them at the same time that, if they exercised their unquestioned right, it would be disastrous for us all, including their fellow-countrymen and the small Sudan frontier forces then under the command of a lifelong friend, Hugh Boustead, fighting with our Ethiopian patriot bands against the Italians who had invaded their country at Kassala and along the Blue Nile.

It was a moment, I said, when the right way would become the wrong way and, with all that I knew of them and their manly history, they would choose of their own free will to help me to deliver our supplies by the only means possible, where they were most needed, if the Italians were not to win.

They listened to this attentively but with growing doubt and an inclination to say no, particularly among the younger men. But I had hardly finished when an old sheik with grey hair asked if he could speak to his countrymen too. He stood on the mound beside me and said quietly, with the confidence of total conviction: "Like you, I have come further than I ever thought I would have to come. Like yours, my feet are hurting and my arms and legs and body are covered with sores. Like you, I am hungry and cold, and my clothes are torn. But standing here, perhaps where no men of Kordofan have ever stood before, unlike you I am old enough to remember things you could never have known. I remember how the Sudan was before the government* came, and if our Effendi† here tells me the government would want us to go on, I and my sons and camels will go on. All I would ask him is to give us a Warraga stating that the government would wish this so." All Sudanese had an unquestioning respect for writing, and this reference to the letter, for which Warraga was the Sudanese, was decisive.

As a result they all came on, bright-eyed and without any holding back within. We took our loads to the top by halving them and then descending again and once more climbing to the top. It took two hard days before we could move on and take our supplies right into battle, where they arrived just in time.

So strange was the appearance of camels in that world that it evoked a feeling of awe among the Amhara of the Gojjam bordering on unbelief. Indeed the first group we met were only prevented from keeping clear of us by the sight of Bill Allen on Prester John. Mules they knew well and treasured, so that one of their most polite forms of greeting was "Bakhlo yet no!" (How is your mule?). Prester John drew them to us, and after a while one bearded Amhara was confident enough to ask me for permission to dart between the legs of the foremost camel from one side to the other. When I asked why, he said simply, "But surely that would be a most remarkable thing to have done". Accordingly he did it, with obvious apprehension but to the unreserved applause of his fellow-men.

The last I saw of my camels and their attendants in their soaked and thin smocks was as they entered the wings of the heavy rain closing in on us, sealing them forever but bravely from our theatre of war. So far had we come from the low wasteland of Kordofan and so high had we climbed that only on the morning of their going did I feel, after shaving, that I had combed the thunder and brushed the rain cloud itself out of

---

* I was to find that all the oldest Sudanese used "when the government came" for the moment British rule was established in the Sudan.
† They always addressed me as "Effendi".

my hair. Doing so on that raised and strangely cleft earth of the Gojjam of Ethiopia, where one's head was in the temple cloud itself one week and one's feet were like lead in a valley below sea level the next, it seemed to me that their achievement could be measured only against a description of the interior by one of the private soldiers in the dashing Napier's nineteenth-century army launched against Magdala in the highlands just ahead of us. Writing to his mother, he said in effect, "We were told that the interior of Abyssinia was a great table-land. But to me it looks more like a table turned upside down." How could I not therefore feel compelled to salute in the only means possible to me these anonymous men of Kordofan as a significant element in facing up to the gathering storm of history, more important to me than even the great who directed them and the rest of the millions driven before it like leaves in a great fall of the seasons as well as of the spirit.

However, the aspect of the whole operation that perhaps mattered most was that I achieved my mission without the loss of a single camel. This may not sound much unless set against the history of the camel trains which followed, without exception, so disastrously in my footsteps. The common and justifiable comment in the Sudan was, "You need no map references or compasses to find 101 Mission in Abyssinia. You only have to follow the smell of dead camels and it will lead you safely in."

Then before that, there had been my journey out by sea to Egypt. I could see now my random past had suddenly acquired a focus even in that. Indeed, the act of preparation could never have been so complete had it not brought me to the sea again to sustain and confirm me as a sort of missionary for my unconverted, unknown and unforeseen future. I had, despite a commission in the South African army, joined as a private in Britain and was not long after sent to Camberley and Aldershot to train for commission as an officer. I was still amazed by the result whenever I thought of it because, although I felt I had done badly in my course, I passed out first and was declared "exceptional in every way" in my final report. I mention this solely because it would seem from this as if my own inner and concealed nature had also been prepared without my knowing what was to follow. But it was the voyage to Egypt that really enlarged and gave substance to what others regarded as 'promise' in me. I travelled in perhaps the most imposing convoy of great ships ever seen on the sea: large P & O, Empress, Royal Mail, Red Star and other luxury passenger liners all crammed with soldiers. There were some fifteen ships in all, and I myself travelled in a remarkable Royal Mail ship which survived the war to become one of the most popular ships ever built. It was called the *Andes*, so new that its real maiden voyage was this round about passage to war. It was especially endeared to me by its pure Union Castle lines because it was designed and built by the same

Harland & Wolff which regularly produced the Castle ships and, though larger, was as of the same maritime family as the *Dunottar Castle* in which I had sailed from Port Natal to Britain for the coronation of King George VI and Queen Elizabeth, in 1937.

As the only unattached officer in the ship, with no knowledge of the why and wherefore of my going, I was ordered by the officer-in-charge to look after the welfare and entertainment of the soldiers on board. I was appalled by the responsibility at first, and yet soon found that it was almost made to measure for my own inclinations and interests. It lead at once to a privileged discovery of the composition and character of a truly democratic British army. I found it unusually imaginative, responsive, sensitive, richly diverse and eager for knowledge, and particularly for knowledge rooted in experience. Accordingly, although I did not know precisely where we were going but was certain that the men could only be destined for Northern Africa and the Mediterranean, I talked to them daily about Africa, of our classical beginnings and a complex of related things. The gatherings rapidly grew and were so swelled by officers that I had to make them twice- and on occasions thrice-daily events. I found others in the ranks who could add to the theme, and I remember in particular a soldier who had been awarded a double first in Classics at Oxford. He talked about Agamemnon's expeditionary force sailing in its black ships to battle on the great Trojan plain, and about Odysseus, Aeneas, Penelope, Dido and all, so that we were made to feel part of a company of men in and yet also out of and beyond space and time, and far greater in numbers than those massed in our convoy of ships or for that matter in all the ships sailing to other beachheads beyond our sight and reckoning. I thought of a poem by Samuel Butler, who fell so much in love with the story of Odysseus that he spent a small fortune trying to prove it was written not by Homer but a woman. The particular line was like a pebble dropped in a still pond of time to raise a ripple of heart and mind widening towards its invisible limits: "We shall meet on the lips of living men."

The imperishable of human endeavour implicit in it was as a mirror to reflect for me how this parade for roll-call of the soldiers of yesteryear on our own lips at sea, brought a known sense of the indestructible into our own voyage despite all the clouds of unknowing that surrounded our destination and fate. I would feel this and also the immense fall-out in all dimensions of my imagination that emerged from it, particularly when I was alone with the ship and sea. I made a point of leaving a comfortable cabin at nights with just a rug and trench coat for a pillow, and would stretch out as I had often done in the *Canada Maru* on the upper boat-deck just behind the funnel of the ship. An exchange of recognition between me and the stars and, in particular, the Southern Cross, rose as leaven in bread for the human spirit, because at its core was an acute re-

remembering of the being into which I had been born, with no fear or feeling of interruption, and a realization how more and more life seemed to be an acting out of a great mystery, so that it too could be increasingly known and consciously lived for its own and our increase on earth. The meaning to come, it would strike me then, had to be lived before it could be known. And as I lay there, thinking that in that sense all life was the living out of an answer to a question of cosmic import, I would look at the Southern Cross and rejoice in its lack of symmetry and the imbalance of light which is reputed to disappoint so many from under the cold polar star. I rejoiced because its lack of perfection was some sort of reassurance from on high that the love which is concerned with a symmetry still to come would never fail and would remain fully employed. It would not vanish in a universe of perfection, for what use could perfection have of love designed to serve our asymmetrical 'now' on behalf of a greater and less imperfect 'not yet', except to make it redundant? It was the beginning of an awareness of belonging to a community such as I had never known, and could not have discovered as a reminder of committed life without that voyage, and all those other sorts and conditions of men who accomplished it with me. They gave me the most profound feeling of being included in a human congregation that I had as yet experienced. That remembrance, dynamic with recognition, was perhaps the most vital part of my preparation for the future. It stood and held a firm centre round which all those pieces of my own experience of life gathered together on that December birthday night, twelve days before Christmas 1941. My odd life in the Interior of Africa; the High Dutch of the Netherlands that I had learned at school and spoken officially; the Malay I had studied in the Cape; my French, and finally Japan and the Japanese language which had long seemed irrelevant if not atrophied through lack of use, suddenly seemed urgently needed, and to make sense in a unified and dynamic pattern. I was no longer haphazard.

So, of course, I went to Java and duly reported to General Wavell, at his Headquarters at Lembang on the slopes of an immense sleeping volcano. It was called Tangkoeboehan-Prauw because it evoked the disturbingly prophetic image of a ship turned upside down, for that is what its Javanese name said it was. I found General Wavell, despite another overwhelming defeat which was not his own fault or contrivance, looking straight out of the one unblinking and far-seeing eye left to him, still a moving and honourable person, dignified and high in the stature of a deeply and poetically involved spirit of obligation to life, which went far beyond the call of office or duty to country. He told me that the main purpose for which he had intended me was no longer a possibility, and he would have to give me some minor but nonetheless urgent tasks. When it came to the last of these, on the night when, sad at heart but undismayed, he and his essential staff flew out of Java, he said: "I will not order you to do this because I do not think it is possible and the chances are you will not survive; I will only ask you, because even the impossible must be tried."

As a result I came to the moment some months later when the random elements I have summarized, above all the Japanese factor, fell into their rightful place in the wider plan of my life. On a mountain called Djaja-Sempoer, the Mountain of the Arrow, where the jungle of Bantam, the turbulent west land of Java consisting mainly of great jungles which swept like a green wave of a typhoon-tossed sea over the summit, I had come to establish a provisional headquarters. It was a setting that once again might have been lifted straight out of Conrad, made from the heart of the earth of tangled trees, footpaths with giant moss growing chin high, volcanoes, mangrove swamps and all, whose outline I had observed with such awe and wonder from the decks of the *Canada Maru*. It was here that the truth of my own past caught up with me, as it did with Conrad's Tuan Jim, in the sense that it too arrived veiled in the manner of an oriental bride at my side.

Going down the mountain one morning and walking along a spur towards the beautiful valley below, called Lebaksembada because, as the name states, it was indeed 'well made', I came to an area where the path widened and formed a small plateau. I had hardly reached it when trees, ridges, the footpath, and the wild, tasselled and tall buffalo grass along the edges shook, bubbled and erupted with Japanese, who charged on me with their bayonets fixed and looking even longer and more pointed than normal. I myself knew only that there was no escape, and that there was nothing I could do to stop them: I did not even have a pistol on me, and every pocket of my jungle-green bush-jacket was stuffed only with M & B tablets and pills of quinine for the wounded and sick. But something else in me knew different. It took over command and called out in a loud and clear voice as of a stranger with special and singular authority over the occasion. It called out moreover in a Japanese that I no longer knew I remembered, and whose nature should be explained in order to make its impact comprehensible. There are many degrees and nuances of polite speech in Japanese, ranging from what passes for good manners among peasants but would be rude if used on others, to the highest which is used only when one dares to utter in the presence of Japan's Emperor and Son-of-Heaven. The normal "Dōzō chotto maté, Kudasai" ("Would you be good enough to wait a little") I remembered and would have used had I still possessed a mind of my own. But it did not seem adequate in the circumstances to this other person in me who rushed into command of my senses, and rounded off the address with the highest of all polite forms, taught to me one day not by my teacher but by Mori himself, as an example of how great and important these degrees of politeness were for the Japanese.

As from afar, this voice, in contrast to my naturally quiet tone, now rang out clearly and loudly, at the same time as my right arm shot up high into the air like a traffic policeman's signal to stop: "Dōzō chotto maté, Kudasai!" As the call caused confusion, I quickly followed it up with Mori's version: "Makotoni osore – irimasu-ga shibaraku omachi Kudasai-ka?". I can only translate it idiomatically as "Would you please excuse me and be so good as to condescend and wait an honourable moment?"

Nothing, it appeared instantly, could have had a greater effect on the Japanese. They had clearly expected all sorts of military responses: rifle and machine-gun fire, hand grenades, explosives and possibly even a mortar or two; but certainly not the highest degree of politeness. Immediately a solar-plexus shout of command broke out of a slight Japanese officer who was running at me sword in hand. His soldiers halted and stood their ground, bayonets at the ready. In a stillness which, in the acute sensitivity with which confrontation with death equips one, sounded loud above the surf of sunlight breaking over the lofty cathedral of trees high above us, the Japanese officer came up to me, fixed the point of his sword on my navel and asked, as if in a daze, whether it was truly Japanese that I had spoken to them. When I said "Yes", he asked where I had learnt it, and when I answered "Japan" his bewilderment gave way to a reassuring amazement. He hissed between his teeth before the inevitable "Ah, so desu ka?" broke from him, to be followed by: "You have been to Japan?"

That fact seemed to conclude the preliminaries for the moment because he rushed on to an imperative "Where are your men?"

Again, that other person in me remembered the almost pathological fear of infection and disease that tended to make the Japanese of my day a nation of hypochondriacs, and I heard the voice declare: "If you look round this mountain you will find many graves but very few people."

"Why graves?" he asked, in an alert voice all a-quiver with suspicion.

"Because there has been a terrible plague of death on this mountain."

"What kind of plague?" he asked, obviously alarmed.

"The deadliest of typhus," the other voice declared.

"Then we go at once," he replied and made as if to turn about. But the other voice in me stopped him and begged his permission to go back to fetch a very badly wounded soldier who was still alive.

"You certain he is only wounded and not ill with pest?" He demanded.

"Absolutely certain, and on my honour as an officer," the voice reassured him.

"Well, you hurry quickly and we wait," he responded, which was all the other voice had wanted and worked to bring about.

I hastened back and up at the double, roused my officers and told them what had happened.

"You are to disperse at once and lie low. I'll lead them away and try to escape later and rejoin you."

Then, taking the wounded Australian soldier, badly in need of expert medical attention, with me, and supporting him with my right arm under his shoulders as he was a small man, I made my way back to rejoin the impatient Japanese.

The authority of this other voice was from then on supreme. Not only had it obviously saved my life but also its bluff had worked and saved the bulk of my officers and men to carry on for months more. It is true that I was never to have a chance to escape and rejoin them. But I led the Japanese without a further searching out of the area for good. From that moment on that other voice was increasingly all that stood between me and death for, as I was taken back in stages towards Japanese Headquarters in central Java, I was passed on to increasingly higher, more organized, complex, angry and politically motivated and fanatical units of command. At every stage the crisis of instant death and a final sentence of execution was repeated, and it was resolved only by this other person in me; until one evening when a Japanese sentry, without whom I might have succumbed from starvation, beating, minor torture, and a combination of long-untreated tropical diseases, told me that I was to be executed in the morning. I was released into an old Dutch prison at a place called Soekaboemi, 'the desired earth', instead. There, ultimately, I was allowed to join a band of soldiers, gunners, airmen, and both Australian and English prisoners who had been captured on their way to try and join me in Bantam rather than accept the surrender which was announced barely a week after the first Japanese invasion, without any real fighting in between. They were all considered the hardest of cases by the Japanese and subjected to extreme severity and yet, having heard about me and fearing for me, on the morning I was suddenly released, still quaking at the knees from the shock of a reprieve from sentence of death which I had never hoped for or imagined even in fantasy, they congratulated me on being set 'at liberty', and gave me my first taste of how relative prison values had already become and were increasingly to be. They were in the charge, as King's Regulation would have it, of a most remarkable officer, Wing Commander W. T. H. Nichols, RAF, who enjoyed authority of rank. I began there and then, as the senior army officer, a relationship which was to last for some three and a half years and to prove one of the most important of all my life. Without him, his sensitive courage, his natural wisdom, physical and spiritual endurance, and the way in which he responded to even the most unlikely promptings of this other voice in me, without him we would, I know, not only have failed to survive as we did but, far more significantly, we would not have achieved an immense emancipation of spirit through incarceration. I am well aware that the full story should one day be recorded. Here I can only say how to this day I marvel at the imperviousness of Staff Command in all services, which need gunfire to make them recognize courage but remain blind to subtler and far greater exercises of this supreme grace of the human spirit in other fields. No airman, soldier or sailor showed greater and more sustained courage, or set a finer example in command in impossible circumstances, than did Nichols. And this went on, day in and day out, through three and a half years of brutal, highly dangerous and, for many, lethal imprisonment. One would have thought that RAF High Command would have examined closely the cause of the remarkable consequences of his prison command but they could not have done so. If they had, they would surely not have been content to recognize Nichols' services by the award of an O.B.E., however worthy, and failed to promote him to the highest command.

I dwell on this matter not just for the lack of imagination it reveals but because of its bearing on what was to follow. For some two months Nichols and I and our little gathering of 'hard cases' went through a kind of crash pilot course of education as to what imprisonment with the Japanese was to mean. It was made all the more realistic and instructive because we were, as I knew from my Japanese guard who remained a contact, under potential sentence of death since we had followed neither the allied Commander-in-Chief's act nor their own edict of surrender. In the process we became one of the most united, finely honed bodies of men, in mind and heart and resolve to endure to the end, which Drake in his famous signal before Cadiz called the 'true glory', that I was ever to know. Moreover we planted among one another the seeds of a prison integrity without which we would not possibly have transformed the vast disordered prisons of alienated and embittered men of all nations and services, into the instruments of self-imposed order and creative imprisonment. When after two months we were released among them, first at Tjimahi and then at Bandoeng, had it not been for their trust, the almost missionary zeal and unwavering support of this compact body of Soekaboemi storm-troops, as I called them, we could not have survived as we did.

But ultimately I believe that without this 'other voice', and the experience out of which it came, we could not even so have saved ourselves from then on as it had saved me. Nichols was a naturally religious person though not in a conventional sense, or he would have become the priest he tried in vain to be many years later. He was unusually open and imaginative, and immediately understood when I told him my own version of the origin and evolution of this other person in me. I told it to him as what we came to call the parable of the 'Two Cups of Coffee'. I did so because I had realized with a shock of lightning revelation how on that cold winter's day in Pretoria, South Africa, in 1926 when I obeyed without hesitation the prompting of instinct and invited Messrs Shirakawa and

Hisatomi to have coffee at my table with me, I was saving my own life in Bantam in 1942; and now, in the same fashion, might be able to save the lives of thousands of others. It raised my own predisposition of character and temperament to be aware of the importance of the small in life into a highly conscious and cardinal imperative. Creation, it seemed to me, was even more active in the apparently humble, often despised, rejected and insignificant detail of the immediate life of the individual, than on the melodramatic and grandiloquent stage on which the establishments and authorities of the world sought it. No life, however humble, the parable stated plainly, was ever without universal importance if it truly followed its own natural gift, even if it was only to plough a straight furrow and plant potatoes well. Out of sight of heart and mind of their world, as prisoners were naturally predisposed to think, and even in their dreams, unconvinced that saving intervention from without was possible, it had a decisive element of reassurance for me and in due course for all. Whatever chance remained of survival, it proclaimed, depended on a creative attention and positive response to the daily trifles of imprisonment, living an imperilled 'now' as if it were a safe and assured 'forever'.

Accordingly I gave myself and all the reason I possessed to this 'other person'. I came to listen to what it had to say with the utmost care. More and more, it was always there like some kind of guardian angel over us, not just to resolve the petty but painful daily confrontations with the Japanese, but to defuse the major crises which convulsed their relationship with prisoners in a rhythm of a monotonous regularity, it seemed to me, with the appearance of every full moon and first few days of its waning, and which could have resulted either in selective killing or indiscriminate massacre. It gradually helped to create an atmosphere of behaviour, of example on our part, and of dealing with Japanese authority, from unpredictable and cruel Korean guards to eccentric Sergeant-Majors and camp commandants, that made the negative marginally less active than the positive on our prison earth. It says something for the Japanese that, totally deluded and collectively and archaically possessed as they were, nonetheless somewhere they remained part of the Japan I had known and loved, to make them not unresponsive to atmosphere of this kind. And in order to maintain this atmosphere, which seemed to me to be of decisive importance because, as I put it metaphorically, it was like the atmosphere in the physical world which made the rain fall, I always started with myself and an assumption of a special responsibility and obligation ultimately invested in me by life for just such an occasion. More and more it seemed that being and doing involved not just myself and this other but a foursome: myself, and the other, and, through the greater awareness created as a result, the finding of two more senses: one of special awareness for my own and our men's sake, and another to all those less aware like our captors. To be aware and to contain the less aware seemed to be the beginning of morality and a true dynamic of change, and the way to meaning and truth of one's own.

It was not always as easy as it might sound. I had no comfort or reassurance in my own appreciation of these intangible feelings. I had, for instance, over many years tried to be obedient to my own intuition; and I knew how necessary it was to begin by learning to separate what was truly objective intuition from the personal fear and hope that so often usurped its role. I tried harder than ever to observe the distinction in prison and often discussed it with Nichols. He had little doubt that the chances of our survival were almost non-existent, and very early on confessed to me that any hope of us seeing our families again was "all Kathleen Mavourneen and a day" – which made his conduct all the more impressive. But I could find nothing definite either way in my hunches. Right to the end my intuition, later re-inforced by the kind of intelligence I described in *The Night of the New Moon*, was that the issue was daily in the balance, and daily we had something of our own to put in the scales of fate to keep the balance even. I often thought in this regard, of course, how a mere 'two cups of coffee' had already kept the scales in balance; and there were further illustrations from the mythologies of Egypt, and from the first people of my native Africa. According to the Egyptians the soul, on leaving the dead human body, was weighed against a white feather for its claim to pass on into life, beyond death. The white feather had been an image of sublime importance for me ever since my old Bushman nurse Koba, whose love and concern were still a living source of encouragement, had told me her stone-age story of how a single feather was all of truth that any single individual hunter could grasp for himself in a life-time of hunting the great mythological white bird of truth. Truth of behaviour, being and thought, in this small daily individual sense, was our only defence against the Japanese: truth about ourselves and, above all, the truth wholly and steadily perceived about them. For the moment they were clearly like the little man on Peach Tree Hill, largely men who no longer knew what they were doing and had forgotten what their own history and culture had taught them of a truth and a seeking greater than themselves. It seemed to me, therefore, of decisive importance that I should not forget it and indeed should remember it constantly on behalf of them.

This was no fantasy, a clear intuition told me, but the sort of imponderable which the inscrutable law of chance looked upon with awesome favour. No matter how dark with primordial shadow was the face that the Japanese – caught in a trap of history – turned to us, I never allow myself to forget the illumination I had found and experienced in Japan and with Japanese. I would often, alone at night in the private world of which my mosquito net had become the frontier, break down the collective, angry, vengeful and destructive mass with which we were confronted, by remembering the individual Japanese who had been my

friends and acquaintances in my own life, as well as their art and
religion: I would also, as an addition to the prayer I had always said to
myself every night since I had first been taught to do so, have a roll call of
Japanese individuals: Mori, Shirakawa, Hisatomi, my 'sensei', Gengo,
Mr Tajima, Teruha, and even our diplomat in plus fours, Mr Fujita and
his little bird, and on to Bashō who would not allow even his articulate
soul another word about a holy place, because it would violate its
mystery of sanctity; the priest in the abstract Zen garden in Kyoto, and
so on to Sesshu, many a face in trains, in Noh drama, and across crowded
rooms, and so on and on even to a kind of mediaeval Japanese Bayard,
sans peur et sans reproche, Taira No Tadanori who, the night before the
battle in which he was killed, added to the poems found in the band of his
helmet:

> *Night falls,*
> *My hospice*
> *A cherry tree,*
> *My host*
> *A flower.*

In the process, a kind of healing came to remove the hurt and negation
of the day and to reinforce me for the next. Without this I know even now
with a certainty greater than ever despite many waters that seek to
quench the memory of long eventful years, we would not have survived.
And more important even than survival, we could not have come out of
prison so much a something *other* than we had been before: an *other* totally
bereft of bitterness and of longing either for revenge or for an exercise of
uncomprehending fundamentalist justice.

At the end Nichols and I could count on the fingers of two hands the
numbers of multi-racial thousands we had served with in prison and who
had re-emerged embittered and diminished. In fact, going back, as I did,
straight from this to more years of war in Indonesia and extraordinary
military duty, I was sustained by hundreds of letters that I received from
fellow prisoners, saying how strange they had found it to be taken to
rehabilitation camps in Britain where they thought the people in need of
rehabilitation were not they themselves but the officers and officials in
charge of the camps. Many of them went further and told me how one
look at the world awaiting them frightened them more than the Japanese
had done. They haunted London and other great cities in groups out of a
fear that dispersal into the life that they saw would take away what they
had earned and cherished together in prison.

As for myself, I could not have succeeded in doing what I had to do
from the moment the Japanese High Command sent for me in Bandoeng
to tell me of Japan's capitulation that August afternoon and evening of
the full moon that blessed our deliverance from what had seemed for so

long certain death, had it not been for this other self and its growth in captivity. In this regard one illustration must speak for many. When the Japanese General and his staff raised their glasses on that afternoon of a light that still vibrates like a powerful spontaneous combustion engine at a window of my mind, and said: "We drink sincerely to your victory", and then added these astonishing phrases: "We Japanese have decided to switch. When we Japanese switch, we switch sincerely", this 'other person' was needed at once. For the General went on to tell me they had been ordered from 'on high' to turn to me. It all happened so fast that I could not ask what he meant by 'on high'. I just accepted that they had no option but to turn to me, and I none but to accept. They told me that they had to turn to me because they were faced with an extremely grave and potentially uncontrollable situation in Java and Sumatra, and did not know what to do. The Indonesians had declared their independence and set up their own government.

I already knew of the declaration and the hoisting of the Indonesian flag at Japanese Naval Headquarters four days before Hiroshima. I knew also, of course, how for some two years the Japanese had begun the selective training of Indonesian soldiers and officers, the 'Hei-Hoes' as they called them. Furthermore, I knew how that scheme some months before had shown signs of becoming more than the Japanese had intended because of a mutiny of Hei-Hoe officers in the interior of Java and a bloody but vain effort at its total suppression. So I was not altogether surprised now to hear that this movement of independence, as well as the news of the Japanese capitulation, had started to spread like a grass fire which could inflame the whole of Indonesia.

Already an Indonesian schoolmaster called Soedirman had appointed himself General, surrounded a Japanese division, cut if off from base, and was demanding all its weaponry in return for a guarantee of safety. More immediate, everywhere in prisons all over Java, the long-suffering Dutch, the moment they heard of the Japanese capitulation, had started to break out of their prisons in twos and threes and were being murdered spontaneously by an indigenous population determined they should never again be subjected to Dutch rule. In vain the Japanese commanders had pleaded with the Dutch men and women, senior officers and leaders and told them of all these dangers. But they were not believed, and the lethal leakage from safety in concentration camps into danger and death outside continued on an increasing scale. What were they to do?

I did not hesitate. I ordered them to send immediately for the senior Dutch Military Officer and a young Dutch civilian leader. I had never met Herman van Karnebeek but knew of him as a brave, totally respected and resourceful person. I would arrange, I told the Japanese, that prisoners stayed in their concentrations provided they had total freedom and command within, were properly fed, and had the immediate medical attention and supplies that they needed. I looked upon the Japanese to ensure the absolute safety of all military and civilian internees. More, I commanded them to see that the insurrection they had told me of did not spread, and that they handed over the administration of all those islands intact to the commanders of the Allied Forces who would already be on their way to accept the Japanese surrender.

The first of my demands they accepted immediately and without reservation. But the second they objected to and said how could they now, in all honour, fight against something which they themselves had encouraged and helped to organize, because that would be the inevitable consequence of the last command. I said that, as someone who had been to Japan, I knew, despite what they had done to us in prison, how much life was an affair of honour to the best of Japanese. What I was requiring of them was precisely addressed to their greater honour. Their obligation of honour to any local independence movement was lost in their incalculable debt of honour to redeem the disaster they had inflicted on millions everywhere, including their own country.

"You have to learn. . . ." I remember the words that fell into me precisely because they were applicable to me as much as to them. "You will have to learn, as I have had to in all these years in your power, how there is a way of losing that can become a way of winning."

At that the hiss of emotion so characteristic of their people when deeply stirred, escaped through their lips like steam from a safety valve. They bowed as one to me and, on coming out of their bow, the General, his eyes wide and moist, exclaimed: "That is a very Japanese thought!"

From that moment, they did all and more than asked of them by and through me, even to the extent of later fighting side by side with the British soldiers of the Fifteenth India Army Corps at Bandoeng and, above all, in the defence of places like Semarang where many thousands of Dutch women and children were concentrated and where the Japanese had to commit a brigade of infantry to help us in one of the fiercest of all our battles in Java.

Meanwhile, I had got Herman van Karnebeek to see that all the many thousands of civilian internees in Batavia did not scatter, while I and the senior Dutch officer visited all the other civilian internee and prisoner-of-war camps elsewhere to make absolutely certain that they too conformed. It says something for the person that I had become that I could take on all this despite a body so weak, and near its end, that a glass of lemon squash made me vomit; and the short journey by Japanese staff car from Bandoeng to Lembang made me violently sick. Yet I did not fall out for a day or even an hour from then on until, on the orders of Britain's Prime Minister, Clement Attlee, I went to talk to the Dutch Cabinet in the Hague. Even then, despite some weeks of regular food and medicine, I was still so weak that the dinner the Dutch Prime Minister, Professor

Schermerhorn, and his cabinet gave for me at Le Vieux Doelen in the Hague – although I ate it sparingly – gave me a spasm of hiccups; and, hiccupping, I had to talk, plead, argue and plead again with them in Dutch for seven hours. At the end I arrived at the British Embassy, where I was staying, so exhausted that the wife of the Ambassador, Portia Bland, had to put me to bed almost like a child in its cradle with a fit of whooping-cough. As a result, on my return to London, I spent four days at 13 Cadogan Street, where Ingaret Giffard had allowed me to share her home since 1938, before hastening back to the East.

I have deliberately not attempted to give a detailed history of all that happened in between these events; nor do I propose one for what followed, because it would fill volumes and must be properly written and supported by verification from all my reports, letters and signals to Lord Mountbatten, whose personal military/political representative I had become. It could not be written from memory as I am writing now. All that needs emphasis here is that the consequences of the parable of the 'two cups of coffee' continued and go on to this day. Also, the halfway-house sense of history and the sea, with which I began, became a full-house through the following years of the strangest of wars in those lovely Indonesian islands, in Java most of all.

I say Java most of all because that island was in a sense the terminal of the thrust of history which had dropped off my ancestors at the Cape. The Governor-General of Java, more than the Lords-Seventeen in Holland, had presided over our beginnings. Our founder was, in fact, a subordinate of his and was not entitled to the label of Governor but merely that of the Commander he was appointed to be. Our first slaves came from Java and its great Malay surroundings. We had Javanese blood in some of our most esteemed families; we even cooked much of our food in the Javanese way, and we evolved a language which carried a profound imprint of the Malay that was the 'Esperanto' of the thousand and one islands of opal and emerald of which Java was the heart, and much else that was Javanese and Malay beside. It was strange, humbling and profoundly inspiring that the four and a half centuries of history that I carried vividly in my imagination should have brought me to a decisive role in its enactment and to a condition which enabled me at long last to fulfil my own personal feeling of a special obligation to history. I say this not boastfully or egotistically but out of the heart of an inordinate and undeserved sense of privilege, and in the belief that it might be of some service to ordinary people who feel a loss of identity and individual meaning in the stifling collectivity which this grey, impersonal world of our day imposes on them. I believe that my own life established some small but undeniable and imperial facts: namely that every life is extraordinary; that the 'average man' is a statistical abstraction and does not exist; and that every single one of us – not excluding the disabled, maimed, blind, deaf, dumb and the bearers of unbearable suffering – matters to a Creation that has barely begun.

I say this with all the more certainty because this growing feeling that I had to 'soldier on' out of an obligation to history, and because I was the only one in that place and at that moment capable of responding to something special in overwhelming necessities, was not fed solely by subjective interpretations of this unlikely prompting from within me. It was confirmed by the reaction and admonition of many others who worked with me and urged me to go on with my soldiering. I shall only mention the first who drew my attention to it because he opened my eyes to what was objective in the inner prompting and helped it to a sense of balance and proportion. It came from Herman van Karnebeek, and was all the more convincing because he was not uncritical of me; and as time went on he and his closest friends, to my great distress, became so critical that they would no longer speak to me.

"It is extraordinary, Overste," he said (using the Dutch for Colonel, as he always did, which was, I thought, truly neighbourly of him), "how lucky the British have always been because, no matter where and in what situation they find themselves, life always throws up someone qualified to serve their unforeseen needs. Look, even here in all this mess where one would have betted for once that luck would have failed them, someone like yourself who knows the Japanese, the Dutch and the Javanese, and has experienced all that we have gone through, is there ready at hand for them to use."

So there it was, confirmed without as it was confirmed within. I had no honourable option but to serve on through another kind of war, far more distasteful than the one behind us.

I do not know how I managed the two years it took for me to persevere until I was convinced that my own mission was fulfilled. The complexities of it by themselves were daunting; and the two main personal targets that I set myself were always in peril. The first target was to prevent Britain from starting a colonial war on behalf of the Dutch who were convinced that they had the right, if not a divine mission, to hang the Indonesian leaders as Japanese collaborators, and then to resume government over a grateful indigenous population exactly where they had left off so abruptly and with such a glaring lack of dignity on the part of many an exalted leader of their Colonial establishment in 1942.

The truth, of course, was that the indigenous population had never longed more to see the last of the Dutch, and were now ardently seizing their best opportunity since the year 1602, when the Dutch began Empire over them, to get rid of them for good. And how, in any case, could one call men 'collaborators' who had been imprisoned without formal charge or trial by law, and then exiled to the outer islands for years, merely because they had advocated forms of independence long since practised in vast parts of the British Empire? We had not, I felt deeply, fought a world war in order to fight an old-fashioned colonial war for others, all the more when we ourselves were in full process of granting independence to Burma and the vast sub-continent of India. It seemed to me to be as immoral and obscene as it was unwise. So I set out to prevent it. Thanks to the hearing granted to me by Attlee, in whom the Major of the First World War was far more alive and in command than people realized; and thanks likewise to Ernest Bevin, Stafford Cripps, and Lord Louis Mountbatten, we achieved just that, despite a mobilization of deviousness and deceit on the part of the Dutch colonial leaders which had to be experienced to be believed. Moreover, we did so with scrupulous moderation, fairness, wisdom and vision that was in the true interests of all.

We did it so successfully, in fact, that there was a moment when the Indonesians under Soetan Sjahrir (who became a real friend and was a great and true little man) were fully prepared to settle for the kind of Dominion Status enjoyed in the British Commonwealth. But the Dutch refused it, and as a result they went on to lose the lot.

In all this I was the only person who could talk to all sides and was constantly coming and going between 'rebel' Indonesian territory, its insurgent armies and my generals and our allies, so much so that I was lucky to survive. But I was far from lucky in another and most painful way. It soon became a common belief in Batavia that I made British policy. As the belief spread, I started losing my Dutch friends until finally I had none left among men with whom I had endured and suffered much. I could only negotiate with the Dutch through a noble Indonesian whom I had, from the beginning, invited to live with me, and whose capable wife, Sri, ran my mess. He was the Raden Abdul Kadir Widjojoatmodjo, 'Dul' as those of us who loved him called him, who was for a brief period heartlessly exploited and set up by van Mook as a cover Governor-General of Indonesia in the belief that it would serve an archaic imperial purpose.

Even the last of our own generals, an able, brave and dashing soldier, but vain and invested with an almost Italian *bella figura* compulsion to an 'issima' degree, ended up by not speaking to me except on points of duty. He was as offensive to me at meetings with the Dutch Governor-General designate as he was attentive and solicitous in front of Lord Mountbatten, who never ceased to support me. His senior staff officers, who loved wining, dining and dancing with Dutch ladies and being entertained at their Governor-General's palace, also took their tone from him and were openly hostile to me in the closing phases of our occupation, with one outstanding exception: the General's principal 'B.G.S.', Victor Campbell, who remained a staunch, loyal and believing ally.

There were two people of my own kind to whom I owed most of all: a remarkable minister sent by the Foreign Office to Batavia, Gilbert

Mackereth, and his even more remarkable, beautiful, true and far-seeing wife, Muriel. Without their love, trust and unwavering support, I am not certain that, after all I had been through in war and prison, I would have been able to withstand the hurt and the loneliness that my concept of what life and history demanded of me had inflicted on me, so deeply that even now, in writing about it, my hand trembles at times with emotions I would have thought permanently tranquillized by the long years which have proved, I believe, that I had not been too far out in my stand.

The second main strand in the complex skein of necessities was a decision not to leave until the eighty thousand or more hopeless Dutch civilians and soldiers who had escaped being murdered by the Indonesians in the early days, only to be re-imprisoned by them later, were safely released and sent home. That, happily, was the part of the mission soonest and most successfully achieved, thanks largely to Soetan Sjahrir, Mohammed Hatta and their own circle, and even Soekarno. But the main task kept me in Java for some two years more. It ceased only when I myself decided that it had to come to an end, and that I would have to leave because the Dutch had come to feel strong enough to fight their own colonial war, convinced in their righteousness, I hasten to add sadly, because it is not love but conviction, supported by the best of apparent reasons, that tends to be blind.

I knew the date and the hour of their attack, warned London accordingly, and flew to Singapore to report in full to Neil Ritchie. He had not long been appointed Commander-in-Chief of Allied Forces in South-East Asia, and I spent my last night talking to him and his admirable Chief-of-Staff, Dixie Redman, until early morning. They wanted me to go back to Java, but for once I refused. I was convinced I could do no more, and one more hour in Java would be to over-stay the welcome of my own history and usefulness. I had left a small but expert military staff in Java who I was confident could do all that was necessary, but, as far as I was concerned, to return to Java would only be a source of confusion and above all the beginning of a compromise of what I had hoped to represent. Neil Ritchie wanted to order me to go back until he could get someone of my seniority from the War Office to take over, and I have never ceased to be grateful to Dixie Redman, who had seen as much of war since 1914 onwards as most, for intervening and saving in that quiet way of his: "You cannot do that, Sir. If anyone has earned the right to decide when it's best for him to go, he has!" And so, sad at heart and not without profound feelings of failure in my larger-mission, I left.

I went by air as far as South Africa since that was the only way possible for someone anxious to leave the tragic scene at once. But from the Cape I went back to report to the War Office in London by sea. I travelled in a Union Castle intermediate ship, the *Llanstephan Castle*, which, like me, had had no rest since the War began and was badly due for an overhaul. The voyage, in fact, took twenty-seven days instead of twenty-one because, somewhere in the Doldrums, the heart of the ship's engines had the mechanical equivalent of a coronary. While the engineers worked at its repair, we drifted on a Coleridge sea without ripple, wind or cloud, martyred by sun during the day but reprieved and blessed by some moonless nights of fathomless compassion and the mystery of healing the injury caused by day. The delay so got on the nerves of the passengers that they quarrelled and did unpredictable, hysterical things, like throwing all of the music of the indifferent ship's orchestra overboard at midnight; but I loved it all, for every hour I spent with that flawless mirror of sea I seemed to find reflection of more aspects of life not visible before. I only wished it could have lasted longer because, as always in the past, the sea helped me to bring up to date and balance the books of the accounts of my own spirit. Without that voyage I could not have arrived with the relative clarity of purpose I was to need in London far beyond any measure that I could foresee.

Clearest in focus of all the many complex considerations was a determination not to return to my pre-war way of life, a determination so firm that even my own vocation of writing had no conscious voice in it. It is true that it had compelled me to write for my fellow prisoners when the relevant words were imperative. In the course of a report on post-war events, which I wrote on Gilbert Mackereth's instruction, I had been amazed how the desire to write again on my own account had been quickened. It did so, I believe, because, defying all precedent, I wrote the report as a creative piece of writing. I was convinced that a report, official in form and formal in shape and phrase, could not possibly convey the volcanic reality that Mackereth and I had had to contain. For instance, wherever an individual had to be mentioned I would describe his person, character and record as one might a character in fiction. As a result, I was told, the report became the longest ambassadorial missive ever written for the Foreign Office. But it had Mackereth's blessing and, within the relevant services, obtained the widest possible circulation and became what he called a 'best-seller'.

I knew as a result how the writer in me lived on, but during that sea voyage it lived out of sight and mind. I was preoccupied by an immense, almost uncontrollable gratitude: gratitude to my parents and ancestors for having lived their lives in such a way that they bequeathed me with a physique strong enough to come out of imprisonment unimpaired and not maimed for life, as so many were; gratitude to life for survival against impossible odds; gratitude to Britain for standing fast even when abandoned and alone in a world that had abdicated, or was abdicating, before evil. I had to do something to express this gratitude, which is so inadequately defined here. Finally I wrote to Major Attlee (for so I always thought of the Labour leader): "I owe Britain so much that I do not want to return to a life of mere private pursuit and personal gain. Is there not something I can go on doing for Britain, something which I shall gladly do on no more than a soldier's pay, perhaps something in Africa?"

I mentioned Africa because with this gratitude went a feeling of immense urgency about my native continent. In all those years in Java, what had most astonished me perhaps and caused me greatest alarm was that a people of such high quality as the Dutch could have lived for some three hundred and fifty years in Indonesia without realizing that, night and day, the millions they ruled wished for nothing more in their secret hearts than to be quit of remote colonial rule. I saw the same imperviousness widespread in my native country by an all-pervasive colour and racial prejudice (which on the whole the Dutch themselves did not have) and blanketing, like a November fog of time, the swift approach of another repetition of a sullen, vengeful and amply discredited pattern of history which I had already attacked in my first book, *In a Province*, and in endless newspaper essays and interviews. That, too, imposed special obligations that I could not ignore since, more and more, the meaning of life seemed to indicate obedience to one's awareness, not only of doing and being, but of one's intimations of greater becoming.

I arrived far more resolved in England than I would have thought possible, but I was briefly confused by my reception at the War Office. My wartime commission had long since been converted in a regular commission, and accordingly I was offered two postings that carried with them flattering promotion and promise of military advance. One I instinctively rejected because it meant a return to the East. I had been so long part of Asia, and so traumatically involved in its history and spirit, that I had dangerously neglected my own special origins in time and culture. I feared that it would not take much to make me incapable forever of a return to my own native context. So a firm and quick 'no' was the only answer. The other offer was of even greater promise, and I only rejected it after weeks of reflection. It involved military administration at a high level, yet I could not see myself being interested for long in administration, however exalted. So inevitably I resigned my commission. So long had I been at war and on active service that I had earned some sixteen months of leave before demobilization. What that final demobilization meant, and how it exemplified the personal conflicts that preceded it, can be conveyed quite simply by the fact that it was one of the saddest days of my life. I was leaving the only true community of men with whom I had ever completely identified, and I was leaving it for a civilian life and what, for fear of the truth, we call peace; which is merely a euphemism for the preliminaries of a life of evasion and unawareness, charged with potential of more and bigger war. War, I knew from my own experience, had been a direct product of the nature of peace; that, and not armaments manufacturers and generals, was the villain of the whole piece which we all augmented with our own little villainies. By comparison with civilian life, the army often seemed a monastic order of unselfish, humane, idealistic and dedicated purpose. The war that had to be won, if another cataclysm of the world was to be prevented, was the insidious war in the mind and spirit disguised in the ample folds of a cloak of peace. So I came to begin all over again.

I went back almost at once to Africa. I went, of course, by sea, as a freelance servant of the Government, undismayed that my subordinates were earning three times the soldier's pay that I had suggested for myself. The going back and what followed is recorded in many books I came to write, and several films. What is not recorded is my work for Britain because I preferred to do it anonymously and unofficially. Only one aspect of it protruded as the tip of an iceberg above the private surface of my life; namely the part I played with David Stirling in promoting the work of the Capricorn Society of Africa which he had just founded. I persuaded myself to join him despite a growing mistrust of any group endeavour. Just then it seemed to be the best instrument available for penetrating the sort of imperviousness of Empire which I have mentioned. For years he, I, and several remarkable men and women who deserve a whole book devoted to their endeavour, did all we could to prevent the recurrence of mistakes of Empire in the past. We wanted to transform it into the instrument we believed it both could and should be; namely, a new order of emancipation and multi-racial integration in Africa. That, too, is another strange story in which our worst and most subtle opponents were not colonial subjects who rallied enthusiastically to us, nor were they Conservative statesmen. They were the radicals and liberals whose natural and invaluable role as instruments of change was increasingly corrupted by a growing inability to contain themselves and the collective and mindless forces that they released. It was a story, moreover, in which David Stirling and I nearly ruined ourselves financially and which, after many years, left David especially rejected and dangerously ill. David, who had fought for this project with even more courage and pertinacity and invention than he had in the war, was dangerously infected with defeat for the first time in his life. I was more protected than he because I did not have his inner sense of defeat. Already at the beginning I had told him: "We must not imagine we shall succeed in terms that the world defines as success. If we do, we shall fail. All we must realize is that at most we can influence events in the right direction."

As a result, when Capricorn 'failed', I escaped his fatigue and almost lethal physical depletion, and carried on the battle as I had begun it at the age of sixteen, in my own natural way. In the course of doing so I travelled repeatedly by sea between Africa and Britain, many times between England and the Americas, and to various other places. Every one of these voyages had the same decisive slot in the evolution of my life as had their predecessors. Indeed at times they even produced a kind of para-physical evidence of being synchronized to a rhythm of universal eventfulness beyond the range of my conscious 'becoming'. So startling was this that I might have become inflated were I not increasingly humbled by it.

For instance, there was a particularly startling voyage from the Cape to Southampton. I had sailed from Africa so shattered by events in my native country, and so full of foreboding and a sense of my own helplessness, that I could not sleep, despite massive doses of sleeping pills. After several days and nights of increasing unrest and sleeplessness, which induced a fear that I would never sleep again, it happened. Alone in the dark in my cabin and the wash of the sea, sometimes so loud that it hurt my ears, I suddenly realized that a calm I had thought to be permanently out of reach was spreading through my being, until at last, on the frontier of sleep, I had a sort of dream-vision. In this vision I was in a deep cleft of a long valley under an overhang of immense peaks, fat with snow so loose that even a whisper could send its avalanches thundering down. Then at the far end of the valley a man appeared on a peak which was still in clear sunlight. I recognized him as C. G. Jung, who had become a close friend. As I saw him he recognized me, and waved at me in a gesture that I knew well because always when my wife and I, or even just I alone, visited him at his home, he would walk to the end of his large garden, lean on the gate and wave a curiously young goodbye just like that. And as he waved, he called out in the schoolboy English that he relished, in the dream-vision as in reality: "I'll be seeing you?". Then he vanished behind the sunlit summit. Immediately, the calm within me assumed absolute command and I fell asleep.

I woke early the next morning feeling that the day was of unusual significance. I stood at my porthole and watched an albatross, the sun making snow of its breast, station itself on quivering wings by it to look at me as if it had something special to say to me. After a while I turned about and picked up beside my bed the Dante that I had tried to read the night before. I saw that I had pencilled beside Dante's "In thy will, my freedom" a strange remark. I had written: "Time is the dream in between when time was not and when time will cease to be." At that there was a knock on my door and the steward came in with some tea, grapes and the ship's newsletter. The first item on the sheet read: "C. G. Jung, the famous Swiss psychologist, died at his home at Küsnacht in Zürich last evening."

Comparing the time differences caused by latitude, it was clear that my dream-vision came at almost the precise moment of Jung's death. Since death had always seemed to me one of the greatest moments of truth, when all that is false and imprecise in life is erased, the comfort of those words "I'll be seeing you", the power of resolution and gift of calm in the avalanche valley of fear and disaster in which we find ourselves, that were at the disposal of such a vision, still remain for me part of the ultimate truth that it was that morning.

In the midst of all those years so crowded that I often thought I could not manage, I was nagged at all sorts of moments by a voice, as of a kind of conscience, reminding me that I still had an account to settle with my experience of Japan. It insisted that I could not leave it with that unpalatable war and brutal imprisonment on my mind if I were to practise what I had preached, namely the significance of the role of the individual to the universe. So I still had to transcend the last instalment of my Japanese experience in a personal way. That way, the nagging process unfailingly suggested, led straight to Mori. I had to make a private and personal peace with him, as the world had already made peace with his country. I had to do it all the more because I believed that, in such a time of disintegration and retrogression, renewal could only come through a kind of recognition of the increased sanctity of friendship and a deepening of all personal relationships.

Mori, I knew already, had survived the war, but there was something else I did not know, until I was told by letter from William Plomer's brother James, who had met Mori with me. When I had seen him last he had been in Port Natal as a boy in grey schoolboy flannels. Now he was in command of a Royal Canadian destroyer on a visit to Yokohama, and he managed to track Mori down. Among other things, he wrote, I ought to know that Mori was extremely angry with me. He had read an indifferent translation of Part One of my story of war, *The Seed and the Sower*. It had been published in a Japanese newspaper. There the title I gave it, 'A Bar of Shadow', was changed to 'Gunzo Hara' (Sergeant Hara), and Mori reacted as if I had unfairly maligned him and all my Japanese friends in my portrayal of Hara. Today, I hasten to add, it and the rest of *The Seed and the Sower*, superbly translated, has no more appreciative a reader than Mori. But for the moment I was profoundly disturbed that someone like Mori, to whom I owed so much, could totally misunderstand what I regarded not as a reflection on the Japanese so much as one of my own and many another human inadequacy. So I wrote to Mori immediately and received a courteous but obviously wounded and wounding reply. And 'that', the nagging voice insisted, is not something you can allow to let pass.

I therefore went back to Japan as soon as I could. I would have liked to go by sea but it was no longer possible. The erosion of passenger traffic and closing preliminaries of the post-Renaissance era, with which so much of this book is concerned, had already gone too far. I had no option but to go by air, and had a most moving and healing meeting not only with Mori but with post-war Japan. The essence of that meeting is all too briefly set out in an illustrated essay I wrote under the title of *Portrait of Japan* and I have only to speak here more directly of Mori as I could not do there, because that is the heart of the matter for me.

Significantly enough, although I was not without apprehensions about visiting Japan, I had none about Mori. My apprehensions seemed to focus on the fact that, although I had spoken Japanese daily in prison, and had continued reading and writing it so that I could then almost think or rather feel in it (since it is a feeling rather than the thinking empiric tongue which, for instance, Chinese is), all memory of it had been erased. That, I thought, suggested a profound area of unresolved hurt in me. And as I fear unconscious injury more than any conscious wounding of the spirit, I was nervous about what re-immersion in the life of Japan could bring back to the surface for me. But as for Mori, I was convinced I had only to tell him how the thought of him had helped me and several thousands of others under my direction and all would be well. Indeed, all was well from the moment we met. And because all was most movingly well between us and, our friendship re-proved and increased, all manner of things were well likewise.

I found him for the moment fallen on hard times materially. I knew how national defeat, even in a war of which he would have disapproved, however loyally, must have made him suffer more than most. That by itself produced inner antagonisms and conflicts, to overcome which would have been almost too much for such a sensitive and single soul. But on top of that he now had to make a living on land, where, after his long years at sea, he was like the albatross on the deck described in Baudelaire's great poem.

At the moment of my arrival, he was facing an almost insoluble struggle with a totally un-Japanese pattern of trade unionism which the new post-war trend, to break absolutely with a discredited past, had imposed on the country. The small stevedoring company he had formed, and which provided him with an extremely modest living even by Japan's austere measure, was being bankrupted by strikes and incomprehensible unrest. Characteristically, he did not mention this to me, and I had to learn it later from Mr Shirakawa who happily had also survived the war. Nor did Mori tell me, when I spoke to him on the telephone and accepted an invitation to dinner, that his home had been burned down only a few hours before. It was deliberately set on fire by a burglar who, enraged to discover that the Mori family were so deprived that there was nothing worth stealing in the fragile building, had set it alight.

I heard this from Mori after we had embraced each other on a suburban railway station near Tokyo. We had only to take one look at each other to know that nothing that had happened could divide us against each other. Like myself, of course, he looked older but still so much the upright, clear-cut person I had known twenty-six years before. The impact of recognition was indeed so immediate and powerful that it released emotions I could hardly control or even wanted to control. But they became almost unbearable when he followed the profound ritual of welcome that his culture so wisely and imaginatively provides for such meetings, with a preparation of myself for the celebration of a welcome from his whole family that was awaiting me. He had to begin this by telling me about the fire. He did so with far more concern for my feelings than his own.

"You would hardly believe it, Post-san," he told me, "but the funniest thing imaginable has just happened. We cannot stop laughing about it because it was so extraordinary and farcical a coincidence. Here I am to welcome you properly after all these years . . . and my house takes it into its silly head to allow itself to be burnt down! Have you ever heard of anything funnier?"

In this tone he told me the whole story in detail, to end up with the remark that funniest of all was the expression they imagined on the face of the burglar when he discovered he had gone to such trouble to break into a house so poor that there was nothing to steal.

Of course, to make light of one's own misfortunes in order not to distress one's friends corresponds to the deep-feeling values underlying Japanese concepts of good manners. And Mori could not have known that he could have chosen no more certain way of upsetting me than by treating it all as a great joke. I knew that he could also laugh genuinely despite regret at such an excess of fate because at heart he was not a Galsworthian equivalent of a man of property transposed in a Japanese idiom. His wealth was always in the texture and quality of the human being's responses to the challenges of circumstances, however undeserved and enigmatic. But my own feelings, and my recollection of what he had bestowed on William and myself when he had some patronage of giving in his power, could not just leave it at that. Fortunately gifts between friends in Japan are not suspect and, since my own financial affairs had begun to recover from Capricorn, I resolved to cable my bank the next morning and arrange for what I had in my current account to be sent immediately to Mori. The resolve helped me enough to prevent the rest of the evening from becoming over-emotional, however full of feeling, but even so the thought which the Mori family, despite their misfortune, had given to it often came near to destroying my control.

Since it was a perfect midsummer's evening, they insisted on welcoming me as if their home were still intact. They served me with a banquet of many courses, while we sat on mats spread over the ashes of their home. All their neighbours had joined in as a matter of personal honour to make it the 'banquet imperial'. Indeed it lives on as one of the feasts of feasts, with a warmth and happiness that many more privileged occasions never equalled. No detail was overlooked that could add to the spirit of welcome, and to my feeling of being back in a place where I belonged. For instance, since it was Sunday and my religion was Christian, some Christian neighbours for some weeks had trained a chorus of children to sing a special hymn as a form of grace before eating.

They were mustered between the ashes of the home of the large Mori clan of father, mother, children and grandchildren, and the bamboo, cherry, plum, peach trees and peonies so white and powerful as to floodlight the rest of the vegetation. A piano-accordion was produced and in clear, Shakŭhachi pure voices, the chorus sang 'Abide with me'. Whatever the musical elite of the world might feel about the quality of this hymn, its power over me, through a lifetime of associations, is almost unrivalled. At every stage of my progression, these associations had been brought up to date even in the war. The way that my apparently doomed little band of Australians sang it one Sunday night, after a number of executions in a Japanese prison, still resounded within me. Even more poignant were the deep bass voices of the gallant Ambonese soldiers of the Dutch colonial army, singing it in the words of the 'Heer Blypt met my' of the Netherlands, despite the efforts of Japanese soldiers to silence them, so determined were they to support three of their number who were marching out to be executed as live targets for bayonet practice by a platoon of Japanese infantry. The consequence was that the anonymous threesome went to their death as if to a wedding. Even more recently my mother had quietly hummed the music to herself as one of the last ordered preliminaries that her unswerving religious nature imposed on her just before she became unconscious and died. As a result this choice of hymn seemed almost supernatural, and perhaps intended as a message of abiding support from the deeps of life even more for Mori and his family than for me; and also an assurance that the ashes on which we feasted were not an end but a new beginning.

This seemed confirmed almost at once by the rising of a full moon. I do not know what it is about the moon in Japan but it seems to rise there with a beauty, and to exercise a power over one, that I have never seen equalled except in the heart of the Kalahari desert from which I had just come. As it rose, even more wonderful than the night in the *Canada Maru* which instantly came to mind, everyone spontaneously stopped eating and did something of their own to welcome it; the smallest children clapping hands, the older ones chanting an ancient welcome, an elder son writing a poem of welcome, and so on. Only I sat inert and bound as if by some merlinesque spell.

Soon, one of the grandchildren noticed my apparent inactivity and protested to Mori: "But grandfather, Mr Foreigner too surely must do something to welcome the august moon!" I had then, of course, to comply. But how? The emotions released in me ran too deep for me to express them in words suitable for the occasion. Almost in a panic, I searched my memory for a suitable poem to recite but none seemed appropriate. For how could Romeo and Juliet's "On such a night as this", or Byron's "We shall go no more a-roving, A-roving in the night, Though the heart be still as loving, And the moon be still as bright", or even Lawrence's "Who has seen the moon, Ah! Who has not seen her rise, from out of the deep", all of which occurred to me simultaneously, how could they be thrown at a family making a palace of an ash-heap?

Then, just as I seemed defeated and was on the verge of confessing my inability to contribute, something surfaced in my mind and ordered me to tell them the stone-age story from the lost world of the Kalahari. Since I had now been schooled in peace as in war no longer to resist this other and greater objective thinking, I obeyed without hesitation. I told them how the Bushmen had a moon story which was most important for us all that evening because it was a testament of unfailing renewal.

In the beginning, the story had it, the moon looked down on the earth and saw that life everywhere was terrified of death. Out of its heart of unfailing love, it chose the swiftest available messenger: the hare.

"Run," it commanded the hare. "Run as you have never run before and tell all men on earth to look at me and know that as I in dying am renewed again, so they in dying will be renewed again."

The hare was in such a hurry that it got the message wrong and told the people on earth: "Unlike the moon who in dying is renewed, you will not be renewed again."

The moon was so angry with the hare for falsifying a vital truth that it bit it on the lip. That is why the hare still carries a split lip, as a sign to all that he bore false witness to a universal promise of renewal that will never fail.

Although neither the Bushmen nor the Japanese at that moment had ever heard of one another, the story went home and produced a silence far more eloquent than words.

It was broken first by the grandson. "Grandfather," he said in a little voice of great awe, "isn't it wonderful what Mr Foreigner has just told us because you will remember the death of the Lord Buddha. All the animals gathered round him weeping as he was dying until he comforted them by saying: 'Animals, do not weep but look at the moon. As the moon in dying is renewed, so I Buddha in dying will be renewed again'."

I do not think the New Testament injunction that we should 'become like little children' had ever had so apt and poignant an illustration for me. It was proof of what I believed: that in the deeps all men are already brothers awaiting a call to become conscious of this in their daily lives; and that, from stone-age flint to electronic chip, all cultures in their greatest seeking are tributaries of one great river which, however enigmatic its flow and tortuous its bends, is seeking the same sea of truth.

Finally, there was a brief return to Moji where I had last been nearly thirty years before. It was of course so little like the Moji in my mind that I almost wished I had not gone. It had been transformed into a modern port, loud with engine noise, the fine pure air polluted, hideous with smoke and a skyline of concrete. I could not retrace my steps in any practical manner, or imagine it once more as a setting for a mediaeval festival with a professional story-teller to bear witness to the immortal pattern of his art; but it was worth it for one reason. I went there for news of Teruha. After much asking around among the formidable 'Madames' who command the world of Geisha, at last I found one whose determined, demanding face softened at the mention of Teruha's name. Yes, indeed she knew of Teruha. No . . . she had not met her and did not know where she was. But she was alive and had brought great honour to her vocation. She had become a revered Buddhist nun and had founded her own little nunnery somewhere, which was much sought after by all those in search of illumination. The word illumination for me had great four-dimensional logic. The name Teruha – the shining one – not only seemed to have been prophetic but to have, as this rumour proved, led to a fulfilment of all the quality that Mori and I had discerned in her. The news made me almost as happy as my meeting with Mori had done, and invested a vulnerable memory with a permanent full-moon glow in the dark of time and distance between our last and final meeting.

From that moment on, my contacts with Mori by letter and exchange of friends were resumed and increased. Not a year has passed wherein someone special has not gone as an 'envoy extraordinary' of friendship from him to me, and me to him. Some of his children have used my home in England as their own; and apart from many letters, I have trunks full of photographs illustrating Mori's rise once again from hardship and the danger of social oblivion to a position of well-being and honour. Every event of meaning to him was photographed and then despatched with heartwarming urgency to me. Piled upon one another I believe the photographs would rise as high as the old Coastguard's tower in Aldeburgh where I do my writing almost in the surf of the North Sea.

One of the last of the series shows Mori at the summit of his well-deserved recognition: namely, his reception by the Crown Prince and Emperor, as a principal performer in a modern Noh drama written to honour the ghosts of all sailors who had been drowned in peace and war at sea. At one point in the theatre, he had to 'roar like a lion', and I have various views in technicolour of him roaring accordingly, as well as a tape-recording to show how uninhibited and loud the roaring was, loud enough to be audible beyond those amber walls of mortal senses that define reality between the now and the undiscovered light in the night beyond.

All the while Mori tried to persuade me to persuade William to come back to Japan to see him and the host of his old pupils who remembered him with love. But William would not heed it. Nor would he heed my constant pleas for him to return to the Africa which had been his first inspiration and briefly given him a heart of flame and made a sword of his imagination. To my regret, I could not accept that William's urbane 'Englishness', and also the poetic, gifted man of letters he had made of himself by a supreme exercise of will, was good enough for him. And his refusal to go back saddened me further. But for all that, William did from then on keep in closer touch in a way that went on to widen the working out of the parable of the two cups of coffee in each of our lives.

For instance, there was a visit to Japan by Benjamin Britten and Peter Pears. They had become close friends in Aldeburgh and, soon after the war, they told me how much they longed to meet William because they admired him so much both as a man and artist. Accordingly I arranged a meeting in our Suffolk home, and was delighted to see how they all three took to one another; so much so that William, who had never taken any interest in music, came to write some of Britten's most eventful librettos. When they told us they were going to Japan, we both begged them not to miss the Noh theatre and above all to see 'Sumida-gawa' which had moved us so in 1926. They saw it not once but three times, and out of this came Britten's first Church Parable, 'Curlew River' that proved to be the pagan and aboriginal stock of his own religious exploration, on which was grafted the biblical flowering of 'The Burning Fiery Furnace' and 'The Prodigal Son'. William naturally wrote the librettos of all three.

Even this fruitful re-connection with Japan did nothing to remove William's resolve not to go back there. But I believe it had much to do with the composition of one of his finest poems, 'Bamboo', which proves, perhaps, how an artist needs neither trains, ships nor aeroplanes to go over his tracks creatively, and how wrong I may have been in wishing other forms of return on him to the end of his days.

And the end came very suddenly. I had seen him only some weeks before, well, laughing and happy at Covent Garden, when suddenly Ben Britten, who himself was on his own dignified way to a much too early death, telephoned me in great distress: "Laurens, you are wanted badly. William has just died of a massive heart attack."

As soon as immediate practicalities allowed, I arranged the dispatch of three full-rate cables: the first to Mori, the second to a Japanese headmaster who had been a favourite pupil of William's, and a third to one of his closest friends in England.

Mori cabled at the most immediate priority: "On receipt of cable went immediately to shrine of ancestors to report tragic news and light two candles to spirit of beloved departed writing."

The favourite pupil cabled almost as soon: "Moon rising when sad news arrived so took my shakŭhachi and played to moon where souls of departed rest on their way beyond."

The third, from the close friend, read: "Regret cannot interrupt holiday in France but will attend memorial service."

And people still wonder why I like the Japanese.

When Mori finally wrote, he blamed himself bitterly for not having come to England as he had intended the year before and chosen instead to 'roar like a lion' before his Emperor. We must all, he said, work harder at our friendships. He proceeded to do so with characteristic energy. Within a few months, although over eighty, he attached himself to a package tour which took him to all sorts of places that he had no desire to see but which, at a price he could afford, gave him three days in England with me. On the very first day he insisted on going to Brighton to pay his respects to William's scattered ashes. I called for him at his hotel in Whitehall and found him in full dress uniform, complete with decorations, and his eldest son Hiroaki, reverently binding a wide length of black crêpe round his right arm. From there we went to visit St Martin-in-the-Fields, where the memorial service for William had been held, and Mori said a prayer. Then we made for Victoria Station by way of St James's and Buckingham Palace. At once, some extraordinary things began to happen. Both at St James's Palace and Buckingham Palace the guards behaved as I have never seen them behave before or since. The moment we went by, they came smartly to attention and presented arms to Mori. At Victoria we found just three first-class places left in a crowded first-class coach of an overcrowded train. Immediately an Englishman by the window stood up, bowed to Mori and in perfect Japanese and with the highest degree of politeness, insisted on his taking the seat by the window so that he could see the country. At the crematorium where William's ashes were scattered, Mori reduced the director to a condition of penitence and contrition bordering on tears because no plaque to commemorate William had yet been placed on the crematorium wall. Moreover he compelled the gardener to give him a cutting of the white lilac that was in a fire of flower under the spring sun where William's ashes lay, to take back and transplant in Japan. Alas, the cutting failed to take root and the failure weighed heavily on Mori, as if it were a direct consequence of 'not having worked harder at his friendships', and his letters contained more and more urgent pleas for another white lilac plant to take the failed cutting's place.

Because of severe Japanese customs regulations I did not get a chance to put the matter right until recently. All my books were either translated or in the process of translation into Japanese. An inspired translation of *The Seed and the Sower* by Professor Yura, who presided over a complete edition of my writing, found its way to the distinguished film-maker, Nagisha Oshima. He had recently won the prize at Cannes for the best film director of the year and, in spite of his relative youth, was regarded by many as potentially a greater and more contemporary artist even than Kurazawa. Unlike the tragic Yukio Mishima, who had become a friend in the course of the seminars that we conducted together at Berkeley and other American colleges, he was by nature radical and determined to emancipate Japanese art from any lingering mediaevalism that prevented a renewal, as he saw it, of national culture. He read *The Seed and the Sower* seven times, came to Professor Yura deeply moved, and told him that he had decided to turn it into a full-length film.

That a Japanese artist could find a story of mine about the war against Japan sufficiently valid to wish to give it new form, was just about the most rewarding thing that has ever happened to me. Asked by Oshima to come to Tokyo to help, I did not hesitate; and, even in my haste to comply, did not forget Mori. Although warned once again by the Japanese embassy in London that the import of plants, particularly plants with roots, was still strictly forbidden in Japan, instinct insisted that I should take a white lilac shoot with me. Accordingly I set out with a robust little tree, its roots washed of all traces of soil, wrapped in masses of wet cotton-wool and stuffed in a plastic Harrods shopping bag. I arrived at Narita airport outside Tokyo in due course and, as I presented my passport, saw on the walls beyond what I had never seen in any airport before: a large black arrow pointed at the ominous legend: 'Plant Quarantine.'

I made for this fearsome place before approaching Customs, and was met by two young officers in immaculate uniform, with discipline written all over their faces. I placed the plastic bag on their desk and explained carefully what it was and why it was there, without infection.

After reflection and grave consultation, they said with immense courtesy and real regret that it was against the law to admit such a plant into Japan. To my amazement that 'other voice', after so many corrosive seasons, was still alive and spoke up in me to ask:

"But do you realise this plant is above the law?"

At once a flicker of totally non-official interest appeared in their composed and formal expression. Above the law? they questioned politely. Would I be so good as to explain how that could be possible?

So I told them the story at length, and as I told it they listened more and more attentively until I believe none of us was, except technically, in a vast official establishment any more. When I finished, I noticed from their eyes and a slight quiver of lips how absorbed and moved they were. We stood in silence as one, perhaps for a minute, as people do before a cenotaph of remembrance. Then they looked at each other briefly before the senior of the two took his stamp, pressed it firmly over the 'Harrods' on the bag and bowed, before saying: "We shall be honoured to receive this flower in Japan."

As a result, it now flourishes in full bloom in the latest photographs of it, with all the Mori clan arranged in impeccable order of protocol of proliferating generations around it. It was preceded by a series of others recording its unimpeded growth, and I like to think of them all as witnesses of a transmigration which carries the parable that began with two cups of coffee as far as it can humanly be carried. I say this all the more because, all the while I was in Tokyo, the story of the White Lilac had to be told and related wherever I went. It has become for great numbers of Japanese what it is for Mori and me, the most eloquent symbol of the transfigurative power of friendship available in our friendless age. It is also final vindication of my belief in the purity and constancy of Mori's love for William.

For me, too, it was a fulfilment that helped me to find on-going meaning in the final crumbling of my own special association with a centuries-long era of history. Perhaps it was this that made my final parting with Mori so moving and so pivotal an event. He organized a farewell gathering at the maritime club in Tokyo. Some three hundred of his friends and acquaintances participated. By individual Japanese standards this was an abnormally large gathering, but it was not a disproportionate measure of the position of honour that Mori had come to occupy in national esteem. Not only was he Japan's senior sea Captain, whose biography had been a bestseller and was now being made into a film, and had had the highest official distinction and recognition from his Emperor bestowed on him, but also, through his own brave merit, unswerving loyalty to his own spirit, upright nature, lively imagination, diversity of interests that were never unfocused, generosity, and above all a child-like simplicity that went hand in hand with the immemorial man in his heart, his friends and acquaintances came from all walks of life: aristocracy, middle classes, workers, writers, universities, water-front, high seas and market-places.

As a result, I saw the occasion not as a tribute to me but as a recognition of the unique quality of the man, and of the boundless extent to which friendship in such a spirit had transformed the world around him. I am compelled to turn again to this element of friendship because that was the theme of Mori's formal introduction of myself to his guests. The attention with which he was heard as he stood there once more, Captain on a new bridge, taking out to sea another and greater ship than ever he had conceived in his most ardent and secret dreams as a naval cadet, was almost ponderable in the vast dining-room. My own reply was far too inadequate in terms of all that I felt, and valuable only as confirmation of the course that this ancient mariner and navigator of unfamiliar seas had steered.

Happily the inadequacy was quickly buried in the superb feast which followed it and, above all, the music of a classical Japanese chamber orchestra which Mori, knowing my love of those antique instruments, had summoned to perform. The ceremony was so appreciated and lasted so long that I had to leave in an unseemly rush in order not to miss my flight to England. I feared that I might spoil all that had been achieved with so remarkable a balance of the abundant and the delicate; but I need not have worried. The Banzais of farewell, which Mori called for, and led, shook the building as it shook my memory. The last time I had heard their sound was in Java on the Mountain of the Arrow in the Valley-that-was-Well-Made as a preliminary to what seemed inevitable death by bayonet. Yet it was the one dark element, the dissonance that even harmony needs to complete the occasion, to draw a full circle around the square of our respective histories, and so to relieve them of the conflicts, tensions and unrealities that had angled them sharply and even disastrously against themselves as against one another. After the Banzais, which still seemed to be echoing in my ears, as I stepped into a waiting car there came the notes of a farewell played on a single shakŭhachi. As in the beginning, some fifty years before in the dark in the *Canada Maru*, it fell on my senses like the first drops of rain on dry earth.

I went back refreshed and reassured to deal with the redirection of my own inner life after the brutally abrupt ending to this special relationship with an era of history and the sea which had overtaken me not long before.

I should, of course, have been fully prepared for such an ending. But when its intimation finally came in a formal and insignificant-looking little slip handed to me by a postman early one morning at my home in Chelsea, I felt overwhelmed with a compound of surprise, unbelief and dismay.

"The *Windsor Castle*," it read, "will sail on August 12 from Southampton outward bound for the Cape and Port Natal, and will return to Southampton on September 19. It is regretted that thereafter the ship and its associate vessels will be permanently withdrawn from service."

Since they were the last passenger vessels plying a regular all-the-year-round service in the world, the ending was irrevocable. To me it was rather like our common reaction to news of death which, for all our knowledge of its inevitability, never loses its power to amaze, distress, and outrage when it strikes someone near to us. After all, the process of elimination of the great ocean-going lines of the world had been going on for a long time.

It was already years since the unbelievable of unbelievables had happened and the last passenger sailing-ship sailed from Britain to Bombay, as well as the last of the Royal Mail Packets from Solent to River Plate, of Canadian Pacific Empresses from Liverpool, and all the teeming Western Ocean traffic like the imperious Cunarders with the largest ships that the world has ever seen, and their American, French, Italian, Greek, German, Dutch, and Swedish rivals, all of which had ploughed the stormiest of waters with an assurance and splendour that their passengers felt as if they were not in ships so much as in some miraculously sea-borne Venetian, Louis Quatorze or Regency palace.

Yet somehow, against all these formidable convoys of evidence, I seemed to have assumed that the sea route whose discovery began it all would be immune to ultimate corrosion, and that this regular passenger service to the ambivalent Cape would provide the one saving exception to the rule of extinction at sea. I could not have been more wrong, and was even more in error in the instant belief that some public outcry of horror would go up to match my own dismay.

The announcement received only the most perfunctory of notices, and the amount of space given over to it by most newspapers was less than the normal obituary of a public servant in *The Times*. Nowhere did I detect any popular inkling that a momentous era stretching back to the earliest and most fateful compulsions of the Renaissance was formally coming to an end. The loss of our sense of history, which for so long has alarmed me as much as it does the psychiatrist when he encounters it in the life of an individual, and diagnoses it with an ominous label such as the 'dissociation of consciousness' which can be the prelude to insanity, could not, it seemed to me, have produced a more blatant symptom. As the days went by and no sign appeared of a demand for some sort of public commemoration of this turning of the last page in the great volume of an era, my own private sense of history persuaded me I could not leave it at that.

Happily, I found that the *Sunday Telegraph* and *The Times* had not lost their historical reflexes as had the rest of their contemporaries. The *Sunday Telegraph* agreed to let me write in haste a brief series of the special history of the sea and ships that was about to end. *The Times*, in a generous measure of its weekend space, accepted an account of the final event itself under the heading: 'Last Liner from the Cape.'

The end by sea, when it came, was like some ambivalent dream event. The August morning when I was driven by some friends from London to Southampton was warm, cloudless and full of a harvest-gold sunlight. The fields were ripe and high with corn, heavy heads of yellow seed already bowed for the sickle, and red poppies everywhere blazing in between as in some sort of remembrance of a vanished summer. They were proclaiming that the time had come for gathering a harvest not only of summer but also of history. England had never looked more English to me, nor the imagery more evocative, until we came to Southampton. There all memory of what it had been, was, and is still intended to become, seemed to have vanished. The great harbour which I had known so well for over fifty years, always crowded with purposeful shipping from the smallest to the greatest, was empty except for one or two outsize freighters and a lone passenger ship. Her black and red funnel, spotless white decks, lilac hull and grace of line were unmistakeably and poignantly evident and revealed her as one of the Union Castle line, with a company of the prettiest and best-dressed ladies that ever sailed the incorruptible sea. But for good measure she flew a thirty-foot paying-off pennant from her main mast, to proclaim what was to come.

She was, of course, the *Windsor Castle*. However, except for the routine attention of local stringers of national newspapers and a perfunctory television camera, she was not at all receiving the courtesy to which good manners of history entitled her. I could not at once believe such casualness on the part of a city and harbour with so old and great a maritime history, and I expected at any moment some special and appropriate demonstration of recognition of the momentous, if not monstrous, amputation of its present from the ship's past that was about to be inflicted on her. But I expected in vain, and when we cast off with the ceremonial that went with it, our ship was outwardly less adorned than I had ever seen her.

I thought of Guillaume Apollinaire's lines about the First World War, wherein he himself found his death. The lines had often come to me before when I was confronted with the indifference of the contemporary world to the Second World War. They did so, I believe, because they made specific in his and my own life the universal theme which had been incomparably expressed, with a profound corrective irony, in Villon's *Testament* with its recurring refrain, "But where are the snows of yesteryear?" This obviously was Apollinaire's inspiration. Now it seemed to me capable of enlargement from the original lines.

> *Où sont-ils ces beaux militaires,*
> *Soldats passés? Où sont les guerres,*
> *Où sont les guerres d'autrefois?*

and I found myself adapting the model to my own view of the fading Southampton:

> *Où sont-ils ces beaux matelots?*
> *Vapeurs passés, où sont les bateaux?*
> *Où les voyages d'autrefois?*

Fortunately for the ship itself, from there to the Cape and back to Southampton again, it was totally different. The ship was crowded, and I had never sailed with so many people, both passengers and crew, that I had known from previous voyages. They had come from all over Britain and beyond, instinctively determined to make this final journey a wedding and not a funeral; an enthronement and not an entombment of history. One example must speak for the complex many.

At the Cape, as soon as disembarkation was complete, a group of passengers was installed in the Mount Nelson Hotel which is, as I have already mentioned, suitably owned by the Union Castle Line. They stayed there far longer than I had ever known any other group do. After some days I had to leave to do some urgent business in the interior of an unhappy and deeply troubled country. When I returned over a fortnight later, as always disturbed and apprehensive by my experience, I was fully expecting to find the group dissolved. In fact it had grown larger and remained so until I left to sail back for Southampton on the very last lap. Indeed, an unexpected recall by air to the Cape three months later found a nucleus of the same group still there, as if paralysed by the shock of the event. But the rest of the passengers sailed back to arrive in a disturbed and unusually emotional state at a Southampton that was, if possible, even more indifferent to this ship's end than it had been to the beginning of her end.

Yet her departure from the Cape, by contrast, had provoked a response to the significance of the meaning of the history that gave birth to it, and which had for so long nourished and contained it. The great harbour at the Cape had been packed with people, and the traditional mail-ship quay, long and wide as it was, was capable of holding only a fraction of the people who wanted to join in homage to this last episode of departure of the 'great mother' which the necessities of traffic and travail in her by sea had been to them. On the quay there was a military band to play the ship out to sea with a sort of music I had first heard in the company of my father when the shipping of the world glowed as in a fable devoted to increase of life and discovery without end. But the crowds on land, on the decks and at the rails of the ship dense with passengers and crew, were unashamedly overwhelmed with new emotions. Most were close to tears, and a great many wept without restraint. Cries that were more like despairing wails of pledges never to forget one another, echoed constantly between ship and quay, and diminished, for once, the dominance of voices of the squadrons of gulls, wheeling in preparatory patterns for their hereditary function as heraldic ushers of the mariner in man to his chosen sea. They were clearly audible, but only at intervals, and then in tones that were acute with alarm rather than the normal sadness of farewell. Coloured streamers were thrown from deck to quay and were held in such quantities that, when the great hawsers and cables that bound this last 'Castle' of history to the land were cast off, it looked as if there were paper enough strung in between shore and ship to hold her in place for good.

Most significantly of all, as the last warning gongs, bells and siren blast sounded, there was a final rush of people from ship to shore and from quay to the overcrowded decks, and all sorts of varied estates of men and women embraced one another. The diversities which normally set us apart were erased for an instant by an eruption from this patient, brooding and ultimately irresistible urge gathering in the unconscious of the world, namely the power to assert one day, and for good, the values of the family of man on our journey ahead, however round about the route – just as our ship was bound for home. That sight of black, coloured, white, Indians, Pakistanis, Afrikaners and English, single and at one because the right occasion was there to reflect this concealed potential, still remains as a candle in the dark of what has followed on since. It remains, too, a reproach to the bankruptcy of a world that seems incapable of producing other and more positive evocations of the meaning of history, to make the manifestation of our latent human singleness not random and spasmodic but permanent.

Finally we broke the last of our paper streamers, and passed a lone Scottish piper in a Black Watch tartan standing on the tip of the last breakwater to grace our going with the kind of lament of which only the Highland soul is capable. We were followed out to sea by hundreds of little ships, and as they and a snow-storm of gulls escorted us out to the dark blue albatross-haunted roadstead of the South Atlantic, the land spontaneously sent us its own special message of hail and farewell. Wherever we looked from city to Lion's Head, Signal Hill to Sea Point, Devil's Peak to Fish Hoek and Simonstown, mirrors – of all shapes and sizes – reflected the levelling afternoon sunlight back at us, as if to say: whatever the rest of an indifferent and even hostile world may feel, we care still and can only reflect our care with light because, in so far as our history was an instrument of light, it was possible only because of the long line of ships of which you are so glittering an 'envoy extraordinary'. This may sound corny and banal. But I have long since learnt that all life is an orchestration of great and constantly recurring platitudes, and that our meaning is to be found in the way we re-orchestrate them from birth on to procreation, death and, through instinct of rebirth, beyond.

Long after the last of the little ships had given up, the gulls had turned about, and our escort of albatross on wings that for me, at least, always have something of heaven about them, had assumed the final office of Admiralty, I remained on deck. I remained there until the Cape itself sank into the sea, because I knew I would never again see it do so in that way. Moreover, I did so not in order to indulge in my emotions, because in a sense I had already had enough of them. I did it as someone who felt bound to bear some sort of objective witness to the last positioning of the future in the vacant place left by our farewell. I felt bound to bear witness to neglected meanings only accessible at such a time and on such an irrevocable course on an ocean stretching without impediment from Antarctic to Lands End, and add my final testimony to an evolution of meaning of which the process we call history is the instrument. We are contracted by life to live out this meaning as a great question before we can become aware of the one answer that can serve to transcend our private and collective conflicts. I seemed to know and to hear it, almost as a voice, in the rising murmur of sea and note of wind, comes to us first as the smallest and most improbable of intuitions. Indeed, such an intuition invariably strikes the established and institutional world either as some absurdity, clamouring clown-like for undeserved attention, or as a criminally subversive impulse. But to the individual who has the courage to let it into the shelter of his awareness, it assumes a clarity and certainty that makes end and beginning one. It came to the individual, I thought, while that moment of personal distress began to assume shape as of an Euclidian supposition capable of mathematical and demonstrable proof; because it would seem as if Life itself ordained that all began where what was to come had to begin, merely as a point in some lonely and questioning heart in which it just had position but no substance or size. It seemed to follow on from there, as if life supported the individual against the collective man because only the individual, and not committees and groups and minds that sought safety and permanence in numbers, could become the carrier of new being and the instrument of true increase. I thought I had the proof in me there as the ambivalent Cape, the clouds of a rising gale covering its descent into the sea, vanished in the twilight and left a trail of blood between the lilac port bow of the ship and the West, as if the event had cut light itself to the heart.

Some four hundred and fifty years before, the great Camões had seen the Cape vanish astern of a ship in which he was one of a handful of survivors of the original crew bound for home. He had sailed on from there over the same swinging water to write in his epic *Lusiad* what that journey and that view of the Cape had done to him. The powerful symbolism of his description, drawn from the kind of intuition of which I was thinking, bore witness to the Pentecostal nature of all art, and shone out as another and greater light than the one just extinguished on my own heaving horizon. He wrote of the Cape personified as the last of the Titans, an image of the great natural forces that had fought the Gods engaged in their campaign to impose order on chaos, and to bring light to aboriginal night. He it was, Camões proclaimed, who had been Admiral of vast and hidden seas, and who had longed, over immemorial ages in his loneliness of power and natural duty, for the love of a white daughter of the Gods of the Sea. One night he saw her by moonlight, a nymph of snow moving in the smoke and spray of a phosphorescent swell, and he strode out with the stride of a giant to embrace her, when the Gods snatched her away. For that and for so natural an impulse, he was turned to stone and became that 'far-flung and much-tormented Cape', that 'Cape of Storms', as Camões significantly renamed it again, rejecting the 'Good Hope' that his countrymen had tried so naïvely to make of it. The accuracy of the presentiment embodied in this imagery, and the significance of intuition that could guide the spirit of man to the centre of its target over a range of four and a half centuries, was for me, in that moment, as awesome as it appeared miraculous. For what could be a more conclusive symbol of all that the mobilization of white empire, in Camões' wake, had done to subjected peoples everywhere, than this imagery which clearly stated that a great natural heart in search of love had been turned to stone because of its denial of love by those who possessed the power to bestow it in abundance?

I stood there on watch troubled, as never before, by what I had just seen of the process of petrification of heart and mind of both rulers and ruled in the land lost below the horizon. But as if this generalisation of intuition were not enough, Camões made it more specific. His Titan goes on to warn the poet's hero, Vasco da Gama, that the time would come when all who followed after him would pay dearly for having broken so brutally into those seas and new worlds which the Cape of Storms had guarded for so long. The reckoning implied, at that very moment was being presented to us all in the approved manner of the Shylock 'accountancy of fate' reserved for all those who mortgage love for power. I myself, in my own small way in South-East Asia and all over Africa, had tried in vain to achieve a more merciful settlement of our debt to life. I had tried for some fifty years through my writing to prevent petrification and judgement according to the letter of archaic law in a court of fate whose appointed officers were executioners without love, and disaster without human bonds. How could men still doubt that disaster and suffering were the terrible physicians summoned by life when all other more gentle means of healing them had failed?

Yet still the incorrigible and fatal drift had gone on and even wrecked the Portuguese Empire which had come first and seemed more immune to retribution than others. Already Angola, Mozambique and Timor had gone and only Macao remained, perhaps because it was too trifling for a people with so long a view of history as the Chinese to be concerned about it. Only South Africa still stood intact, apparently more powerful than ever in its history. But from what I had just seen, I had no doubt, as I had had no doubt at the age of sixteen, that its summons before the court of disaster was inevitable. The final note would soon be delivered at its door unless the men of stone in power were swiftly disarmed, and power, through a change of heart and mind, was made subordinate to the law of love with which another and greater poet than Camões had tried to protect the spirit of Renaissance from the hubris of another fatal round of the absolutes of might. Indeed, at any moment now the final knock of the bailiffs of fate might resound on all our doors. The odds against it happening in time to prevent another round of suffering and disaster at that moment seemed greater than ever. I had first learnt as a boy that love was not just a kiss against a technicolour sunset but also a call to battle in the only campaign worth fighting in life.

I had always felt that, from the beginning, the history of southern Africa was one of growing psychological isolation from Europe, and that ultimately no solution of its problems was possible unless the process were reversed. I could not see it coming about soon enough in a world so sick that even the church (whose task it was to heal, as its New Testament master had come to heal) was itself as sick as the rest of the institutions charged with the duty of nourishing the spirit of man. It had utterly forgotten that it should draw as close, if not closer, to the sinner as to the sinned against. Instead of diminishing the distance between men, as charged by its own inner calling, it was, or so it seemed to me, widening it in the most important dimension of all; and also abetting from without the process of isolation of which this last liner from the Cape just then was the final melancholy symbol. The fall of night which became absolute at that moment would have been unbearable, so much did it coincide with the blackness that I was experiencing within myself then, had it not been for the appearance of the stars in a cloudless sky.

I had been brought up to sleep whenever possible outside the walls of houses, on the earth and under the stars. Even when life in the cities of the world shut me in as in a prison against them, I was let out for long periods, as on a parole of the soul, to sleep again underneath the stars. They were, moreover, the stars of the Southern Hemisphere, and I do not think I am guilty of some form of star 'chauvinism' but am merely being subjectively-objective when I say that I believe there are no other stars like them. And on this sombre evening the Southern sky seemed so great and packed with stars that there was hardly room for another in it. Such a sky had never failed me in the past, and it did not fail me now. It sent all my associations winging over the course of my improbable life that had brought me safely there. And despite all the dark waters in between, I found the associations not changed at all but re-confirmed. As I watched, the foremast of the *Windsor Castle* reached up to stir the stars around in the same eager way that the slender mast of the *Canada Maru* had stirred them about on the first night of my voyage from a half-way to a full-house of history and of self. They were still the witnesses that no cross-examination by pure reason could shake, of ancient and imperishable certainties of a universal love that could not fail, however desperate a defeat may be inflicted on it in the immediate now. It could and would still lose many a specific battle; but never the ultimate campaign. This was a certainty that came so convincingly to me by way of another Pentecostal intuition straight from the heart of Dante. It was he, of course, to whom I alluded as the other poet greater by far even than the Camões whose intuitive foreboding, known and proved at last, was almost substantial enough to stand there on the heaving deck like the dark figure of an ancient mariner beside me. Ever since I set eyes on the face of a young girl whose name I was never to know, Dante fell into a place that seemed specially designed for it in my spirit. From the moment I saw her face across a dusty street in the Interior of Africa, it made my dusty little native village a Florence for me, and was to guide my own seeking, as Virgil and Beatrice between them had guided Dante down into the depths of Hell and up again towards reunion with a transfigured feminine value which was his intermediary between Hell and Heaven. I thought of him because his intuition carried on where that of Camões had left off, on the battlefield between Titans and Gods and all that they symbolized for men in his, and our own, confused and retrogressive day. It took Dante unerringly on to a world of doing, being, and becoming, which is beyond the stars where freedom was the will of creation, and the ultimate law a love on which the universe, sun and all the stars, turned and moved onward as if in a wheel whose movement nothing could mar.

At that moment I seemed in some way to discover a still centre in the storm of painful emotions clamouring within me. It may be true that I would never again see the Cape of one last Good Hope sinking into the sea, and that its loss would never lose its power to hurt. But the hurt had acquired a meaning, and once suffering had a meaning whose imagery could be recognised and named, it could, I believed, be made welcome and endured. And this new meaning was not only inaccessible to rational and verbal expression but superseded it through the power and glory of the symbols and images which had served its predecessors in the wounded and vanished past. Just as the great Cape which once had evoked so much had sunk, superannuated, into the ocean at last, so all symbols and images that seek, night and morning, to enlarge our day-time awareness and to open wide the gates that bar the way to the future, once they have served their turn they too sink back into the dreaming unconscious of man, where all is refreshed, transfigured and restored to new life. Though many a symbol inevitably had to vanish and give way, the progression of meaning that they served never ceased, and produced other forms and shapes to take their place and continue the campaign in a more contemporary manner against the loss of meaning which would deprive life itself of purpose and leave it defenceless. The wood may burn out but the fire flames on, and somewhere beyond those dancing stars, floodlit as for a night of universal celebration, there was a universe and a life that was profoundly seasonal and needed both fall and winter in the spirit of man and its symbols for the great resurrections of spring and summer, for another thrust of meaning.

This, then, was a certainty which has never left me since that moment: life was greater than non-life, meaning more powerful than meaningless-ness, and all were bound irresistibly from a half-way to a full house of love. This certainty already seemed to propose an image of compensation for the loss of the Cape. All the time that I had been watching its slow and irrevocable sinking into sea and night, another vision had begun to rise like a new star and to trouble my imagination with a strange demand for overall authority in my intimations of the future. It was strange because I could not, at first, see any connection between it and what I had just lost. It was composed around a picture I had once seen on a master television screen at the Space Centre in the United States of America: it was a superb and unbelievably detailed view of the earth as seen moments before man first set foot on the moon. Ever since, at all sorts of unlikely moments when I was preoccupied with things totally unconnected with it, that vision had intruded and demanded to be looked at, before it withdrew again and left my imagination free to resume its prescribed course. But never before had it intruded with the force and authority that it did now. And even more important, never before had it so overtaken all that I had with a nostalgia that was not just my own – that was the least of it – but as something imposed on me by life and its course towards the future. For a moment I was almost in a panic, so great were the feelings of a boundless home-sickness in my heart and mind, and I might not have been able to contain so much had it not been for a falling star which suddenly dropped across my vision of the sea and the night. It came down solemnly in a great, grave and awesome arc between the stern of the ship and the Southern Cross. That one burnt-out morsel of light re-directed me to the stars which had first inducted me to this mood, and at once I seemed to know.

That view of the earth from the moon of renewal meant so much to me because, in the new age of travel and exploration that was upon us, it would be to a new breed of explorers and voyagers what the Cape had been to all of those who had preceded them over the five centuries on which the gates had just been finally closed. The time would come, I knew, when from somewhere out in space, on a sea of the greatest darkness, they would look back and see this little speck of dust of ours gleaming like a brave jewel in the night, and they would feel about it as men had done for centuries about the Cape. They would feel it all the more profoundly because the earth, rehabilitated and replenished with a new reverence for life, would be at one and no longer fragmented and ambivalent. My own emotions were so enlarged because they were overflowing already with news of valediction and nostalgia for the future with men made whole on earth, at last, for exploration and discovery in the service of all creation.

I thought here of how I had begun this account with a belief which was idiomatic to me and which held that no new thrust in the human spirit ever occurred without an equal and opposite thrust into the physical world around us. I even elaborated as to how the thrust into the spirit of man that had revealed the existence of a great unexplored universe within his spirit, which the empiric psychologist in depth called our 'collective uncon-scious', had already joined forces with the physicist in his exploration of the universe in the nucleus of the atom; and how all this must inevitably lead to a greater thrust than man had ever known into both the world without and the world within, and so be joined inevitably and irresistibly into a world without end. This thrust was ultimately irresistible because it had at its command the greatest energies of which the human being is capable: energies streaming out of a master pattern in this mysterious dimension of a 'collective unconscious' where already all life and men are one.

I hesitated to give a name to this pattern even then, although I carried within me a description of this great unknown which had become, for me, one of the most moving statements of ultimate truth ever written. And its impact was all the more profound and conclusive because it was contemporary, and abolished the spurious divide which the slanted, compartmentalised concepts of over-rationalized ages had sought to establish between the scientists' devout search for truth and the quest of religion to give their search its relevant meaning. It was something that Jung had written to me just before his death which, in a dream-vision in a 'Castle' ship in these very waters, coincided precisely with it, and contained the promise that he would be seeing me.

"I cannot define for you what God is," he wrote in reply to a desperate plea of mine. "I can only tell you that my work as a natural scientist has established empirically that the pattern which men call God exists in every man, and that this pattern has at its disposal the greatest transformative energies of life."

There was more, but this much was enough of truth and life for me, and was reassurance that the heroic quest of man was far from over, indeed had barely begun, and may even have been resumed, as implied by this view of 'earth-rise' seen from the margins of a moon that was, from the beginning, an image and promise of renewal of life-without-end for all men.

It is impossible to convey, and I shall not be so foolish as to attempt it, how moved I was by that view of so tiny a speck of dust as the 'earth' taking on itself the mystery of the universe; and, despite all its own divisions, conflicts and suffering, compelling men to combine and set out once again to make the great 'unknown' that begot it, and us who inhabit it, known, as already we were known to it. Indeed, the evocation of the heroic and invincible in the spirit of man setting out into the dark around him from so tiny, precarious and threatened a base, raised a storm of emotions that could have been unbearable except that, small as was the pull of earth from which his craft was projected into space, it had a one-ness, seen from such a height, that was not visible to eyes blurred in the heat and dust raised by competitive feet still pounding races that had long since run their course. For the first time in recorded history, all the sense of challenge between one man and another, one race, one nation against others, which had seemed so necessary to raise the human spirit to new endeavour, had served its purpose and sunk into insignificance, as had that Cape into the water behind me. In this master pattern of the dreaming 'collective unconscious' of man, confirmed fifty years before in Kenya in the year of the *Canada Maru*, it seemed to me that a great shift in design and re-grouping of energies had taken place, and already provided man with a physical and tangible perspective of the essential unity of his native earth; and how in the travel and exploration and resumption of the quest for wholeness and life once symbolised in a legend of a Holy Grail, he would see his departure not from a specific place, town or country but as from an earth at one, and the unique, beloved and infinitely heroic home of the life and family of a man whose frontier was no longer to be found on any man-made map or chart, but beyond that undiscovered beach on which the swell of the seas of the night was breaking above the ship's mast into the foam and spray of the 'milky way'. In so far as he still had obligations and feelings of citizenship, I felt that it would be as a citizen of earth; certain that this master-pattern within him would always have need of him, never leave his spirit unemployed, but give it increasingly work of ever greater meaning to do.

I thought of my grandmother, long before I was born, sitting in the dark of winter in the Interior of Africa and listening to the carrion of the night. She remarked that there – outside and beyond the night – all was a preamble of preparation that made her afraid. I had no doubt then that this immense act of preparation, and this transfigurative process which had haunted me as a boy, was now on its way to consummation all around me, and redeeming all the negations which the finality of this last voyage, in the last liner from the Cape, had been trying to inflict on me. Suddenly, far from being afraid, I was restored to the fullness of hope, and to the certainty that such an end had to come in order to release us for our journey to the stars, and to restore our wounded and troubled earth to its lawful place in their convoy, as they, and all the unapprehended life beyond, wheeled to conform to the love which the Pentecostal soul of the poet, the prophet and the dreamer already knew, and which I myself believe is all that there is of life and meaning that is worth serving. Had it not been for all that, I was sure, I would not have done the little of living in the way that I have done, or been encouraged not only to add my mite of evidence to all that of the great pioneers and explorers of this dimension before me, but also to be seen doing it by taking passage in the last ship of a long meaningful era.

As we hastened home over the swollen Atlantic water, the star-light came down like the dew which falls so large and heavy before the dawn in the desert I have known, both within and without, as a balm on wounds inflicted by burn of the sun, and sends the land all upon it, healed, back into battle in the coming day. The Cape may have been ambivalent, but the meaning of which it had for so long been a dominant image did not change. The thing itself stood fast. Only the images changed in order to inform us of new and deeper levels of discovery and re-discovery. However great its power of evocation of life, and its enlargement of our awareness of the paradox of the truth which alone can transform and set free the explorer in the spirit from the explored, this beautiful and beloved Cape had served its purpose, and the time was upon it to vanish in the sea of time behind us as other loyal images had done.

But as it vanished, there moved into the vacant place entire this jewel of earth that contained it, glowing in the dark night of soul and time and space, no longer ambivalent but single at last, a star of unfailing and universal good hope.